Max Dehn
Polyphonic Portrait

HISTORY OF MATHEMATICS VOL **46**

Max Dehn
Polyphonic Portrait

Jemma Lorenat, John McCleary, Volker R. Remmert, David E. Rowe, and Marjorie Senechal, Editors

AMERICAN MATHEMATICAL SOCIETY
Providence, Rhode Island

EDITORIAL COMMITTEE

Sloan Evans Despeaux (Chair)
Fernando Q. Gouvêa
Colva M. Roney-Dougal
William Thomas Ross

2020 *Mathematics Subject Classification.* Primary 00A66, 01A60, 01A70, 01A73, 20F10, 20F65, 52B45, 57K10, 57K30.

The image of Max Dehn on the cover was taken around 1930, and it is printed with permission, JoAnna Dehn Beresford Private Collection.

For additional information and updates on this book, visit
www.ams.org/bookpages/hmath-46

Library of Congress Cataloging-in-Publication Data

Names: Dehn, Max, 1878–1952, honoree. | Lorenat, Jemma, 1987– editor. | McCleary, John, 1952– editor. | Remmert, Volker R., 1966– editor. | Rowe, David E., 1950– editor. | Senechal, Marjorie, editor.
Title: Max Dehn : Polyphonic portrait / Jemma Lorenat, John McCleary, Volker R. Remmert, David E. Rowe, Marjorie Senechal, editors.
Description: Providence, Rhode Island : American Mathematical Society, [2024] | Series: History of mathematics, 0899-2428 ; 46 | Includes bibliographical references.
Identifiers: LCCN 2024021474 | ISBN 9781470461065 (paperback) | ISBN 9781470478353 (ebook)
Subjects: LCSH: Dehn, Max, 1878–1952. | Mathematicians–Germany–Biography. | Mathematicians–North Carolina–Black Mountain–Biography. | Dehn family. | AMS: General – General and miscellaneous specific topics – Mathematics and visual arts, visualization. | History and biography – History of mathematics and mathematicians – 20th century. | History and biography – History of mathematics and mathematicians – Biographies, obituaries, personalia, bibliographies. | History and biography – History of mathematics and mathematicians – Universities. | Group theory and generalizations – Special aspects of infinite or finite groups – Word problems, other decision problems, connections with logic and automata. | Group theory and generalizations – Special aspects of infinite or finite groups – Geometric group theory. | Convex and discrete geometry – Polytopes and polyhedra – Dissections and valuations (Hilbert's third problem, etc.).
Classification: LCC QA29.D446 M39 2024 | DDC 510.92 [B]—dc23/eng/20240513
LC record available at https://lccn.loc.gov/2024021474

History of Mathematics ISSN: 0899-2428 (print); 2472-4785 (online)
DOI: https://doi.org/10.1090/hmath/46

Copying and reprinting. Individual readers of this publication, and nonprofit libraries acting for them, are permitted to make fair use of the material, such as to copy select pages for use in teaching or research. Permission is granted to quote brief passages from this publication in reviews, provided the customary acknowledgment of the source is given.

Republication, systematic copying, or multiple reproduction of any material in this publication is permitted only under license from the American Mathematical Society. Requests for permission to reuse portions of AMS publication content are handled by the Copyright Clearance Center. For more information, please visit www.ams.org/publications/pubpermissions.

Send requests for translation rights and licensed reprints to reprint-permission@ams.org.

© 2024 by the American Mathematical Society. All rights reserved.
The American Mathematical Society retains all rights
except those granted to the United States Government.
Printed in the United States of America.

∞ The paper used in this book is acid-free and falls within the guidelines
established to ensure permanence and durability.
Visit the AMS home page at https://www.ams.org/

10 9 8 7 6 5 4 3 2 1   29 28 27 26 25 24

Contents

Preface	vii
Photo and Figure Credits	xiii
Chapter 1. Max Dehn's Family: A Brief History BRENDA DANILOWITZ AND PHILIP ORDING	1
Chapter 2. Max Dehn as Hilbert's First Star Pupil DAVID E. ROWE	17
Chapter 3. Dehn's Early Mathematics JEREMY J. GRAY AND JOHN MCCLEARY	53
Chapter 4. Dehn's Early Work in Topology CAMERON MCA. GORDON AND DAVID E. ROWE	83
Chapter 5. Golden Years in Frankfurt DAVID E. ROWE	119
Chapter 6. Three Students of Max Dehn JOHN STILLWELL	153
Chapter 7. Max Dehn and the Word Problem JAMES W. CANNON	167
Chapter 8. Max Dehn, Axel Thue, and the Undecidable STEFAN MÜLLER-STACH	181
Chapter 9. Mathematics under the Sign of the Swastika DAVID E. ROWE	197
Chapter 10. Max Dehn's Long Journey MARJORIE SENECHAL	219
Chapter 11. Max Dehn's American Students MARJORIE SENECHAL	237
Chapter 12. Toward a Happy Life: Max Dehn at Black Mountain College BRENDA DANILOWITZ AND PHILIP ORDING	249

Preface

Max Dehn's name is known to mathematicians today mostly as an adjective (Dehn surgery, Dehn invariant, etc.). He is also remembered as the first mathematician to solve one of Hilbert's famous problems (the third) as well as for pioneering work in combinatorial topology. Beyond his accomplishments as an eminent mathematician, however, he was a remarkable scholar and teacher whose influence extended far beyond research mathematics. He also led a remarkable life, parts of which are described in this book along with some of his most important contributions to mathematics and their afterlife. This book's three-part structure is based on three distinct periods in Dehn's life. The first centers on his early career as one of David Hilbert's most gifted students. During the second phase following the First World War, he emerged as the leader of an intimate community of mathematicians in Frankfurt, an idyllic era that quickly darkened once the Nazi government seized power in 1933. After fleeing to Scandinavia in 1938, the Dehns had to undertake an arduous journey that brought them to the United States. The third part takes up his initial wanderings in the US before focusing on his years at Black Mountain College, where he found a final home.

As the title of this book suggests, many voices contributed to the chapters. The central events in Dehn's life are presented in chapters 1, 2, 5, and 9 through 12. These are interspersed with four other chapters (3, 4, 7, and 8) that focus on mathematical themes that are found in Dehn's work and his influence on their development. Chapter 6 presents a mixture of biography and mathematical exposition in a discussion of the work of Dehn's three most prominent students. A great deal of the information found in all twelve chapters has never been presented before. The various perspectives and interests are often original as well. The book, however, remains a collection of essays; it does not pretend to offer a comprehensive picture – a full-scale scientific biography of Max Dehn remains to be written. That being so, we nevertheless feel confident that this book contributes significantly to the literature and readers of all sorts will find something new and enriching.

Max Dehn was born in Hamburg, Germany, on 13 November 1878, the fourth of eight children of Maximilian Dehn, a medical doctor, and his wife Berta, née Raf. The family belonged to Hamburg's vibrant middle-class Jewish community, which was mainly comprised of secular rather than religious Jews. Dehn's family background is described in Chapter 1.[1] After spending a single semester in Freiburg, he entered Göttingen University in the summer of 1897 (Chapter 2). It took him altogether only six semesters to complete his studies, so he had not yet reached his twenty-first birthday when in November 1899 he passed his doctoral examination

---

[1] Further information about the eight children and their relatives can be found in Matthias Brandis, *Meines Großvaters Geige. Das Schicksal der Hamburger jüdischen Familien Wohlwill und Dehn*, Leipzig: Hentrich & Hentrich.

and took his degree *summa cum laude*. Soon thereafter, his star was already clearly visible in the mathematical firmament, illuminated by his noteworthy contributions to the foundations of geometry (Chapters 2 and 3).

FIGURE 1. Max Dehn, second from right in top row, with the Göttingen Mathematical Society, 1899. To Dehn's left stands Harvard's William Fogg Osgood, a former student of Felix Klein (seated in the front row, third from left, next to David Hilbert, second from left). In front of Max Dehn, to his right, is Alexander Ziwet from the University of Michigan. Ziwet, a native of Breslau, is occasionally remembered for having performed the service of preparing the written text based on Klein's *Evanston Colloquium Lectures* from 1893.

During those years, American and other foreign mathematicians flocked to Göttingen to study or visit, if only during the summertime, since the universities in Germany normally remained in session until early August. The photo of the young, innocent-looking Max Dehn (Figure 1) standing next to much older mathematicians was probably taken in the summer of 1899, possibly in June during the celebrations at the time of the unveiling of the Gauss-Weber Monument. That ceremony had special significance for Dehn, since he had been intensely following Hilbert's lectures on the "Elements of Euclidean Geometry" during the previous winter semester. Thus, no one was in a better position to appreciate the importance of Hilbert's contribution to the *Gauss-Weber Festschrift*, the original version of his *Grundlagen* *der Geometrie.*² In later years, Dehn would help his former mentor refine and improve that original edition.

The mathematician Francisco José Craveiro de Carvalho described the young man in that photograph poetically:

> This is the summer season
> of the quiet year of 1899
> in Göttingen.
> In the background there is a river
> or a stage-setting that creates a river.
>
> Only we still know
> that the young Max Dehn
> will have to rush
> out of the photograph
> and get rid of his coat
> high-collared shirt and former life
> to try finding a new one for himself
> on a strange and distant shore.

(Translated from the Portuguese by Manuel Portela)³

Max Dehn's years as a student seemed to set him off on a fast track. Yet, as the biographical chapters in our portrait show, his academic career was a long uphill climb, not without setbacks along the way (Chapters 2 and 5). Indeed, seen from a longer term perspective, he faced many of the same obstacles as other young men from this era. The religious confession of candidates was at times a definite factor, though ethnicity rather than religious affiliation played an even larger role during the first phase of his career. After spending more than a decade as a private lecturer (*Privatdozent*) in Münster, Dehn received his first appointment as an associate professor in Kiel, where his candidacy for a post as full professor was passed over. Shortly thereafter, however, he gained that promotion in 1913 with an appointment to the Technical University in Breslau, where his work was interrupted by three years of war service.

Dehn made his major contributions to mathematics roughly between 1900 and 1920. He was attracted by the newest currents of research, beginning with his thesis work on the foundations of geometry. Riding on the wave of interest generated by Hilbert's work, he became a leading authority in this field (Chapters 2 and 3). His interest in the geometry of polyhedra led to his solution of Hilbert's third Paris problem, and his attempt to solve the Poincaré conjecture took him into the vast realm of 3-dimensional manifolds in topology (Chapters 3 and 4). In all this work, Dehn fashioned powerful tools that took on lives of their own. For example, his invariant for polyhedra involved a geometric property of rectangles filled by sub-rectangles – a subsequent paper of Dehn's has been recast and reproved

---

²There are significant differences between this text and the many editions that followed it. A reprint of the *Urtext* along with commentary on it and subsequent editions can be found in Klaus Volkert, Hrsg., *David Hilbert, Grundlagen der Geometrie (Festschrift 1899)*, Heidelberg: Springer, 2015.

³The original Portuguese version of this poem was first published in *Gazeta de Matematica*, # 149, July 2005.

by many geometers since its publication (Chapter 3). His work on 3-manifolds required a deeper understanding of the fundamental groups of surfaces, for which he developed what became the foundation of geometric group theory, a thriving field today (Chapter 7). Dehn's pioneering studies in combinatorial group theory also opened the door to research on the word problem, as described in Chapter 8.

After the war, Dehn returned to Breslau, but his career entered a new phase in 1921 with his appointment in Frankfurt, where he soon became head of its mathematics program (Chapter 5). Today this is remembered as a "golden age" for mathematics in Frankfurt, whose faculty was unusually harmonious and dedicated. Dehn's former Göttingen teacher, Arthur Schoenflies, headed the original faculty, acting as Dean in 1920/21. Max Dehn's appointment not only elevated his own career, it also led to the appointments of Hungarian Otto Szász and Paul Epstein as associate professors is 1921. At the outset of the new republic, the Prussian Ministry supported the promotion of scholars (including many Jews) who had been held back under the monarchy. Schoenflies retired one year later, opening the way for young Carl Ludwig Siegel to assume his position. A key figure during the 1920s and early 30s was Ernst Hellinger, whom Hilbert had chosen earlier as his assistant. Hellinger came to Frankfurt University at the time of its founding in 1914, and during the Dehn era he took charge in organizing the course work and advising students.

Siegel would later memorialize this unique community in a moving lecture, delivered to members of the new faculty in 1966.[4] He warmly remembered the atmosphere created by these selfless colleagues, but especially their weekly historical seminar, led by Dehn, in which they read classic mathematical texts in their original languages (Chapter 5). He also described how this brilliant era came to an abrupt end once the Nazis came to power in 1933. The Dehn children thereafter found refuge in England and the United States, leaving their parents behind in Frankfurt, where Dehn's position was terminated in 1935. Then came the terrifying pogroms during the "Night of Crystal Glass" in November 1938, after which Dehn and his wife Toni fled to Scandinavia (Chapter 9).

They first went to Copenhagen in January 1939 before moving on to Trondheim in Norway, where Dehn took over the post of a vacationing colleague at the Technical University. With the German invasion in 1940, however, hopes of remaining in Norway suddenly vanished. Thanks to support from friends in the United States, the Dehns were able to leave Oslo in late October 1940. Chapter 10 describes their long journey across Siberia, reaching San Francisco at the very beginning of 1941, followed by four years of temporary teaching jobs across the country, a story recounted in Chapters 10 and 11. Max Dehn was remembered by many of his students in the US as a man of unusually broad interests.

Dehn finally found refuge, and a setting for his many talents, at Black Mountain College (BMC) in North Carolina (Chapter 12), a unique though short-lived experiment in higher education situated in the mountains of western North Carolina. During the last eight years of his life (1945–1952), he taught mathematics, philosophy, Greek, and Italian at BMC. The curriculum at Black Mountain College focused on arts and crafts, and BMC is celebrated today for having served as a

---

[4]C.L. Siegel, On the History of the Frankfurt Mathematics Seminar, *Mathematical Intelligencer* 1(4): 223–230. Wilhelm Magnus, Max Dehn, The Mathematical Intelligencer, 1(3), 132–143, 1978.

catalyst for several renowned artists. In addition to teaching elementary mathematics and projective geometry and tutoring a few advanced mathematics students at BMC, Dehn also taught a very popular course called "Geometry for Artists," while working in close association with others on the BMC faculty. Love of nature had always been a central part of his life, and he felt especially at home in these surroundings, far from what others would call civilization.

When he died, "Dehn" was not yet the adjective familiar to mathematicians today. As his former student Wilhelm Magnus remarked at the Max Dehn centennial in 1978, "his name is probably more widely known and certainly more widely-used today than it was at the time of his death in 1952." In present-day nomenclature we find the Dehn invariant, Dehn twists, Dehn's Lemma (about which, see Chapter 4), Dehn's algorithm, Dehn filling, and Dehn surgery. Terms that carry names honor (sometimes incorrectly) the work of the mathematician who introduced them. Dehn's influence can clearly be seen in the work of three of his most distinguished students (see Chapter 6): Jakob Nielsen worked on the topology of surfaces, Wilhelm Magnus studied groups through geometry and topology, and Ruth Moufang took up the foundations of geometry in the Hilbertian manner by building bridges between geometry and algebra. Thus, while this book focuses on Dehn's life, the temporal range extends to the end of the twentieth century through the afterlife of his works and the research of his students.

To provide an overview, we here list various milestones in Max Dehn's career, framed by the structure of the book.

Part 1: From Hamburg to Breslau (1878–1921)

- 1878–1896 Early years in Hamburg
- 1896–1900 Student in Freiburg and Göttingen
- 1900 PhD Göttingen under David Hilbert (Legendre's Theorems on the Sum of the Angles in a Triangle, published in *Mathematische Annalen* 53(1900): 404–439)
- 1900/01 Assistent of Friedrich Schur at the Technische Hochschule Karlsruhe
- 1901 Habilitation in Münster under Wilhelm Killing (*Habilitationschrift* on Hilbert's Third Paris Problem, "Über den Rauminhalt", published in *Mathematische Annalen* 55(1902): 465–478)
- 1901–1911 Privatdozent in Münster
- 1911–1913 Extraordinarius in Kiel
- 1913–1921 Ordinarius at the Technische Hochschule Breslau
- 1915–1918 Service in World War I on the Western Front

Part 2: Frankfurt Years (1921–1938)

- 1921–1935 Ordinarius in Frankfurt/Main
- 1922 Founding of the Frankfurt Mathematics Seminar
- 1935 Dismissed on formal grounds immediately prior to implementation of the Nazi racial laws (he receives regular pension payments until 1940)
- 1938/January to April teaches at the boarding school attended by Dehn's children in Kent/England

Part 3: From Frankfurt to the USA (1939–1952)

- 1939 Emigration (Copenhagen and Oslo)
- 1939/40 Visiting professor in Trondheim/Norway (substituting for Viggo Brun)
- 1940/October 30 The Dehns leave Oslo for San Francisco.
- 1941/January Arrival in USA (via Sweden, Finland, Soviet Union, Japan)
- 1941/42 Asst. Prof. of mathematics and philosophy, University of Idaho–Southern Branch (today Idaho State University)
- 1942/43 Visiting Prof. of mathematics, Illinois Institute of Technology, Chicago
- 1943/44 Tutor, St. John's College, Annapolis, Maryland
- 1945–1952 Prof. of mathematics and philosophy, Black Mountain College Lecturing at the University of Wisconsin, Madison 1946/47 (fall semester), 1948/49 (fall and spring semesters), and Notre Dame 1949 (summer)
- 1952 Prof. em., Black Mountain College

Acknowledgments

We are grateful to the Mathematisches Forschungsinstitut Oberwolfach for helping us launch this work through a four-person Research in Pairs grant in 2015 and a mini-workshop the following year. (See Mathematisches Forschungsinstitut Oberwolfach Report No. 59/2016 DOI: 10.4171/OWR/2016/59 Mini-Workshop: Max Dehn: his Life, Work, and Influence.)

The editors

Photo and Figure Credits

The Editors and Authors gratefully acknowledge the kindness of the following individuals and institutions for granting permissions for the use of archival and photographic material.

Courtesy of the Josef and Anni Albers Foundation
Preface, figure 1.
Chapter 12, figures 3 and 5.

Courtesy of the American Mathematical Society
Chapter 11, figure 2 (from the *Notices of the AMS*, July 1949).

Courtesy of Antonio Alcalá
Chapter 1, figure 2.

Courtesy of Ruth Asawa Lanier, Inc., Dept. of Special Collections and University Archives, Stanford University Libraries, Stanford, California (Asawa Papers)
Chapter 12, figures 7 and 9.

Courtesy of JoAnna Dehn Beresford Private Collection
Cover image of Max Dehn
Chapter 3, figure 1.

Courtesy of the Western Regional Archives, State Archives of North Carolina. (Archives of Black Mountain College)
Chapter 11, figure 3.
Chapter 12, figures 1, 6 (Gift of Mary Emma Harris), 8, 10, and 11.

Courtesy of Dr. Matthias Brandis
Chapter 1, figures 1 and 3.

Courtesy of Peter Spencer
Chapter 1, figures 4 and 5.

Courtesy of the Dolph Briscoe Center for American History, University of Texas, Austin (Dehn Papers)
Chapter 4, figures 8 and 9.
Chapter 5, figure 4.
Chapter 12, figure 4.

Courtesy of the Dean's Office, St. John's College, Annapolis, MD
Chapter 10, figures 3 and 4.

Courtesy of the Universitätarchiv Frankfurt am Main
Chapter 5, figure 3.

Courtesy of Google Maps
Chapter 10, figure 2.

Courtesy of the New York Public Library, Emergency Committee in Aid of Displaced Foreign Scholars records. Manuscripts and Archives Division. The New York Public Library. Astor, Lenox, and Tilden Foundations.

Chapter 10, figure 1.

Courtesy of Manuel Portela

Preface, the translation of the poem on pages viii and ix.

Courtesy of Springer Nature

Chapter 4, figures 7, 10, and 11 from Johansson, I. Über singuläre Elementarflächen und das Dehnsche Lemma, Math. Ann., **110**(1935), 312–330.

Chapter 6, figure 5 from Wilhelm Magnus. Über Automorphismen von Fundamentalgruppen beranderter Flächen, Math. Ann. **109**(1934), 617–646.

Courtesy of Taylor and Francis Group

Chapter 12, figure 2 from Max Dehn, Mathematics, 200 B.C.–600 A.D., The American Mathematical Monthly, **51**(1944), no. 3, 149–157, reprinted by permission of Taylor & Francis Ltd. on behalf of the Mathematical Association of America.

Courtesy of the University of Wisconsin Archives

Chapter 11, figure 1.

Courtesy of Wikimedia Commons, CC-by-3.0

Chapter 5, figure 1 (by G.F. Hund). https://commons.wikimedia.org/w/index.php?curid=11968671

Figures in the public domain

Chapter 3, figures 4 and 5 are taken from Max Dehn, Die Legendre'schen Sätze über die Winkelsumme im Dreieck, Math. Ann., **53**(1900), 404–439.

Chapter 3, figures 13, 14 and 15 from Max Dehn, Über den Rauminhalt, Math. Ann., **55**(1902), 465–478.

Chapter 3, figure 16 from Max Dehn, Ueber raumgleiche Polyeder, Nachr. Ges. Wiss. Göttingen, Math.–Phys. Kl., 1900, 345–354.

Chapter 3, figure 17 based on C. Juel, Om endelig ligestore Polyedre (On finitely equal polyhedra), Nyt Tidss. for Math., **14**(1903), 53–63.

Chapter 4, figure 1 from Max Dehn and Poul Heegaard, Analysis situs, Enzyklopädie der Mathematischen Wissenschaften Ill, AB 3 (1907), Leipzig: Teubner, 153–220.

Chapter 4, figures 2, 3, 4, and 5 from Max Dehn, Über die Topologie des dreidimensionalen Raumes, Math. Ann. **69**(1910), 137–168.

Chapter 4, figure 6 from F. Klein, *Gesammelte Mathematische Abhandlungen*, volume 2, p. 111 (from 1922).

Chapter 5, figure 2 (by Karsten Ratzke). https://commons.wikimedia.org/w/index.php?curid=33764343

Chapter 6, figure 2 from J. Nielsen. Untersuchungen zur Topologie der geschlossenen zweiseitigen Flächen, Acta Math., **50**(1927), 189–358.

Chapter 6, figure 3 from R. Fricke and C.F. Klein, *Vorlesungen über die Theorie der automorphen Funktionen*, Vol. 1, Teubner, Leipiz 1890.

Figures drawn by the authors

Chapter 3, figures 2, 3, 6, 7, 8, 9, 10, 11, and 12.

Chapter 6, figures 1, 4, 6, 7, and 8.

All figures in Chapters 7 and 8.

CHAPTER 1

Max Dehn's Family: A Brief History

Brenda Danilowitz and Philip Ording

On April 15, 1925, Berta Raf Dehn celebrated her eightieth birthday. Her eight children with their spouses and her twenty-four grandchildren gathered at 40 Heilwigstrasse, Hamburg, the home of her eldest daughter, Elisabeth, where they took turns being photographed grouped around Berta on the front porch (see Figure 1). Their father and grandfather, the physician Dr. Maximilian Moses Dehn had died at the age of fifty-six almost three decades earlier in 1897, the year after Max, their fourth child and second son, graduated high school. Like his siblings, Max Dehn spent his school years in Hamburg, where he received his *Abitur* from the Wilhelm Gymnasium in September 1896.

FIGURE 1. Berta Raf Dehn's 80th birthday, April 15th, 1925. From left, front row: Berta Dehn, Berta Raf Dehn, Grete Schenkel Dehn; second row: Marie Dehn Mayer, Alice Sussman Dehn, Elisabeth Dehn Goldschmidt; third row: Toni Landau Dehn, Rudolph Dehn, Hedwig Dehn Wohlwill, Max Dehn, Ruth Omega Dehn, Eduard Goldschmidt, Georg Dehn, Heinrich Mayer, Karl Dehn, Heinrich Wohlwill.

©2024 by the authors

> gras und sende leise meinen blick na
> ch oben. auf flügeln   des gesanges
> herzliebchen trag'ich dich fort, fort,zu de
> n ufern des ganges, da weiss ich den
> schönsten ort. ihr naht euch wieder,sc
> hwankende gestalten, die früh sich eins
> t dem m trüben blick gezeigt,versuch ich
>   wohl euch diesmal festzuh

FIGURE 2. Toni Dehn. Lettering exercises from her classes at the Staatliches Kunstgewerbeschule Hamburg 1911-1912.

By the time his mother celebrated her eightieth birthday, Max had been married for twelve years. He and his young wife, Antonie (Toni) and their three children, Helmut born in 1914, Maria in 1915, and Eva in 1919, were living in Frankfurt where, in 1921, Max had been appointed Professor at Frankfurt University and chair of the Frankfurt Mathematics Seminar.

Toni Landau Dehn was born in 1893 and grew up in Berlin where her father Isidor Landau was a theater critic for the daily newspaper the *Berliner Borsen-Courier* from 1877 until 1912. Her mother Luise Lowenthal Landau was a literary translator from English and French into German. Toni and Max met in Hamburg when he was visiting from Kiel where he was an *Extraordinarius* from 1911 to 1913. Toni had enrolled at Hamburg's Staatliche Kunstgewerbeschule in October 1911. According to school records, she boarded with a Frau Ludwig Mauen at Mittelweg 117, and took classes in lettering (Figure 2), geometric drawing, anatomy, nude drawing, material studies, the history of style, and art history over the 1911/1912 winter semester and the summer semester of 1912, the year she and Max were married.[1]

Years later Dehn's daughter, Maria Dehn Peters, would recall family visits to Hamburg as "visits to heaven for us ... our dream-town" where "the special occasions were something else again!...

> For Grandmother Dehn's eightieth birthday everyone came, even Uncle Karl from Seattle with his beautiful American wife and little son. Father's eldest sister had sent bales of pink crêpe de chine to all the mothers to have dresses made for all the little granddaughters. For part of the day the youngest grandchildren had to come as the birds grandmother fed each day. We Frankfurt kids had to be "mere" house sparrows, but mother consoled us by making beautiful costumes. [Peters, 1972, p. 3]

Before Frankfurt, the young family had lived in Breslau, where Max held a professorship at the Technische Hochschule from 1913 to 1921, during which time

---

[1] Archive of the University of Fine Arts Hamburg. See Chapter 5 for more on their whirlwind courtship and marriage.

he served in the German Army [Magnus and Moufang, 1954, p. 225]. Throughout World War I, Toni and the two children "stayed with dear little Grossmutter Dehn" where "[w]e were happy among our Hamburg cousins and aunts and uncles. When father came on leave, he took us up the water tower in the Stadtpark: I remember from my tots-eye-view the hem of his long military coat flapping against the steps" [Peters, n.d., p. 9].

Returning to Breslau at the end of the war, the family lived in Grüneiche on the outskirts of town. Despite the hard times and food shortages, Max provided daily excitement and distractions for his young family. "Father took Helmut and me to kindergarten all the way across the woods. He took us by the hand, one on each side of him and ran through the woods with us; we were practically birds, practically flying! How lucky we were" [Peters, n.d., pp. 10-11]. Later in Frankfurt, besides weekend mountain hikes, there were visits on rainy weekends to the Städel Museum where "Father's favorites were the Dutch painters, but [Helmut] and I enjoyed going down to the 'moderns' admiring [Franz] Marc's horses, [Edvard] Munch's man with a duck, [and Pierre-Auguste] Renoir's breakfast" [Peters, n.d., p. 16].

Max Dehn, ever the historian, later noted that he could trace his family back 300 years [Dehn, 1940]. In fact, the Dehn family can be traced back to a line of Polish rabbis, the Pollak-Tiktins, who took their family name from their native country and their hometown Tiktin (Tykocin) northeast of Warsaw that was founded in the 11th century. Its synagogue, built in 1642 is one of the best preserved in Poland and its Jewish cemetery one of the oldest.[2] In the eighteenth century the family migrated to Copenhagen and the name Pollak-Tiktin most likely changed to Dehn at that time. Some family members settled in the Netherlands where the name became Deen or van Deen.[3]

At the start of the fifteenth century, Hamburg was a prominent member of the Hanseatic League — a center of commerce since the Middle Ages. The League, a loose trade alliance of "free cities" dominated trade along Europe's North Sea coastline from Bruges to Bremen, Hamburg, and Lubeck, and on to the Baltic coast from Danzig and the coast of Latvia to Estonia and as far north as Novgorod. The League stretched westwards into Denmark, east to southern Sweden and inland to Cologne and Halle. According to Ferguson, "Along with London, Liverpool, Rotterdam, Antwerp and New York, the city of Hamburg was one of the elites of great Atlantic entrepôts which flourished in the century between the Battle of Waterloo and the battle of the Marne — the British age of free trade, industrialization, imperialism, and the gold standard. Never before had goods, people, information, and capital moved so freely across oceans ..." Not only did Hamburg have close business ties to Britain at the height of its imperial prosperity, but as trade between Europe and America grew, Hamburg led in creating outposts "between Boston and Brazil; and as the industrialization of western and central Germany gathered pace, so Hamburg's free port became 'the German Reich's gateway to the world"

---

[2]The Jewish community was wiped out by the Nazis in August 1941 [Spector and Wigoder, 2001, p. 1353].

[3]According to [Hanks and Lenarčič, 2022] possible derivations of the family name Dehn, might be from a short form of the German personal name Degenhar(d)t; from the German or Danish name Daniel; the ethnic name of someone from Denmark; or from the Middle Low German and Danish Dene = Dane; or a topographic name from the Middle Low German Denne meaning a hollow or shallow valley.

[Ferguson, 2002, p. 32-33]. Three hundred years later, the Pollak-Tiktin-Dehn family's westward migration drew them towards this prosperous and flourishing part of the continent.

By the late eighteenth century, Max Dehn's great grandfather, Abraham Isaak Dehn, born in Copenhagen in 1764, had become a prosperous merchant. After the great fire of Copenhagen in 1795 the family moved west to Burgsteinfurt, an island of Lutheran Protestantism in the strongly Catholic Münsterland of North-Rhine-Westphalia not far from the Dutch border.[4] Burgsteinfurt boasted the oldest Westphalian university, founded in 1591. Exceptionally, from the late seventeenth century, the city was tolerant of Jews and their religious practices — a policy that encouraged the growth of its Jewish community. Abraham Isaak married twice, had seven children, studied to become a rabbi, and by the time of his death in 1821 he had become the chief rabbi of Groningen in the Netherlands.

Abraham Isaak's sixth child was Max Dehn's grandfather, Bernhard Adolf Dehn, born in Burgsteinfurt in 1808. In about 1838 Bernard Adolf married Hanne-Anna Melchior, member of a prominent and extensive Danish merchant and shipping family. Hanne-Anna was twelve years younger than Bernhard Adolf and probably in her late teens when they married, since her date of birth is given as 1821 and their first child, Arnold, was born in 1839 when she would have been eighteen years old. Two more children, Hanne Hannah (b. 1840) and Max Dehn's father, Maximilian Moses (b. 1841), followed in quick succession. Early in 1843 Hanne-Anna died, leaving Bernard Adolf with the three young children.[5]

Together with his brother-in-law, Sally Salomon Melchior, Bernhard Adolf Dehn went on to found the firm of B.A. Dehn and Melchior in Hamburg. The Melchior family had significant business and trade interests with the Danish West Indies, and it is likely that the firm B.A. Dehn and Melchior was similarly occupied. Bernhard Adolf soon married again, and his second wife, Merle Marianne Goldschmidt (born in 1825) produced five more children, half-siblings to Max Dehn's father.

Max Dehn's mother, Berta Raf Dehn, was also descended from rabbis. On her father's side her roots in Hamburg ran deep. Her great grandfather Menachem Mendel Frankfurter (1742–1823) had been the rabbi of Altona and founder of the Talmud Torah school in Hamburg. Berta's grandfather Raphael Aryeh Hirsch was a prominent Hamburg merchant as was her father, Harry Raphael Hirsch. His business took him to Paris and in 1854 to Valparaiso, Chile where Berta Raf Dehn spent part of her childhood [Peters, n.d., p. 9a]. Such a move was not uncommon among the Hamburg merchant class that carried on significant trade with, and in, South American countries.[6]

Intermarriage between European Jewish families was not unusual and persisted into the twentieth century. Among the extended families of Hamburg's upper bourgeoisie, Berta Raf Dehn was a member of the prominent Hamburg Goldschmidt banking family. Merle Marianne Goldschmidt was not only Bernard Adolf Dehn's second wife, she was also Berta Raf Dehn's aunt (Berta's mother's sister). Berta's mother's brother, Meyer Martin Goldschmidt, was the father of banker Eduard

---

[4] Burgsteinfurt was renamed Steinfurt in 1975.
[5] Genealogy, unless otherwise stated, is from https://www.geni.com.
[6] Harry Raphael's brother, Berta Dehn's Hamburg-born uncle, Samson Raphael Hirsch, on the other hand, became a leading rabbi, scholar and writer, and a vocal opponent of reform Judaism. His writings have been translated into English and remain in print.

FIGURE 3. The eight Dehn siblings – Berta, Max, Rudolph, Elisabeth, Hedwig, Karl Arnold, Marie and, in front, Georg – undated ca. 1892.

Martin Goldschmidt who married Max Dehn's sister Elisabeth (see Figure 3). Thus, the networks and interconnections between large families had long, wide, and tangled reaches. By the time Max and Toni Dehn arrived in the US in 1941, the Goldschmidts had established themselves in Los Angeles.[7]

Unlike most states in the patchwork that made up the German lands in the eighteenth and nineteenth centuries, Hamburg had no royal court or aristocratic patrons. It was a "self-governing state [with] a constitution, an eighteen-man senate elected for life ... and a Citizen's Assembly ... administered largely by elites" [Levine, 2013, pp. 61-62]. Among Hamburg's prosperous and ambitious grand bourgeoisie in the nineteenth century the family was "a large and highly ramified institution" [Blackbourn and Evans, 1991, p. 120]. While the wealthiest among the ruling elites were largely Lutheran protestant Hamburg families, similar family relationships existed among the upper middle class Jewish community, as relationships between the Dehn, Melchior, and Goldschmidt clans demonstrate. "Family ties" Evans writes "helped ease the emergence of big business, they softened the contours of competitive capitalism, [and] they drew together the disparate strands of the professional, mercantile, financial, and industrial grand bourgeoisie." Furthermore, "The social ties that bound the Hamburg grand bourgeoisie together were pulled tighter by the influence of neighborhood" [Blackbourn and Evans, 1991, pp. 129-130].

The Hamburg home of Elisabeth Dehn Goldschmidt and Eduard Goldschmidt at Heilwigstrasse 40 was located on the west side of the Alster Lake at the edge of the Harvesthude and Rotherbaum districts, among the elegant villas of the city's wealthy bourgeoisie. Max Dehn's older brother Rudolph and his wife Alice lived in a large home four kilometers to the north, right around the corner from their second

---

[7]Eduard and Elisabeth Goldschmidt's second daughter, Gertrude Goldschmidt, trained as an architect in Germany. She was unable to get entry to the US and emigrated to Caracas, Venezuela where she abbreviated her name to Gego and became a renowned artist whose reputation continues to grow.

FIGURE 4. Rudolph and Alice Dehn's home on Bebelallee, Hamburg.

FIGURE 5. Rudolph and Alice Dehn's home on Bebelallee, Hamburg (interior).

sister Hedwig and her husband Heinrich Wohlwill (see Figures 4 and 5). The home of their younger sister Marie and her husband Heinrich Mayer was little more than a kilometer from the Goldschmidts. A few streets further south, at Mittelweg 17, was the large Warburg mansion first occupied by Moritz and Charlotte Warburg in 1864 and later by their son Fritz and his wife Anna-Beate. Evans notes that by 1905 half of the city's practicing Jews and many more converts, lived in Harvesthude and Rotherbaum [Evans, 2005, pp. 392-3].

In 1904 Aby and Mary Warburg moved to a large house at 114 Heilwigstrasse.[8] When the house proved inadequate for art historian Aby's rapidly growing library, the Warburgs purchased the adjoining property at number 116, where the newly constructed Warburg Library opened in 1926. With a separate façade, it connected internally to the family home at 114. In 1933 the library was shut down and the contents and staff moved to London. (In 1993, 116 Heilwigstrasse was purchased by the City of Hamburg and reopened in 1995 as the Warburg Haus, an interdisciplinary forum for art history and cultural sciences administered by the University of Hamburg.[9]) In October 1938, as the National Socialists continued to tighten their grip on the Jewish population, and almost all the Hamburg Warburg family had left Germany, the brothers Max and Fritz Warburg set up the Warburg Secretariat in the Mittelweg 17 mansion. The Secretariat's purpose was to oversee the transfer of remaining assets to members of the Warburg family in England, Palestine, and the US. The office became a center of social and cultural activities for Jews in Hamburg, offering assistance in matters of emigration and relations with Nazi authorities [Solmitz, 1975]. The Warburgs appointed Robert Solmitz, husband of Herta Goldschmidt Solmitz, Edouard and Elisabeth Dehn Goldschmidt's eldest daughter, to supervise this enterprise.

In one of the bitter ironies that history throws up, a mere four kilometers further out from these luxury villas of Hamburg's wealthiest Jewish families is Fuhlsbüttel, now the Hamburg Airport. In September 1933 the Gestapo commandeered a prison on the Fuhlsbüttel site as a Nazi concentration camp that became the first point of incarceration for prisoners deported to larger camps further east.[10]

"Chamber music would be played, quadrilles danced, skits put on to tease, and love and laughter filled the rooms." That is how Maria Dehn Peters described the Hamburg family home.[11] The Dehn family musicians included not only a professional violinist in Max's younger sister Berta, but also Max's brother-in-law Heinrich Wohlwill, a talented amateur violinist, and Max himself, a polished cello player. "Father played the cello well enough to make me cry with it being so, so beautiful," Maria wrote, "[In Frankfurt] they had a string quartett [sic], sometimes trio, at various peoples' homes" [Peters, 1972, p. 1]. Music was a fundamental part of the social glue that held these families together. Correspondence in the Warburg family archives includes details of such social events. In a letter to his son and daughter-in-law Aby and Mary Warburg, Moritz Warburg enclosed the guest

---

[8]This account is based on [Levine, 2013, p. 77]. [Chernow, 2016, p. 353] dates the move to 1909.

[9]For the Warburg Haus see https://www.warburg-haus.de/, accessed January 17, 2023.

[10]In 1942 the Nazis deported Max's two older sisters, Marie Dehn Mayer and Hedwig Dehn Wohlwill and their spouses to concentration camps. Marie Dehn Mayer and her spouse Heinrich Mayer were sent to Theresienstadt on July 19, 1942 where Heinrich, aged seventy-six, died on December 2. On May 15, 1944, Marie Dehn Mayer was transferred to Auschwitz where she was killed. The fate of Hedwig Dehn Wohlwill and her husband Heinrich Wohlwill, deported to Theresienstadt on the same day as the Mayers, was little different. Heinrich Wohlwill, who suffered from vascular disease, died at Theresienstadt on January 31, 1943 aged sixty-nine. His death was communicated to their daughter from Theresienstadt in a postcard that Hedwig Dehn Wohlwill wrote from the camp. Hedwig survived the camps but did not recover her health, and died in Hamburg on July 3, 1948 aged seventy-one. https://www.stolpersteine-hamburg.de, accessed November 11, 2021 and Brandis 2020, pp. 46-47 and 77.

[11][Peters, 1972, p. 1]. Max and Toni Dehn lived in Breslau from 1913 to 1921 and then in Frankfurt from 1921 to 1935.

list for a dance party at Mittelweg 17 on January 31, 1900, noting that "only two people out of 60 have sent apologies due to the flu." Among those who accepted were Marie and Hedwig Dehn, Rudolph Dehn, Max's older brother and lawyer to the Warburgs, and Gretchen, Anna, Heinrich, and Paul Wohlwill.[12]

Max Dehn's father, Dr. Maximilian Moses Dehn (1841–1897), practiced as a physician at the Israelite Hospital (*Israelitisches Krankenhaus*) in Hamburg. Funded partly by the banker Salomon Heine (uncle and patron of the poet Heinrich Heine) the hospital opened in September 1843 in Hamburg's working-class dockland neighborhood of St. Pauli. It was open to all Hamburg residents irrespective of religious affiliation and was known to have served the area's local sailors and prostitutes [Grenville, 2012, p. 12]. There is scant documented information about Dr. Dehn's life, university training, or career. We can however get a sense of his personality at a distance, through later family reminiscences.

Maria Dehn Peters observed that her grandfather "seems to have been as close to his children as our father was to us." In 1972 she recounted an anecdote that provides an unforgettable and powerful glimpse into the relationship between Dr. Maximilian Dehn and his children:

> One day, my brother [Helmut] relates, grandfather Dehn was walking in the Stadpark with Rudolph and Max. Rudolph (b. 1874) was trying to put little Max (b 1878) in his place by asking him a difficult mathematical question. Max hung his head, crushed at not knowing the answer. So, quickly, grandpa said to Rudolph "If you are so clever, tell me what you know about Apapuriucasiquinichiquinaqua." It was Rudolph's turn to be embarrassed, admit ignorance, and ask for an answer. "Why," grandpa improvised, "that's the country where the cows give whipped cream instead of mere milk." [Peters, 1972]

Helmut would have heard this parable directly from his father, Max, and related it to *his* sister, Maria. Beyond that, however, it is evidence of Max's burgeoning commitment to mathematics as a young boy, and forecasts his similarly exacting, but affectionate, ways with his own children — and likely also with his students. This is borne out when we find Max decades later invoking the extraordinary word in his unpublished investigation into the processes by which the human psyche experiences the "mysterious results and wonderful fertility" of mathematics. "Counting," he observed, begins with the memorization of a long word, "one two three ... thirteen fourteen" that is as arbitrary as the alphabet or the "famous, completely meaningless Immermannian word Apapurindcasiquinitschiquisaqua" [Dehn, n.d., p. 20].[13] Dr. Dehn's parental admonition and the lightness and humor with which he challenged his oldest child, is illuminating both as a lesson to Rudolph and an assurance to the young Max. It provides a rare example of the physician's character and a model of the "closeness" to his children that Max inherited. Indeed we might speculate that Max inherited a great deal more from what appears to

---

[12]The Warburg Institute Archive online database, https://wi-calm.sas.ac.uk/CalmView/, Ref No WIA GC/27198, accessed 14 January 2023.

[13]The puzzling term "Immermanian" may be a reference to the writer, satirist and friend of Heinrich Heine, Karl Immermann (1796–1840). According to Maria Dehn Peters, Max Dehn's study in their Frankfurt apartment on Wöhlerstrasse contained "some 8,000 books from floor to ceiling" and both Max and Toni were avid readers [Peters, n.d., p. 14].

have been an intimate and fond relationship with his father. Emphasizing Dehn's humanity, Wilhelm Magnus, reviewing his doctoral advisor's life and work in 1978, Dehn's centennial, reflected on Dehn's "exceptional situation" as a mathematician whose work "was an essential part of his personality" that "influenced also his very well founded and deep interest in the humanities, art, and nature" [Magnus, 1978, p. 132].

To get some sense of Hamburg's cultural and scientific milieu in the last years of the nineteenth century, one can turn to the life of Maximillian Moses Dehn's near contemporary Emil Wohlwill (1835–1912) about whom a great deal is known. Their families became closely linked soon after Maximilian Moses' death in 1897 when his daughter Hedwig married Emil Wohlwill's son Heinrich. After studying chemistry and physics at Heidelberg, Göttingen, and Berlin universities, Emil Wohlwill returned to Hamburg in 1860. There he taught physics at the Polytechnische Fortbildungsanstalt and the Bauschule, published an important paper on crystallography, and was a practicing industrial chemist at the prominent Hamburg refinery, the Norddeutsche Affinerie, where, in 1874 he developed a process for the electrolysis of gold and other non-ferrous metals, immortalized as the Wohlwill Process.[14]

Although at the time it was barely recognized as a discipline, Emil Wohlwill is most remembered today as a pioneer in the history of science and, along with Antonio Favaro, the "most important Galileo specialist of the late nineteenth century" [Schütt, 2000, p. 642]. Indeed, while it was not his profession, the Galileo study became a lifelong obsession for Wohlwill. Wohlwill's biographer, Hans-Werner Schütt, has suggested that "[Wohlwill] sensed a spiritual relationship to the man who had to struggle [against the church] for the freedom of his spirit" [Schütt, 2000, p. 643]. According to Wohlwill's great grandson Matthias Brandis, "For Emil Wohlwill, Galileo's life exemplified the struggle between love of truth on the one hand and obscurantism on the other." In his history of the Dehn and Wohlwill families, Brandis writes that Wohlwill received substantial financial support from the Warburg family — from his uncle, Simon Ruben Warburg, who funded Emil's university studies, and from art historian Aby Warburg, who supported the posthumous publication of the second volume of the Galileo biography in 1926 [Brandis, 2020, p. 33].

The ideal of freedom was central to Emil Wohlwill's life, cultivated through the example of his father who had been an early pioneer of Jewish reform, his own adoption of the Enlightenment ideals of self-cultivation and *Bildung*, and his determination to be recognized culturally as a German while acknowledging his Jewish roots. His attempts, at first unsuccessful, to be accepted as a citizen of Hamburg after he resigned from the Jewish religion but refused to become a Lutheran declaring himself a "freethinker," led to him fighting, and eventually winning, a legal battle for recognition [Grenville, 2012, p. 14].

From the late seventeenth and into the mid-eighteenth century, the ideal of *Bildung*, as a process of self-formation developed by philosophers Georg Wilhelm Friedrich Hegel, Wilhelm von Humboldt, and their contemporaries, coincided with possibilities of emancipation for Jews. Humboldt, in particular, as a product of eighteenth-century German Enlightenment (*Aufklärung*), propounded his theory of *Bildung* as a means of creating political and social harmony in the modern state. His model was ancient Greece imagined as a harmonious society in which

---

[14]https://en.wikipedia.org/wiki/Wohlwill_process, accessed December 3, 2022.

the individual embodied citizenship, and truth was equated with beauty. *Bildung* (literally "education") aimed at encouraging individual attainment of knowledge through self-cultivation and self-formation in a context that stressed morality. This, Humboldt proposed, would lead to the identification of the individual with the state [Sorkin, 1983, p. 60]. Educational reform was a first step to enlightenment, and in the German states "[e]ducating the Jews became a substitute, as well as a preparation, for full emancipation" [Kober, 1954, pp. 7-8].

As described by historian George L. Mosse, *Bildung*, as an inner process and a continuous one, was an "ideal ready-made for Jewish assimilation because it transcended all differences of nationality and religion through the unfolding of the individual personality" [Mosse, 1985, p. 3].[15] Mosse suggests that, because Jews did not have strong roots in a class-based system, they readily accepted the ideal of *Bildung* as a new, secular faith. Furthermore "The Jews ... reached for *Bildung* in order to integrate themselves into German society" [Mosse, 1985, p. 4].

Dieter Langwiesche's writing on the nature of German liberalism pointed out that between 1789 and 1848, for German liberals who comprised the political elite, "Timely reform which would make revolutions unnecessary was the liberal therapy to immunize society against revolution. In place of revolutionary upheavals [the liberal political elites] proposed to achieve change through parliamentary resolutions" [Langwiesche, 1992, p. 98]. Emancipation, as reform that held out promise of full political participation for Jews, however, was not a clear-cut issue. Emancipatory laws and proclamations differed from one German state to another, and their tacit purpose was largely to secure the status quo among the political class. Reforms that comprised Jewish emancipation were thus frequently dependent on the politics of the day and were practiced selectively. Jews remained excluded from certain careers, both in politics and higher education, and with few exceptions, academics, writers, musicians, and artists were marginalized from the upper layers of German culture. As Mosse observed, "Jews had been excluded from academic life and from the civil service at the very beginning of their emancipation in spite of the promise of equal rights" [Mosse, 1985, p. 13]. From roughly 1782 until the 1870s, Jewish secular schooling was subject to a series of wide-ranging initiatives that were linked to an array of shifting emancipation orders in the German lands [Kober, 1954, pp. 3-23]. Thus, in the course of the 19th century, the idea of *Bildung* — always a difficult and overwhelmingly intellectual concept — faded into a new and growing nationalism and "the individualism which had provided an anchor for Jews reaching out to Germany was in danger of being cut loose" [Mosse, 1985].

Academic and professional career opportunities for Jews in the early nineteenth century were limited, and Jews were barred from holding political office or positions in the civil service. Medicine, however, was one professional career path that was open to Jews as a so-called "free" profession in the sense that physicians were not public servants. Jewish physicians had studied at German universities as early as the late seventeenth century, and had been the "chief representatives of secular learning in the Jewish communities" [Kober, 1954, pp. 7-8]. However, without professional organization or uniform requirements among different states, standards of training varied greatly. Charles E. McClelland (citing [Finkenrath, 1928, p. 3]) observed that "the concept of a German physician was unknown during much of the nineteenth century. Most officially recognized 'medical personnel' were in fact not

---

[15]For a succinct discussion of Humboldt and *Bildung*, see also [Konrad, 2012, p. 109].

licensed doctors (*approbierte Ärzte*) but 'surgeons' (*Wundärtzte*) of many different classes and with little or no academic training" [McClelland, 1991, p. 38]. In his analysis of Hamburg's cholera epidemic of 1872, Richard J. Evans examined the complicated social position and power of physicians in nineteenth century Hamburg and the attempts to professionalize medicine over the course of the century [Evans, 2005, pp. 211-212]. Professionalization was complicated by attitudes to *Bildung*. For physicians, *Bildung* entered a contentious debate between theory and practice in the medical profession, in which *Bildung* was viewed as the product of the university educated student's engagement with *Wissenschaft*, or scholarship, acquired through the active pursuit of knowledge and erudition. University educated physicians thus enjoyed an elevated social status. "To distinguish themselves from surgeons and mere careerists, physicians began adopting the argument ... that a proper university education conferred a unique form of character development, almost a form of spiritual transcendence. They referred to this cultivation of personal character with nearly mystical reverence as *Bildung*" [Broman, 1995, p. 864].

Without knowledge of Dr. Med. Maximilian Dehn's schooling or university education, Emil Wohlwill's schooling in Hamburg and his embrace of *Bildung* can serve as a pattern. His father, Immanuel Wohlwill, was born Joel Wolf in 1799. He became a legendary Jewish educator and an early and ardent champion of reform Judaism. In 1820-22 Immanuel Wohlwill studied at Berlin University where, two years earlier, Hegel had been appointed to the chair of philosophy. In Berlin Immanuel met likeminded Jewish students including the poet Heinrich Heine, Heine's close friend and confidant Moses Moser, and Leopold Zunz, all active in the *Verein für Cultur und Wissenchaft der Juden* which debated the nature of Jewish faith in a modern world, inspired by Hegel's lectures on the phenomenology of the spirit and his own formulation of *Bildung*.

On a personal level, Immanuel Wohlwill's response was to change his name from Joel Wolf to the more German sounding Immanuel Wohlwill.[16] He did not, however, follow Heine who converted to Lutheranism in 1825, but continued to focus on reforming the practice of Judaism. In 1825 Immanuel Wohlwill settled in Hamburg where he taught at the Israelite Free School, founded in 1815. The school followed the model of the earlier Jewish Free School of Berlin founded in 1778, as a privately funded secular secondary school or *Realschule*. German and French language as well as Hebrew were taught and both Jewish and Christian students were admitted. Those who could not afford the fees were given free tuition. In 1831 Immanuel Wohlwill married Friederike Reichel Warburg, and in 1838 the family moved from Hamburg to Seesen, where Immanuel became director of the progressive Jacobson School that embraced religious pluralism, where Jewish and Christian children were educated together, and where Emil received his early education. After Immanuel's death in 1847, the young family returned to Hamburg where Emil attended the elite Johanneum (*Gelehrtenschule des Johanneums*), Hamburg's oldest school, where the classics remain an essential part of education. It was founded in 1528 and named after its founder Johannes Bugenhagen, a Lutheran priest and close friend and ally of Martin Luther. Emil completed his high school studies at

---

[16]Wohlwill is from the German "wohlwollend" meaning sympathetic or benevolent. See Andrew Kleinert, Karl Sudhoff Memorial Lecture "Emil Wohlwill 1935-1912."

the *Akademisches Gymnasium* (forerunner of the University of Hamburg) and went on to study chemistry at the universities of Heidelberg, Göttingen, and Berlin.

Unlike Prussia and other German states that instituted, or at least had intentions of instituting, state sponsored elementary schooling in the eighteenth century, and where no less a public intellectual than Humboldt himself was charged with implementing the reforms, Hamburg came late to the idea of public education.[17] There was no political interest in state sponsored education and one of the results was that "[e]xcellent schools – the Johanneum and the Jewish Foundation School of 1815 – coexisted with one-room schoolhouses ... as individual communities established a few exemplary private schools and numerous substandard ones."[18]

In 1870 three years after Hamburg joined the North German Confederation of states and then became enfolded into a unified German Federation, the Hamburg Senate finally introduced large-scale reforms, establishing a public school system, and in 1872, a new teacher training seminar under a new administrative office, the *Oberschulbehörde* [Jenkins, 2003, p. 154]. The following year a new Elementary School Teachers' Association (*Verein Hamburger Volksschullehrer*) was launched, followed in 1877 by its own journal, *Pädagogische Reform*, through which Hamburg teachers campaigned for and introduced "new vitality into education by making instruction as creative, concrete, and child-centered as possible."[19] We do not know whether Max Dehn received his elementary education through home, private, or public schooling; in any case, it likely started in 1883 or 1884 when he was five or six years old and was undoubtedly shaped by the progressive winds of the period. The new school legislation also encompassed secondary school initiatives, and in 1881 Hamburg could finally boast an academic public school to rival the *Johanneum*. This *Neue Gerlehrtenschule* was renamed the *Kaiser Wilhelm Gymnasium* in 1883.[20] The educational program embraced the classical humanist tradition of the enlightenment and nurtured Max's deep and abiding interest in Greek and Latin language and philosophy.

Hamburg's senate and politicians had shown little enthusiasm for reform before 1870. Yet there was an active movement in the city for Jewish religious reform throughout the nineteenth and early twentieth centuries [Brämer, 2003]. That turn away from orthodox practices and observance was not, however, accompanied by large numbers of Jewish families actively repudiating their faith and converting to Protestantism. For Jewish scholars seeking teaching positions at German universities, however, conversion had long been a necessity.[21] As Annette Vogt has shown,

---

[17] Karl A. Schleunes (1979) provides a detailed overview of these developments: "In 1794 the principle of state control over schools, as well as compulsory attendance, was anchored in the new general Civil Code," (p. 320) and "[b]y 1837, eighty percent of Prussia's school age children (6-14) were believed to be receiving systematic elementary instruction" (p. 317).

[18] [Jenkins, 2003, p. 153]. Jenkins refers to the Israelite Free School as the "Jewish Foundation School."

[19] [Jenkins, 2003, p. 156]. *Pädagogische Reform* (1877-1921) became "one of the most respected publications in the field in Germany" and progressive education in Hamburg elementary schools continued into the 1920s and in some cases even later. See [Mayer, 2014].

[20] After Germany's defeat in World War I, and the end of the monarchy, the royal reference was removed, and the school became simply the Wilhelm Gymnasium.

[21] The case of Heinrich Heine, who reluctantly converted to Lutheranism in the hope that conversion would be his ticket to admission into German culture and a university career in law, proved this assumption could not always be relied on. Heine, unable to obtain an academic appointment, left Germany in 1831 and spent the rest of his life in Paris. See Chapter 2 for a discussion of how Dehn ran up against Anti-Semitism in Kiel.

for Jewish mathematicians even junior professorships were out of the question [Vogt, 2012, pp. 22-23].

Max Dehn, who "was not inclined to attach much significance to the events in his own life" as David Rowe observes in the following chapter, is not known to have commented on his reasons for converting to Lutheranism at some point after he received his *Abitur* from the Wilhelm Gymnasium in September 1896. Writing in response to Constance Reid, the biographer of Hilbert and Courant, who may have been contemplating a similar biography or an article on Dehn, his daughter Maria confirmed Max's conversion and offered this account of the family's relationship to religion: "My mother was Jewish (non-practicing). We three children were all baptized in the Lutheran Church. At school we were enrolled in the religious classes of protestant children. There was no emphasis on 'church' in our family or in the families of our and our parents' friends. But it never occurred to me to think of us as 'Jewish,' as in the pre-Hitler days 'Jewish' was considered a religion, just as '[C]atholic' and 'Protestant' were. Of course, all that changed in 1933, when 'Jewish' became a matter of race and according to Hitler, a crime deserving the death penalty" [Peters, 1972, pp. 9-10].[22]

In the same letter Maria wrote "[Max's] father had expected him to become a physician, like himself, but after his father's death (I am told) he made his final decision to study mathematics." Although the source for this momentous decision "I am told" cannot be guessed, we do know that Dr. Med. Maximilian Moses Dehn died in April 1897, by which time Max would have been well into his first semester of study in Freiburg. We might speculate that it was his father who encouraged him to enroll at Freiburg University, which had a venerable medical faculty founded at the same time as the university in 1457. Dehn spent his first university semester enrolled in the mathematics department at Freiburg, then, in 1897, moved to Göttingen, the renowned center of mathematics since the end of the eighteenth century.[23]

References

David Blackbourn and Richard J. Evans. *The German Bourgeoisie: Essays on the Social History of the German Middle Class from the Late Eighteenth to the Early Twentieth Century.* Routledge, London and New York, 1991.

Andreas Brämer. The dialectics of religious reform: The Hamburger Israelitische tempel in its local context 1817–1938. *The Leo Baeck Institute Yearbook*, 48 (1): 25–37, 2003.

Matthias Brandis. *Meines Grossvaters Geige: Das Schicksal der Hamburger judischen Familien Wohlwill und Dehn.* Hentrich & Hentrich, Berlin, 2020.

---

[22] According to Matthias Brandis, after Dr. Maximilian Moses Dehn's death, "The family no longer followed the Jewish religion, but lived in an enlightened [*aufgeklärten*] atmosphere without strong connections to a religion" [Brandis, 2020, p. 119]. For a detailed account of the period after 1933 see Chapter 9.

[23] We wish to thank Marjorie Senechal for inviting us to the Mathematisches Forschungsinstitut Oberwolfach Mini-Workshop: Max Dehn: his Life, Work, and Influence which prompted this collaboration. The second author received support for this research while a visitor to the Program in Interdisciplinary Studies at the Institute for Advanced Study in Princeton, NJ. We are grateful to the following members of the Dehn family: Antonio Alcala, Joanna Beresford Dehn, Dr. Matthias Brandis, Max Dehn Jr., Enrique Mayer, Renata Mayer Millones, Maria Mayer Scurrah, and Peter Spencer.

Thomas Broman. Rethinking professionalization: Theory, practice, and professional ideology in eighteenth-century german medicine. *The Journal of Modern History*, 67 (4): 835–872, 1995.

Ron Chernow. *The Warburgs: The Twentieth-Century Odyssey of a Remarkable Jewish Family*. Vintage, New York, 2016.

Max Dehn, *Unpublished notes, n.d.a. Personal notes*. Written by Max Dehn, circa 1940, and possibly transcribed by Toni Dehn. In Dehn family papers, collection of Joanna Dehn Beresford and Max Dehn, Jr.

Max Dehn. *Manuscripts 'animi sui complicatum'* [sic], typed ms and carbon copy. each with annotations. ca 60 pages total—on psychology of mathematics, undated, n.d. Max Dehn Papers, Archives of American Mathematics, The Dolph Briscoe Center for American History, The University of Texas at Austin.

Richard J Evans. *Death in Hamburg: Society and Politics in the Cholera Years 1830-1910*. Penguin, New York and London, 2005.

Niall Ferguson. *Paper and iron: Hamburg business and German politics in the era of inflation, 1897-1927*. Cambridge University Press, 2002.

Kurt Finkenrath. *Die Organization der deutschen Ärztschaft*. Eine Einführung in die Geschichte und den gegenwaertigen Aufbau, Berlin, 1928.

John Ashley Soames Grenville. *The Jews and Germans of Hamburg: The Destruction of a Civilization 1790-1945*. Routledge, London and New York, 2012.

Patrick Hanks and Simon Lenarčič, editors. *Dictionary of American Family Names*. Oxford University Press, 2nd edition, 2022.

Jennifer Jenkins. *Provincial Modernity: Local Culture & Liberal Politics in Fin-de-Siècle Hamburg*. Cornell University Press, Ithaca and London, 2003.

Adolf Kober. Emancipation's impact on the education and vocational training of German Jewry. *Jewish Social Studies*, 16 (1): 3–32, 1954.

Franz-Michael Konrad. Wilhelm von Humboldt's contribution to a theory of Bildung. In *Theories of Bildung and Growth: Connections and Controversies Between Continental Educational Thinking and American Pragmatism*, pages 106–124. Sense Publishers, Rotterdam, Boston and Taipei, 2012.

Dieter Langwiesche. The nature of German liberalism. In Gordon Martel, editor, *Modern Germany Reconsidered: 1870-1945*. Routledge, London, 1992.

Emily J. Levine. *Dreamland of humanists: Warburg, Cassirer, Panofsky, and the Hamburg school*. University of Chicago Press, 2013.

Wilhelm Magnus. Max Dehn. *The Mathematical Intelligencer*, 1 (3): 132–143, 1978.

Wilhelm Magnus and Ruth Moufang. Max Dehn zum Gedächtnis. *Mathematische Annalen*, 127 (1): 215–227, 1954.

Christine Mayer. The experimental and community schools in Hamburg (1919–1933): an introduction. *Paedagogica Historica: International Journal of the History of Education*, 50 (5): 561–570, 2014.

Charles E. McClelland. *The German experience of professionalization: Modern learned professions and their organizations from the early nineteenth century to the Hitler era*. Cambridge University Press, New York and London, 1991.

George L Mosse. *German Jews beyond Judaism*. Indiana University Press and Cincinnati: Hebrew Union College Press, Bloomington, 1985.

Maria Dehn Peters. Letter to Constance Reid, August 17, 1972. Max Dehn Papers, Archives of American Mathematics, Dolph Briscoe Center for American History, The University of Texas at Austin.

Maria Dehn Peters. Letter to Joanna Dehn Beresford, n.d. Partially dated November 1, 1997, and June 8, 1998, Courtesy Dehn Family Collection.

Hans-Werner Schütt. Emil Wohlwill, Galileo and his battle for the Copernican system. *Science in Context*, 13 (3-4): 641–643, 2000.

Robert Solmitz. Das sekretariat Warburg : Eine oase fuer die Juden in Hamburg, Oktober 1938 bis Juni 1941., 1975. catalog description `http://search2.cjh.org:1701/permalink/f/1o7aamh/CJH_ALEPH004617457` accessed 22 January 2023.

David Sorkin. Wilhelm von Humboldt: The theory and practice of self-formation (Bildung), 1791-1810. *Journal of the History of Ideas*, 44 (1): 55–73, 1983.

Shmuel Spector and Geoffrey Wigoder, editors. *The Encyclopedia of Jewish Life Before and During the Holocaust*, volume 3. New York University Press, New York, 2001.

Annette Vogt. From exclusion to acceptance, from acceptance to persecution. In Birgit Bergmann, Moritz Epple, and Ruti Unger, editors, *Transcending Tradition: Jewish Mathematicians in German-speaking Academic Culture*, pages 12–31. Springer Verlag, Berlin and Heidelberg, 2012.

CHAPTER 2

Max Dehn as Hilbert's First Star Pupil

David E. Rowe

Making His Mark in Göttingen

After graduating from Hamburg's Wilhelm-Gymnasium in 1896, Max Dehn began his studies in Freiburg, an idyllic city in the southwestern state of Baden. Why he chose such a distant provincial university, rather than the much nearer one in Kiel, remains a mystery. It seems unlikely that his decision had anything to do with its program in mathematics, since Freiburg had only two professors, Jacob Lüroth and Ludwig Stickelberger, neither of whom was particularly prominent. Their careers are nevertheless interesting from the standpoint of academic networking, which would also play an important role in Dehn's career.[1]

When Dehn arrived in Freiburg, Lüroth was in his mid-50s and Stickelberger about ten years younger. Jacob Lüroth was, like his compatriot of the same age Max Noether, an algebraic geometer. Lüroth and Noether both grew up in Mannheim, where they became good friends, sharing a strong interest in astronomy. Later, they studied in nearby Heidelberg, only 20 km from Mannheim. Eventually, they gravitated to the school of Alfred Clebsch, who taught briefly in Giessen, before assuming Riemann's chair in Göttingen in 1868. The Clebsch-network continued to play a major role in German mathematics long after the master's sudden death at age 39 in November 1872. Clebsch had been a student of Otto Hesse, whose final academic station was the technical college in Munich, where he taught up until his death in 1874. Hesse's professorship was then transformed into two chairs that were offered to Felix Klein and Alexander von Brill, both former protégés of Clebsch. In 1880, Klein took a new professorship in Leipzig, and Lüroth then gained the vacant chair in Munich. A few years later, another former Clebsch pupil, Ferdinand Lindemann, left Freiburg for Königsberg, which opened the way for Lüroth to succeed him.

As these examples illustrate, personal connections played a major role in this small German world of higher mathematics. Had Alfred Clebsch lived to a ripe old age in Göttingen, Max Dehn would have found a very different atmosphere there. As it happened, though, by the 1890s Klein had emerged as the central figure for mathematics in Göttingen, supported by David Hilbert, the rising new

---

©2024 by the author

[1] Stickelberger was a Swiss mathematician, who had studied in Heidelberg and then Berlin, where he took his doctorate under Karl Weierstrass in 1874. When he first came to Germany in 1867, he crossed paths with Lüroth, who had just begun his career as a private lecturer in Heidelberg, so undoubtedly they got to know one another during the next two years.

star who arrived in 1895 from Königsberg.² Lacking any documentary sources from this time in Dehn's life, one can only guess why he chose to leave Freiburg after only one semester and then continued his studies in Göttingen. Dehn's daughter Maria recalled hearing that her grandfather, Dr. Maximilian Dehn, had wanted Max to follow in his footsteps by studying medicine. If so, Max senior may well have had some connections in Freiburg and, if so, perhaps he urged his son to study there.³ Were that the case, then Max perhaps tested those waters and soon decided he wanted to change his course of study, presumably with his father's permission, assuming of course that this is what happened. All that can be said with certainty, however, is that Max decided to transfer to Göttingen shortly before his father died in April 1897. If he had gotten to know the two mathematics professors in Freiburg, he may well have spoken with Lüroth about his future ambitions. Assuming such a scenario, one can easily imagine that Dehn would have received a friendly tip, namely, that he should pack his bags and take up studies in Göttingen, where Klein had been teaching since 1886.

By the late 1890s, however, Klein's career as a creative mathematician was already behind him. Not even his lecture courses and seminars, which had long attracted students from both near and far to Göttingen, took on great importance next to his other activities. After acquiring Hilbert, a native of Königsberg who was Germany's most promising young talent, Klein became increasingly involved in organizational projects, in particular reform movements in technology and pedagogy [Tobies, 2021]. Students continued to flock to Göttingen, but the more gifted or ambitious among them were drawn to Hilbert rather than Klein. When Dehn arrived in Göttingen in the summer semester of 1897, Hilbert was 35 years old and on the cusp of new breakthroughs that would make him famous in the world of mathematics. Klein was offering an introductory course on differential equations. He and Hilbert were also teaching a joint seminar on function theory, and Hilbert was also giving a course on algebraic invariant theory.

This was a rare instance when Hilbert returned to the field that had dominated his research interests from the late 1880s right up until 1893. Whether Dehn attended remains unknown, but the contents of the course are known from an elaboration (*Ausarbeitung*) prepared by Sophus Marxsen, later published in English translation [Hilbert, 1993]. Marxsen went on to write his dissertation under Hilbert on a topic in invariant theory, graduating only a few months after Max Dehn. It was common practice in Göttingen for Klein and Hilbert to choose one or more students to prepare an "official" set of lecture notes, which other students could then consult in the *Lesezimmer*, the special library for mathematics that Klein installed when he arrived in 1886. To be so chosen was considered a high honor, as Max Born recalled in his autobiography [Born, 1975, 81] – he was given the task of writing up Hilbert's lectures from his 1904 course on the number concept and quadrature of the circle.⁴ Although, Dehn never wrote about his student days – he was not inclined to attach much significance to the events in his own life – we do know that he was tapped to prepare the official elaboration for Klein's lectures on differential equations.

---

²Hilbert and Hermann Minkowski studied in Königsberg under Lindemann, who took his doctorate under Klein shortly after Clebsch's death.

³Maria Peters Dehn to Constance Reid, August 17, 1972, Texas University Archives.

⁴Born also prepared the *Ausarbeitung* for Hilbert's lecture course from 1905 on "Logische Prinzipien des mathematischen Denkens."

A much later source offers a hint as to what probably happened during Dehn's very first semester in Göttingen. One of the older students at this time was Otto Blumenthal, who would go on to become managing editor of the journal *Mathematishe Annalen* and a major figure within the web of academic power that Klein and Hilbert had begun to spin [Rowe, 2018b]. Blumenthal came from a Jewish family in Frankurt and was a close friend of the astronomer Karl Schwarzschild, whom Klein would bring to Göttingen in 1901. Klein and Hilbert always had their eyes out for new talent, but many times they tasked their assistants and protégés with finding bright and ambitious young men (and later women) among the students. In Blumenthal's case, though he came to be known as Hilbert's first doctoral student, he was actually discovered by Arnold Sommerfeld, who was Klein's assistant during the mid-1890s. Like Hilbert, Sommerfeld was a native of Königsberg, but his years as a *Privatdozent* in Göttingen were decisive for launching his spectacularly successful career. His first professorial appointment came in 1897, when he left Göttingen to join the faculty at the Mining Academy in Clausthal, located in the nearby Harz Mountains region. Hilbert then inherited Blumenthal, whose interests had been strongly formed by Sommerfeld, namely applications of mathematics to engineering problems.

Thus, when Dehn arrived in Göttingen in the summer of 1897, Sommerfeld was on his way to nearby Clausthal as Klein's leading disciple. Nevertheless, Sommerfeld continued to teach courses in Göttingen during that summer and the following winter semester; the topics were ordinary and partial differential equations and the calculus of variations. It appears, however, that Dehn never studied with Sommerfeld. Blumenthal, who spent the summer semester of 1896 in Munich studying under Alfred Pringsheim, returned to Göttingen and began assuming his new role as a "senior" student, and he enjoyed the growing recognition. He enrolled in Klein's course, alongside Dehn, whom he must have befriended from the start.

In any event, it was surely on Blumenthal's advice that Klein asked Dehn to prepare the official *Ausarbeitung* for his lecture course. How difficult this proved to be no one can say, but Blumenthal later wrote about the problems he had trying to work out the connections between Klein's geometric approach, based heavily on Sophus Lie's work on differential equations that admit infinitesimal transformations, and the formal analysis needed to represent these ideas [Rowe, 2018b, 308]. Shortly before Max Dehn's sixtieth birthday, Otto Blumenthal briefly recalled how their friendship began. He wrote on 11 November 1938, "I cannot claim to have much instinct, but back then I had a good one when I picked you out as my lead fox (*Leitfuchs*)" [Rowe/Felsch, 2019, 494]. Presumably Klein had asked Blumenthal to scout for an appropriate student in the course, and the latter chose Dehn as his "lead fox" at that time, more than forty years earlier.

Within the culture of the German universities, the *Fuchs* already had a long tradition, though this term eventually came to be identified with a novice among the older members of a fraternity. In an older guidebook to student life in Göttingen, one finds the following explanation:

> Fuchs is the name of a student in his first half year. The name is not at all inappropriate, because the young person who arrives with highly exalted ideas about a university and with the good teachings and rules of life of his concerned parents feels worried among the students; believes he sees a celebrity in everyone who

> meets him; feels noticed by all people; consequently expresses anxiety in posture, gait, and expression – in fact has many similarities with a fox. After a few weeks these anxious creatures disappeared, and towards the end of the six months usually behave in the opposite way; they often want to fly before they have wings and so fall into a different sort of ridiculousness. Regardless of this, one recognizes them for foxes. [Wallis, 1813/1981, 102].

Blumenthal surely had nothing at all like this in mind, but he knew from personal experience that a new arrival could profit immensely from the friendly help of older students. Unless Dehn came from Freiburg with personal recommendations from Lüroth and/or Stickelberger (which was not entirely unlikely), then meeting Otto Blumenthal was a tremendous stroke of luck.

Very few sources survive that offer clues like Blumenthal's "lead fox", although a post-doc from those early days, Ernst Zermelo, also befriended Dehn. Zermelo later visited Dehn and his family in Hamburg in 1900, and he afterward wrote an interesting letter expressing his thanks for that experience. Putting these pieces together, it seems that Dehn came to the attention of Klein through Blumenthal, who was about to come under Hilbert's wing. We have no detailed knowledge of the courses Dehn took with Hilbert and others, so piecing together a picture of his mathematical education involves a good deal of guesswork. What can be asserted with assurance is that Dehn's course work in Göttingen focused on mathematics, physics, and chemistry, the three fields he chose for his doctoral examination. In mathematics, he took courses offered by Hilbert, Klein, and Arthur Schoenflies, associate professor of geometry. His instructors in physics were Woldemar Voigt, Eduard Riecke, and Eugen Meyer, from whom he studied theoretical, experimental, and technical physics, respectively. The chemistry courses he took were taught by Otto Wallach and Wilhelm Kerp.

Regarding the general atmosphere in Göttingen during the five semesters Dehn studied there, we are fortunate that Otto Blumenthal recorded his own vivid recollections of that time [Rowe, 2018b, 307–310], parts of which surely applied to Dehn's situation. Blumenthal noted that back then the number of younger students, especially those who planned to become teachers, was relatively small, a factor that benefited the older ones, many of whom held doctoral degrees or came from foreign countries. In Blumenthal's opinion, "the lectures in Göttingen were ideally suited for highly motivated young people at the time. Those students without higher aspirations were less likely to get their money's worth, and they were also not well respected." Not surprisingly, it helped to be bright and ambitious, as the competition was ferocious. For some, this had an unhappy ending, because "whoever wanted to be recognized needed to reach higher than they would by their natural inclination," and Blumenthal recalled "quite a number of people who let themselves get carried away and later, in practice, fell down" [Rowe, 2018b, 309]. Blumenthal's professors were nearly identical with those under whom Dehn studied, with the exception of Blumenthal's first mentor, Arnold Sommerfeld. Both of them, for example, took courses with Göttingen's senior physicist, Woldemar Voigt, an experience Blumenthal recalled with real regret:

> It was characteristic of our generation that we only thought about the exams at the last moment; initially, we only wanted

to learn as much mathematics as we possibly could. I believe that, compared with the lecture courses, the exercise sessions at that time were decisively inferior. This was especially the case for students who did not manage to grasp that practical experience was a necessary thing. This was particularly apparent in physics, which is the only point where I see an actual, and purely self-inflicted, shortcoming in my own training. We all took the physics tutorial taught by Voigt, but we treated this so marginally that nothing came of it. It was just seen as a counterpart to the exercises in descriptive geometry, even though Voigt took the matter very seriously and exerted considerable effort and patience with us. [Rowe, 2018b, 309]

Beginning students were offered problem solving sessions in the proseminars, which were generally taught by Schoenflies. Dehn probably took at least one proseminar, though this is uncertain. However, he definitely did some coursework with Schoenflies, who taught the lion's share of the courses in geometry. Thus, Dehn may well have gotten his first exposure to non-Euclidean geometry from a 2-hour course offered by Schoenflies during Dehn's very first semester in Göttingen. The semester following, he had the opportunity to take a course with him on space curves and curved surfaces, whereas on Saturday mornings Schoenflies conducted exercises in geometrical drawing. In the summer semester of 1898, Schoenflies offered a proseminar on set theory, a novelty at the German universities. During his many years in Halle, Georg Cantor never taught a lecture course on set theory. The first such course, taught by Dehn's friend Ernst Zermelo, was only offered in the winter semester of 1900/01. Zermelo was then a *Privatdozent* in Göttingen working closely with Hilbert.

During Max Dehn's student days, Klein and Hilbert co-taught the seminars. Over the course of Klein's long career, his seminars served as a training ground for many aspiring young mathematicians. As was customary at the German universities, the students rather than the instructors did the lecturing, and Klein kept protocol books of all the presentations in his seminars going back to his first semester as a professor in Erlangen in 1872 (when he was 23 years old!). During Dehn's five semesters, Klein and Hilbert often held a seminar together. In the summer semester of 1897 the subject was function theory, whereas for the two semesters following the seminars dealt with topics meant to complement the material covered in Klein's year-long 4-hour lecture course on mechanics. Since testing and grades played no role at the German universities, the success of this system depended on its stringent selectivity – only a very small percentage of young males graduated from a classical Gymnasium – as well as a deeply ingrained work ethic. Seminars were, in general, quite open-ended, but Klein typically presented a set of topics at the beginning of the semester and assigned these to the participants at their first meeting (or even beforehand). Blumenthal remembered these seminar topics as extremely difficult, in fact far too difficult for the participants, as he wrote:

> For practical exercises in mathematics there was, besides the proseminar, only the seminar by Klein and Hilbert. I learned a tremendous amount from the lectures I presented there. Never did I have to work so hard and with such intensity as when preparing these lectures. I have the feeling that the topics were

consistently too difficult. This was particularly evident when listening to the lectures of others. Only on rare occasions did I actually understand something. I don't think these seminars helped steer students' own work onto the right track. One had too much to do to learn about the topics, which were usually too difficult for someone to accomplish something on their own. [Rowe, 2018b, 309]

Blumenthal's mathematical training was heavily influenced by Sommerfeld, Klein, and to some extent by Schoenflies. After Sommerfeld's departure, he attended Hilbert's lecture course on number theory and joined in on Hilbert's memorable "number fields outings" (*Zahlkörperspaziergänge*). These took place soon after publication of his "Zahlbericht" [Hilbert, 1998]. Blumenthal's mathematical education was thus very stimulating and diverse, but hardly systematic, and he later realized some of the more glaring gaps in his knowledge of the fundamentals.

I am very astonished that during the entire course of my studies I never developed a sense for mathematical rigor. That this elementary knowledge came to me so late was certainly due to an overemphasis on receptive learning. In other cases (irrational numbers) the difficulties never became clear to me psychologically; it took me a very long time to recognize the necessity of Dedekind cuts. Whether this was a personal weakness, or perhaps due to the overly deductive presentation of the elementary material, I can no longer say. [Rowe, 2018b, 309–310]

In reflecting back on what drove him to mathematics, Blumenthal dropped a number of important hints, starting with the sheer abundance of courses offered and the extraordinarily varied nature of the subject matter. Through Sommerfeld, he also gained a smooth transition to physics and the excitement of learning about all the things people everywhere were working on. Still, the biggest single factor, he thought, was "due to the reading room and the friendly atmosphere that prevailed there. There people got infected by what each of them was working on."

After completing his doctorate in 1898, Blumenthal took the customary precaution of preparing for and passing the state examination (*Staatsexamen*) for teachers, which qualified him to teach mathematics, physics, and chemistry at the *Gymnasien*. Many of the teachers at these demanding secondary schools held doctorates, as higher education was esteemed very highly in those days. Moreover, the number of university positions was so few and the normal waiting time to attain one so long that few families wanted to risk betting their sons would eventually be called "Herr Professor." Even Hilbert, who had to wait seven years before gaining his first appointment, took no chances and passed the *Staatsexamen* immediately after taking his Ph.D. in Königsberg. The next step for Otto Blumenthal was the *Habilitation*, which normally involved writing a second thesis and then submitting this to a receptive university faculty. Few took this step without the prior support of a faculty sponsor, in this case Hilbert, who encouraged his first doctoral student to spend the academic year 1899/1900 in Paris. He did not need to prod him, though, as Blumenthal spoke French nearly perfectly and was exceedingly fond of the French capital city. By the time he returned to habilitate in Göttingen, his friend Max Dehn had already moved on.

Dehn surely shared many similar experiences as a student in Göttingen, and yet his career took off on a very different track than those taken by Sommerfeld and Blumenthal, who began as fellow colleagues at the Institute of Technology in Aachen. Dehn would also later teach at a *Technische Hochschule* (TH); from 1913 to 1921 he held a full professorship (*Ordinariat*) at the TH Breslau, though he hardly felt at home there. Teaching mathematics to engineers was rarely a labor of love for someone like Dehn, whose life revolved around pure mathematics. Like others from this time, though, he had very broad training and scientific competences. Alongside mathematics, physics, and chemistry, he had also mastered several foreign languages. Over the course of his career, Dehn taught a wide range of mathematics courses, and some of his doctoral students wrote dissertations on topics far removed from their mentor's favorite research fields. This breadth of knowledge has to be kept in mind when considering Max Dehn's mathematical works, which reflect his burning interests as a young mathematician. Hilbert's own courses covered an incredibly wide range of topics, which accorded with Klein's vision that students should be exposed to the full gamut of mathematical knowledge [Tobies, 2021]. Before announcing their courses, the mathematicians would meet to discuss their plans for the coming semester, but the only real purpose was to achieve some balance in the content and level of the offerings. In later years, when the number of students was far greater, more attention was given to standard course offerings, though even these could vary a great deal depending on the whims and tastes of the instructor.

During the five semesters Dehn spent in Göttingen, Hilbert taught invariant theory and complex function theory (SS 1897); number theory, confocal curves and surfaces, and a special course on the number concept and quadrature of the circle (WS 1897/98); differential equations, Fourier series, and number theory (SS 1898); mechanics, theory of determinants, elements of Euclidean geometry (WS 1898/99); differential calculus, group theory, and calculus of variations (SS 1898). Although few documents have survived that shed any light on what Dehn learned from Hilbert, he arrived at the very time when his mentor was riding the cusp of major new developments in the foundations of geometry, a research field he transformed in a dramatic fashion after 1899. The events of the next few years would soon catapult Dehn into a role he surely never imagined for himself. He became Hilbert's star pupil, though in a field far removed from invariant theory and algebraic numbers, the areas in which Hilbert had first staked his reputation. How this came about was largely a matter of unforeseen circumstances, luck, and of course, talent.

Hilbert's "Grundlagen der Geometrie"

Before this time, Hilbert had firmly established his reputation as the era's leading authority in both invariant theory and the theory of algebraic number fields, two formerly distinct disciplines that had now been brought together through his work. But then came a most unexpected turn of events. Many years later, in [Blumenthal, 1935, 402], Otto Blumenthal recalled the buzzing chatter among the students when they read Hilbert's announcement for a 2-hour course on "Grundlagen der Euklidischen Geometrie" [Hilbert, 2004, 185–406], which he offered during the winter semester of 1898-99. Blumenthal and the older students, those who had been accompanying Hilbert on weekly walks, had never heard him talk about geometry, only number fields. Little did they realize that Hilbert had been contemplating the foundations of geometry ever since his years as a Privatdozent in Königsberg, as evidenced by recent historical studies.[5]

Hilbert's lecture course that semester surprised them even more, for in it he sought to lay out the fundamental structures underlying Euclidean geometry as few before had ever imagined. In his classic *La Géométrie*, Descartes had found a simple way to arithmetize geometry in the plane by introducing a unit of length. One can then add, subtract, multiply, and divide line segments by appealing to elementary properties of proportional lengths in similar triangles. Hilbert referred to Cartesian geometry as a segment arithmetic based on the real numbers (whereas Descartes had only a vague notion of number systems). Unlike Descartes' approach to arithmetization, Hilbert based his theory on a set of axioms for abstract number systems. In this way, he was able to derive segment arithmetics based on two central theorems of projective geometry, the theorems of Pappus and Desargues. Dehn very likely took not only this course but also the parallel 4-hour lecture course on projective and descriptive geometry taught by Schoenflies. If so, he surely would have encountered these theorems in their more familiar form. During the twilight of his career, Max Dehn took pleasure in teaching them to his students at Black Mountain College (see Chapter 12).

The following spring, acting on a request from Klein, Hilbert revised this material and presented it in the original version of his famous "Grundlagen der Geometrie" [Hilbert, 1899/2015], published in a two-part *Festschrift* commemorating the unveiling of the Gauss-Weber monument. In Göttingen the partnership between the mathematician Carl Friedrich Gauss and the physicist Wilhelm Weber was already legendary in June 1891, when Weber died at age 86. In fact, the collaboration between these two famous scientists had only lasted for six years, as Weber lost his position in 1837. He had the misfortune (or belated honor) of belonging to the "Göttingen Seven" – a group of professors who had the temerity to protest the annulment that year of the liberal constitution in Hanover. Soon after his death, plans began for erecting a monument commemorating this celebrated duo, the Gauss-Weber Denkmal, which was unveiled on 17 June 1899.

Two of those who came to Göttingen to attend the ceremony on that day were Georg Cantor and Hermann Minkowski. Cantor was curious to learn more about the status of the "arithmetical axioms" (his quotation marks) in Hilbert's *Festschrift* [Meschkowski and Nilson, 1991, 399]. What he may have heard about these no one can say, but Hilbert definitely spoke about this very topic with Minkowski, who alluded to it in a thank-you letter, written one week after the festivities in Göttingen:

> Dear friend,
>
> Now that I've returned to the reality of Zurich, the wonderful days in Göttingen seem today like a dream to me, and yet one can as little doubt their existence as that of your $18 = 17 + 1$ axioms of arithmetic. I felt especially comfortable in your warm home, and I've been reporting here repeatedly with pleasure about the exciting time I spent there with you. ...

---

[5]See, in particular, [Toepell, 1986], [Hilbert, 2004], and the commentaries by Klaus Volkert in [Hilbert, 1899/2015].

> Anyone who experienced these days in Göttingen will hardly get over their astonishment as to the liveliness in the Göttingen mathematical circle, and at the moment this is entirely due to you. Spending time in such air gives a person higher ambitions and an impulse to more intensive creativity. [Minkowski, 1973, 116–117]

Hilbert's approach to Euclidean geometry was by no means an altogether original vision. Italian geometers had been promoting axiomatic methods well before him (he himself alluded to the work of Giuseppe Veronese), and the German geometers Hermann Wiener and Friedrich Schur sought to establish an approach to the foundations of geometry that did not depend on continuity principles [Wiener, 1891], [Wiener, 1893]. Max Dehn's first and most lasting contribution to this program was entirely in that same spirit, as were some of his later publications in topology. In their more systematic investigations, Hilbert and Dehn sought to exhaust the substantive results that could be derived from more elementary axioms before invoking continuity principles or properties of the full real number continuum. Richard Dedekind had famously derived the latter in *Stetigkeit und irrationale Zahlen* [Dedekind, 1872]. He did so by using so-called "Dedekind cuts" to complete the rational numbers by using a kind of transcendental construction for introducing irrational numbers on a rigorous basis. Hilbert sought to avoid such a construction, but the goal he announced in his public lecture "On the Number Concept" [Hilbert, 1900a] was essentially the same. By introducing his completeness axiom, he hoped to give a direct proof that the standard properties of the real numbers were free from contradictions.

One might perhaps wonder why Hilbert chose to avoid Dedekind's principle as a more intuitive way to establish continuity. Part of the answer seems to be that he strove for a simple *non-constructive* approach to this problem. In [Hilbert, 1900a] he referred to genetic methods for grounding arithmetical systems, which he set in contrast to the axiomatic method he found preferable. One should also mention Hilbert's longstanding interest in the Axiom of Archimedes, in particular in demonstrating its independence by way of non-Archimedean geometries.[6] It was precisely in this direction that Max Dehn made an important new contribution to clarifying the foundations of geometry. As Hilbert's tenth student,[7] Dehn belonged to a small group who wrote on a topic in geometry, though only Dehn's dissertation [Dehn, 1900a] had a lasting importance. In it he took a significant step beyond Hilbert by exploiting his model for a non-Archimedean geometry.[8]

---

[6] As Volkert points out, had Hilbert decided to invoke Dedekind's principle as an axiom, he could have simply derived the Axiom of Archimedes from it [Hilbert, 1899/2015, 179–180]. In the seventh edition of *Foundations of Geometry* (*Grundlagen der Geometrie*, 1930), Paul Bernays introduced a weaker form of Axiom V.2, namely linear completeness. Based on this, one can then prove the old completeness axiom as a theorem. In Chapter 2, Bernays then proves the relative consistency of Cartesian geometry by means of the model for $\mathbb{R}$ based on Dedekind cuts [Hilbert, 1971, 31–32].

[7] According to MacTutor, Hilbert had 71 doctoral students over the course of his career in Göttingen, a number that easily dwarfs that of any other mathematician of the period, even Klein, who mentored some fifty.

[8] Dehn's axiom system was identical to the one in the original *Festschrift* [Hilbert, 1899/2015], where the congruence axioms appear in group 4 and the parallel postulate forms group 3; continuity (group 5) consists of the Archimedean axiom alone. Beginning with the second edition, [Hilbert, 1903], groups 3 and 4 appear inverted and the completeness axiom is added to group 5.

For Dehn's dissertation topic, Hilbert posed the question whether one could draw any conclusions regarding the sum of the angles in a triangle without invoking either the parallel postulate or the axiom of Archimedes. Dehn was able to derive two important new cases that differ sharply from the three classical ones: in elliptic, hyperbolic, and Euclidean geometries, the angle sum in triangles is, respectively, greater than, smaller than, or equal to two right angles. In what Dehn called a non-Legendrean geometry, there are infinitely many parallel lines through a given point to a given line, but unlike hyperbolic geometry, the sum of the angles in a triangle *exceeds* two right angles. Likewise, he uncovered the case of a semi-Euclidean geometry, where the angle sum is the same as in a Euclidean geometry. Thus, within the scope of the axioms in the original groups 1, 2, and 4 of Hilbert's *Festschrift*, Dehn was able to unveil two new types of plane geometries (for details, see Chapter 3).

Hilbert was elated by these new results. Three days before Dehn's oral exam, he wrote to his former mentor Adolf Hurwitz about Dehn's work: "Since you have some interest also in the foundations of geometry, I'd like to inform you of a dissertation by one of my best students Herr Dehn, whose results have delighted me" [Epple, 1999a, 230]. In 1902 a French translation of his *Grundlagen der Geometrie* came out, to which Hilbert appended a fairly lengthy account of Dehn's dissertation results.[9] When he later found time to modify his *Festschrift* for the second edition, Hilbert again used the opportunity to underscore the importance of Dehn's results [Hilbert, 1903, 23–24].

Max Dehn had not yet reached his twenty-first birthday when on November 8, 1899 he passed his doctoral examination in Göttingen. On the long list of those who wrote their dissertations under Hilbert, two others had already written on topics in foundations of geometry. These were Michael Feldblum from Warsaw and Anne Lucy Bosworth, a native of Rhode Island who held a master's degree from the University of Chicago; both passed their oral exams in July 1899. Doctoral candidates in mathematics typically chose their secondary fields from among astronomy, physics, or chemistry. In Dehn's case, he was examined by the theoretical physicist Woldemar Voigt and the chemist Otto Wallach.[10] Each probed his knowledge for 30 minutes, whereas Hilbert spent a full hour asking questions on a wide range of topics in geometry and foundations of mathematics. The protocol lists these areas: elements of descriptive geometry, algebraic surfaces and curves, line- and sphere-geometry, differential geometry: surface curvature, geodesic lines, then axioms of arithmetic, and finally elements of set theory. Hilbert's verdict read: "The knowledge of the candidate proved to be reliable and very versatile and is satisfactory in every respect." This may not sound like lofty praise, but in those days German professors rarely deviated from sober language. In fact, Hilbert's verdict was clear, as Dehn took his degree *summa cum laude*.

Hilbert's Paris Lecture

In late 1899 Hilbert received an invitation from the organizing committee to deliver a plenary address[11] at the forthcoming international congress in Paris. Soon

---

[9]Universitätsarchiv Göttingen (UAG) Phil.I.185.b.

[10]The protocol from Dehn's oral exam can be found in UAG Phil.I.185.b.

[11]In the final program, Hilbert was not among the plenary lecturers due to organizational confusion.

thereafter, he wrote to Minkowski asking for his advice about an appropriate theme for his talk. "Most alluring," his friend thought, "would be the attempt to look into the future, in other words, a characterization of the problems to which the mathematicians should turn in the future. With this, you might conceivably have people talking about your speech even decades from now. Of course, prophecy is indeed a difficult thing" [Minkowski, 1973, 119–120]. Minkowski's letters to Hilbert sparkle with witty remarks and telling observations, but at no time did he strike a more resounding chord than here.

Hilbert's first two problems were to become particularly famous, far more than the third, which had captured Max Dehn's attention even before the Paris ICM opened. The first concerned a conjecture raised by Georg Cantor, who claimed that the cardinality of the real number continuum was precisely the same as that of the smallest uncountable well-ordered set. Hilbert raised this as an open question, but his remarks made clear that he believed Cantor's continuum hypothesis was both true and also provable. His second Paris problem stemmed from his own quite different approach to the continuum, which he sought to characterize axiomatically. One year earlier, when he unveiled his new axiom system for the real numbers in [Hilbert, 1900a], he claimed one could prove its consistency directly. He now called for someone to provide that proof. Hilbert's third problem called for a proof that the cut and paste methods used to prove the equality of plane rectilinear figures do not suffice for handling polyhedra in space (see Chapter 3 for details).

Considering that Hilbert's speech would later become famous as one of the most influential ever delivered, it is fascinating to read an eyewitness account written around the time of the event itself. Charlotte Angas Scott from Bryn Mawr College was one of seven women who attended the Paris Congress. Still, she had an important role to play, namely to report on what transpired for the American Mathematical Society. Her detailed and yet colorful and opinionated account in [Scott, 1900] focused on more important talks and neglected those of lesser interest.[12] Charlotte Scott had attended the first ICM in Zurich, which she found far more successful than the second from an organizational standpoint. Under normal circumstances, Paris would have provided an excellent venue, but in 1900 this meant hosting a scientific meeting in the middle of a world's fair. The *Exposition Universelle*, held between April and November, attracted some 50 million visitors; little wonder that the mathematicians had difficulty finding one another. Scott had plenty of advice to impart when it came to making improvements for the future, and she was hopeful that the next ICM, tentatively planned for 1904 in Baden-Baden, would be more successful than the last. Indeed, she emphasized how important such gatherings could be just for the sake of someone's mental health during an era when mathematicians otherwise had so few opportunities to meet. In that connection, her general remarks about what such an event should aim to achieve are well worth quoting:

> The mathematician who is in any degree a specialist is in general rather solitary in the average college – he would have been better

---

[12]Scott had studied at Girton College, Cambridge from 1876 to 1880, and afterward taught there for four years while studying under Arthur Cayley. Since Cambridge did not grant doctoral degrees to women, Scott took hers from the University of London in 1885. That same year she came to the US, where she became one of the eight founding members of Bryn Mawr's faculty. By 1900 she had established herself as a prominent figure within the American mathematical community.

off in Noah's Ark, for at the worst there would have been two of a kind. For his mind's health it is well that he should occasionally be thrown with those of kindred interests; it is well too that he should be made to feel the unity of mathematics. [Scott, 1900, 75–76]

Scott also noted that unlike Zurich, where about half the talks were delivered in either French or German (English and Italian were also permitted, but seldom spoken), in Paris nearly all the lectures were delivered in French; Hilbert was one of the few who spoke in German.

When she came to that lecture, Scott wrote more about Hilbert's general remarks than the individual problems he presented. The most striking among these was also the most significant, namely, his claim that in mathematics there is no *ignorabimus*. In Hilbert's opinion, every well-posed problem in mathematics was capable of being solved, whether in the positive or negative sense. This optimistic assertion marks the high point in his address, and one might well say he took this with him to his grave. Indeed, his gravestone carries the famous motto: "Wir müssen wissen, wir werden wissen" (We must know, we will know). In Paris, he recounted how mathematicians had recently resolved two of the fundamental problems bequeathed upon them by the ancient Greeks: the status of the parallel postulate in Euclidean geometry and the possibility of squaring the circle. These nicely illustrated what he considered to be characteristic of mathematical problems in general: each and every question was capable of being answered with finality. For Hilbert, this belief offered not only a strong psychological support; it even had the quality of a moral imperative: "the conviction in the solvability of every mathematical problem is a powerful incentive to the worker. We hear within us the perpetual call: there is the problem, seek its solution. You can find it by pure reason, for in mathematics there is no ignorabimus" [Hilbert, 1900, 298].

Judging from Scott's report as a whole – and she was an astute observer – we surely should not imagine that the audience to whom Hilbert spoke was on the edge of their seats. Nothing she wrote could be read as confirming Minkowski's prediction that Hilbert's address would turn out to be the event of the congress, though his text with its famous list of 23 problems would gain many avid readers in the years ahead. In 1902 the text of "Mathematische Probleme" became available in a French translation produced by Léonce Laugel.

Max Dehn did not attend the Paris ICM, but he may well have read Hilbert's text before it came out in print. In any event, his correspondence with Hilbert makes clear that they had discussed the third Paris problem, which Dehn set about to solve before the ICM convened. Little is known about the events in Dehn's life for many months after he took his doctorate in November 1899, but in all likelihood he lived at home in Hamburg and spent much of his time there working on Hilbert's third problem while preparing to complete the *Staatsexamen*. In the fall of 1900 he accepted an offer tendered by the geometer Friedrich Schur, who taught at the Karlruhe Institute of Technology, a *Technische Hochschule* (TH) in Baden. Dehn became his new assistant during the next academic year 1900/01.

One can gain a glimpse of Dehn's life after graduation from a letter written by his friend Ernst Zermelo, whose name is today synonymous with axiomatic set theory (the Zermelo-Fraenkel system, often abbreviated ZF). During the semester break from mid-March to mid-April 1900, Zermelo visited Dehn and his family in

Hamburg and he later wrote to thank Max for the delightful days he spent with them.[13] From his letter, dated 25 September 1900, we learn that Dehn spent some time traveling in Norway that summer; Zermelo was curious to know how far northward he had gone on his trip. Ernst Zermelo was seven years older than Max Dehn. A native Berliner, he took his doctorate under Hermann Amandus Schwarz with a dissertation that extended Weierstrass's approach to the calculus of variations.[14] After visiting Dehn in Hamburg, Zermelo began his teaching career in the summer semester with a 4-hour course on elliptic functions.[15] In his letter, he wrote very openly to Dehn (with whom he was on the familiar "Du" basis) about the loneliness he felt in Göttingen and how much he missed his own relatives. Those feelings came to him after experiencing the warm atmosphere in the Dehns' home in Hamburg, a city he got to know for the first time.

Zermelo remarked that the Hilberts were quite delighted with their sojourn in Paris, no doubt in large part because they stayed in the same hotel with the Minkowskis. This was the conveniently located Hôtel St. Pétersbourg on the rue Caumartin, where Minkowski's oldest brother Max often stayed. As a scientific event, though, Hilbert found the Paris ICM to be very disappointing. Zermelo also reported on various other encounters with mutual Göttingen friends, including Otto Blumenthal. Several were returning from their summer vacations, preparing for the new semester, which would commence in mid-October. Zermelo complained about the time he needed to prepare for his lectures, though without mentioning that he had decided to teach a 2-hour course on set theory. Since he had gravitated into Hilbert's circle, one can easily imagine how Zermelo now felt inspired to tackle the first Paris problem, Cantor's continuum hypothesis. In the meantime, Dehn was hard at work on Hilbert's third problem.

Dehn Cracks Hilbert's Third Paris Problem

One month after the Paris ICM ended, the DMV held its annual meeting in Aachen with Hilbert as chair. He reported on German participation in Paris and noted that the DMV should begin work on preparations for the next ICM, which would take place somewhere in Germany in 1904. This Aachen conference was otherwise fairly uneventful with only sixteen lectures on the program. Max Dehn attended, but he was not on the program of speakers. Having missed the chance to speak with Hilbert in Aachen, Dehn wrote to him from Hamburg on 24 September to ask whether Hilbert had any news about potential positions for the coming academic year. In all likelihood, Hilbert had promised to discuss this with Sommerfeld in Aachen, as Dehn referred to the latter in his letter. The matter was now urgent, as he had only one week to decide whether to accept an offer from the TH Karlsruhe. This was a post-doctoral position offered by the professor of geometry Friedrich Schur, which meant teaching the tutorial that supplemented the lecture course in descriptive geometry. Schur, who had been a protégé of Felix Klein in Leipzig

---

[13]Dehn Papers, Briscoe Center for American History, University of Texas at Austin.

[14]In 1894 Zermelo became Max Planck's assistant and began pursuing a career in mathematical physics. So he was heavily immersed in physics when he came to Göttingen in 1897, thus around the time Dehn took up studies there. Two years later he submitted his habilitation thesis on vortex motions on the surface of a sphere.

[15]Zermelo also expressed disappointment that his work was going so slowly, in particular his efforts to finalize his paper on vortex motion, which was eventually published in [Zermelo, 1902], [Zermelo, 2010/2013, II: 300-463].

during the 1880s, was very impressed with Dehn's dissertation (he mentioned this in a letter to his friend Friedrich Engel).[16]

Apart from this more immediate concern, Dehn's letter to Hilbert also contained a very noteworthy piece of information. In it, he briefly described his partial solution to the third Paris problem, which he almost surely knew about from the time of Hilbert's lecture course. Moreover, Dehn wrote about this as if it were a progress report, referring to "my tetrahedron work." This letter would thus appear to confirm that he and Hilbert had already spoken about this, either before or during the time Hilbert was composing the final text for his Paris lecture. Hilbert's third problem aimed to show that, unlike areas of plane rectilinear figures, the theory of volume for polyhedra requires continuity assumptions, such as the so-called principle of exhaustion employed in Euclid's *Elements* (for details, see Chapter 3). In his formulation of this challenge, Hilbert referred directly to an impossibility proof based on properties of tetrahedra. The concluding passage notes that the problem would be solved if "we succeeded in specifying two tetrahedra of equal bases and equal altitudes which can in no way be split up into congruent tetrahedra, and which cannot be combined with congruent tetrahedra to form two polyhedra which themselves could be split up into congruent tetrahedra" [Hilbert, 1900, 302].

Dehn's initial result concerned decomposition alone, and in his letter to Hilbert he mentioned two examples of tetrahedra that cannot be decomposed into a finite number of congruent pieces and then used to construct a prism of equal volume. This, he stated, was the case both for a regular tetrahedron as well as one with a dihedral angle formed by three perpendicular planes. One month later, on 27 October, Hilbert submitted this paper for publication in the *Nachrichten* of the Göttingen Scientific Society [Dehn, 1900b]. No sooner had Dehn returned the corrected proofs in early November than he realized he could modify his argument to prove the general result allowing for the addition of congruent pieces as well (he called this *egalité par soustraction*, though why he preferred this French term seems unclear). Writing from Karlsruhe on 12 November 1900, he felt somewhat annoyed with himself that he had already sent off the proofs: "I'm sorry that I didn't finish this matter two weeks earlier." He then asked Hilbert whether he might write this up as a supplement to his first paper or simply include it in the projected paper he planned for *Mathematische Annalen* [Dehn, 1900b]. Hilbert evidently preferred the second alternative, but a note appears at the end of [Dehn, 1900b, 354] indicating that the impossibility of *egalité par soustraction* would appear in the forthcoming paper in the *Annalen*.

Dehn was probably hoping to habilitate at a German university, rather than starting his career as a geometer at a *Technische Hochschule*, even one as esteemed as the Karlsruhe TH. Although one of the most prestigious in Germany, the Karlsruhe Technical College followed the general developmental pattern of many others. Founded in 1825 by the Grand Duke of Baden as a polytechnical school modeled in some ways on the École Polytechnqiue in Paris, it only gained the full structure and status of a *Technische Hochschule* in 1885. Although the institution had already since 1865 gained the right to habilitate new faculty members, this was seldom exercised in mathematics before 1900. In 1899, Kaiser Wilhelm II, acting as King of Prussia, decreed that henceforth all *Technische Hochschulen* in Prussia would have the right to confer doctoral degrees, a major step in elevating their status to

---

[16]Friedrich Schur to Friedrich Engel, 27 October 1900, Engel Nachlass, Giessen.

a level comparable with the universities. Baden then followed suit the very same year.

Geometry had a particularly strong tradition in Karlsruhe, owing to the long tenure of Christian Wiener, who was professor of geometry from 1852 to 1896, succeeded thereafter by Friedrich Schur. Many positions in geometry were established at the polytechnical institutes, where descriptive geometry was taught to large classes of engineers and architects. Schur's previous assistant, Martin Distelli, was particularly adept in this field, having studied under Wilhelm Fiedler at the ETH in Zurich. Fiedler's drawing courses were famous, and Distelli became known as a leading expert on kinematic geometry, including gearing mechanisms. This hands-on approach to geometry had great appeal for Felix Klein, whereas Hilbert was far more interested in abstract studies based on axiom systems. Both aspects had a strong appeal for Max Dehn, however. In any event, no suitable alternative position was available, and so Dehn accepted the offer from Karlsruhe, where he spent just one year before transferring to Münster. Since he was looking for the opportunity to habilitate, an unlikely possibility in Karlsruhe, this first appointment led to a dead end.

Dehn's First Years in Münster

Dehn's decision to transfer to the Philosophical and Theological Academy in Münster was without doubt due to the support of its senior mathematician Wilhelm Killing. Although Killing, like Dehn, was a leading expert on foundations of geometry, today he is remembered above all for his pioneering work on Lie algebras [Hawkins, 2000]. Killing studied in Berlin under Ernst Eduard Kummer, Karl Weierstrass, and Hermann von Helmholtz. He was a deeply religious Catholic and, like Weierstrass (who recommended him for the post), he taught for a good decade at the seminary Collegium Hosianum in Braunsberg (today Braniewo).[17]

Killing's decade in Braunsberg was followed in 1892 by his appointment as professor of Mathematics in Münster, a city famous for a singular event in European history. It was in Münster's Old City Hall (*Rathaus*) in 1648 that delegations met to sign the Peace of Westphalia ending the Thirty Years War.[18] The traditionally Catholic city of Münster was also the site of the last university founded within the Holy Roman Empire, which was already in its death throes when Napoleon dissolved it in 1806. A decade later, after the Congress of Vienna redrew the map of Europe, the Catholics living in the Rhineland became citizens of Prussia, whose rulers were Lutherans. The Prussian government founded a new university in Bonn in 1818, while transforming the university in Münster to a Philosophical-Theological Academy with just ten professors. This institution gradually grew, however, and during Dehn's years there it regained its former status as the Westfälische Wilhelms-Universität, its present-day name.

---

[17]During this time, Killing and his wife entered the Third Order of Franciscans. The Collegium Hosianum was originally founded in 1565 as a Jesuit college by Polish Cardinal Stanislaus Hosius and soon thereafter became one of the most important centers of the Counter-Reformation in Europe. After the partition of Poland in 1772, Braunsberg became a part of the Protestant Kingdom of Prussia, which suppressed the Society of Jesus. The Prussian government closed the Collegium and converted it into a Gymnasium Academicum, later called Lyceum Hosianum.

[18]The adversaries once again recognized a form of religious tolerance enshrined in the earlier Peace of Augsburg of 1555, according to which each prince had the right to determine the religion of his own state, whether it be Catholicism, Lutheranism, or now some form of Calvinism.

Max Dehn was the first mathematician to habilitate in Münster. During his first years there, the Prussian ministry elevated the academy to a university. The normal procedure when applying for habilitation was for candidates to submit a post-doctoral thesis to the faculty along with other documents attesting to their scientific attainments and potential. One or more faculty members would then report on the candidate's qualifications, and if these were judged positively the formal procedure could commence. This usually required submitting three topics for the colloquium, which began with a lecture on the topic chosen by the faculty, followed by discussion, during which candidates might be questioned on quite wide-ranging mathematical matters. In Dehn's case, he submitted his forthcoming paper on the third Paris problem [Dehn, 1902] as his post-doctoral thesis, which was evaluated along with his earlier work by Wilhelm Killing.

As a true expert on foundations of geometry, Killing clearly appreciated the significance of Dehn's accomplishments like few others among his contemporaries. In his report, he briefly alluded to the classical theories of Euclid for treating the sizes of plane rectilinear figures and polyhedra, highlighting how the latter required the so-called method of exhaustion based on continuity properties. Killing further noted that a number of geometers had discussed the need to prove that the theory of volume for stereometric figures required infinitesimal methods, though no significant progress had been made on the problem, despite recent efforts. This state of affairs had then led Hilbert to call attention to this outstanding problem in his noteworthy speech at the ICM in Paris, about which Killing commented:

> Thus the mathematical world was very much astonished when only a few months later a young mathematician, Dr. Max Dehn, who had recently graduated summa cum laude in Göttingen with an outstanding dissertation, solved this problem. His fundamental ideas for proving this were communicated to the Göttingen [Scientific] Society by Hilbert himself, along with a number of examples of polyhedra having equal volume but which cannot be decomposed into congruent pieces.
>
> This work [Dehn, 1900b] also shows, however, that the proof is by no means simple and that the author did not come upon it merely by a lucky accident. On the contrary, one must recognize that it took true geometrical talent, along with goal-oriented discipline and energetic determination to put this problem to rest.

Killing then went on to discuss how Dehn's post-doctoral thesis (*Habilitationsschrift*) [Dehn, 1902], which was about to be published, went beyond [Dehn, 1900b], in particular by demonstrating that one cannot obtain an elementary theory of volume even by extending two given polyhedra of equal volume, i.e., by attaching equi-decomposable polyhedra to them. This rules out a theory of volume analogous with the theory of area developed by Hilbert in [Hilbert, 1899/2015]. The concluding passage in Killing's report then gave this assessment of the new work:

> One will rarely be in the position to commend the work of a young mathematician to the same degree as the present treatise

of Dr. Dehn, not only by virtue of the importance of the topic
but also the excellence of the solution.[19]

A similar enthusiasm for the candidate can be found in Killing's report on the colloquium, which took place on 31 October 1901. The topic for Dehn's lecture was set theory, a field of research that only a handful of specialists would have felt competent to speak about. As mentioned above, Ernst Zermelo had only begun to study Cantor's theory of sets around this time. The main purpose of such a colloquium lecture was to gauge whether the candidate could present an understandable account of a research field somewhat outside their own special interests. In this case, the chosen topic was even far beyond the ken of most mathematicians, whereas Dehn had to address a general academic audience. Nevertheless, Killing came away most impressed by the way Dr. Dehn

> knew how to pick out the most important results from the multi-layered material on set theory as found in numerous publications that have been strewn about and which lack all systematic order, ordering these by way of a transparent unifying principle, and presented in a clear and elegant spoken form. Accordingly, the lecture demonstrated complete command of the subject matter in this restricted, but important branch of mathematics, thereby making evident an outstanding talent for teaching, as was unanimously acknowledged afterward in the private meeting among the faculty.

Killing opened the public discussion following Dehn's lecture by asking him to describe the historical background that led Cantor to formulate the fundamental ideas of set theory. In the course of answering, Dehn sketched the proof for one of Cantor's fundamental results. After this, he was asked to compare the pros and cons of the two main approaches to complex function theory due to Riemann and Weierstrass, as well as to comment in this connection on the theorem of Mittag-Leffler. Killing then asked Dehn about the justification for complex numbers as well as other hypercomplex systems, before inquiring about the different methods for proving the fundamental theorem of algebra. Finally, he wanted Dehn to discuss different approaches to synthetic geometry. The candidate then characterized Poncelet's methods before giving a detailed comparison of the methodological differences between Steiner's works and those of Staudt. In all of these areas of pure mathematics, Killing attested that Dehn possessed "not only broad and firm knowledge but rather, for his age, an altogether astonishing penetration of the subject matter." Still, the colloquium was not yet over: another colleague asked about the significance of Euclidean geometry, and two physicists wanted to know about the significance of quaternions for their discipline. These questions, too, Dehn answered to the satisfaction of all concerned.

One should add that a candidate for habilitation who got as far as the colloquium stage could virtually count on gaining the *venia legendi*, the official title allowing one to teach at a German university. It almost never happened that a

---

[19] Copy of Killing's report from the petition of 5 November 1901 submitted by the Dean of the Philosophical Faculty to the Minister of Education, in Dehn's personnel record, Westfälische Wilhelms-Universität Münster.

colloquium proved so disastrous that this right was denied.[20] Nevertheless, the performance of a candidate could seriously affect their standing in the faculty. In this respect, Max Dehn clearly got off to a remarkably good start. Probably few would have imagined, least of all Wilhelm Killing, that a decade later he would still be a lowly *Privatdozent* in Münster. It would be a long and difficult waiting game.

As mentioned above, Hilbert appended a fairly lengthy account of Dehn's dissertation results to the French translation of his *Grundlagen der Geometrie*, which in the meantime had elicited several different reactions (see the commentary by Klaus Volkert in Chapter 5 of [Hilbert, 1899/2015]). By early 1903, Hilbert was preparing the second German edition, which would be the first to appear as an independent publication. Not surprisingly, he enlisted Dehn's support for the proofreading, as he would do in the future.

Like the original *Festschrift*, this book was published by Teubner in Leipzig, which employed a large staff. Proofreading, typesetting, and printing were all closely linked, and it was not uncommon for papers and books to go through several rounds of proof correction before publication. Thus Dehn received the galley proofs one sheet (*Bogen*) at a time, each corresponding to 16 printed pages. These technical details help explain what had happened when Dehn wrote from Münster to Hilbert on 5 February 1903 about a problem in the text concerning the completeness axiom, which first appeared in the French edition. Dehn explained that he had already sent back the first *Korrekturbogen*, so he no longer had the precise wording in front of him. He thus paraphrased it (correctly) from memory as asserting that one could not add new elements to the system without contradicting one or more of the prior axioms. Dehn then wrote:

> I now believe that this formulation could lead to misunderstandings. It should evidently mean that one cannot add anything to the system without invalidating one of the earlier axioms, *insofar as one does not alter the existing order or congruence relations*: a point lying between two others before the extension remains so after the extension, segments and angles that were earlier congruent to others, remain congruent. If this stipulation is not made, then it is easy to see that no geometry satisfies the completeness axiom.[21]

Dehn went on to describe how one can easily extend a given geometrical system by changing the order and congruence relations. He then continued:

> I do not know if this condition is immediately clear or whether it is perhaps on account of this open to formal attack. If one were to introduce it explicitly, then the axiom loses its philosophical character, in that it now requires that completely definite geometrical properties (congruence and order) must be fulfilled. Beyond that, the condition is too stringent: one only needs to require that order and parallelism are not disturbed. Then one

---

[20]During the Nazi era, on the other hand, political factors loomed very large. Max Deuring, a favorite pupil of Emmy Noether, was not allowed to teach in Göttingen, and the same fate befell Dehn's former student, Ruth Moufang, in Frankfurt. In such cases, the candidates were allowed to habilitate but were nevertheless denied the *venia legendi* as they were deemed unfit to teach German students.

[21]Dehn to Hilbert, 5 February 1903, Nachlass Hilbert 67, SUB Göttingen.

sees very clearly the close relationship between this axiom and that of Dedekind.

Hilbert afterward added an explicit remark explaining that the extensions ruled out by the completeness axiom were those that maintained order and congruence; the passage is a nearly verbatim rendering of the quotation above from Dehn's letter [Hilbert, 1903, 17]. Regarding Dedekind's method of achieving continuity by means of cuts in the set of rational numbers, Hilbert made no mention of this in the text, though he did emphasize that the Axiom of Archimedes, V.1 in his system, was a linear axiom, like Dedekind's, whereas VI.2, Hilbert's completeness axiom, was not. Paul Bernays would later prove, however, that only a linear version of completeness is needed in order to derive the full completeness axiom as a theorem. His proof was published in the seventh edition [Hilbert, 1930, 31-32]. This new version of the theory focused on linear properties, which led back to Dedekind, and in fact Hilbert explicitly introduced Dedekind cuts in order to prove that his system of axioms with linear completeness was relatively consistent by appealing to Cartesian analytic geometry.

Dehn would continue to advise Hilbert on matters concerning foundations of geometry, but in the years ahead he took his first steps into a related, but far larger realm, namely the geometry of space. The path he followed is described in greater detail in Chapter 4, whereas here we take up key threads that go back to Hilbert's first two Paris problems, both of which surfaced in dramatic fashion at the International Congress that convened in Heidelberg in 1904. There, Max Dehn had a front-row seat as a witness to a major turning point in the history of set theory and foundations.

ICM in Heidelberg 1904

Hilbert's lecture at the Heidelberg ICM was on the foundations of logic and arithmetic. Although nowhere near as grandiose as the address he delivered in Paris four years earlier, he nevertheless adopted the same self-assured tone. Speaking not only for himself but for all experts in the field, he picked up where he had left off with his second Paris problem:

> While we are essentially in agreement today as to the paths to be taken and the goals to be sought when we are engaged in research on the foundations of geometry, the situation is quite different with regard to the inquiry into the foundations of arithmetic; here investigators still hold a wide variety of sharply conflicting opinions. [Hilbert, 1904, 174]

This remark barely hinted at the more recent controversies that had arisen in connection with certain paradoxes in naive set theory. Hilbert and Zermelo were well aware of these, though both were confident that the problems they raised could be overcome. In the meantime, Hilbert proposed to establish the consistency of the axioms for arithmetic by simultaneously developing the laws of logic and arithmetic. This approach has often been regarded as the initial step toward Hilbert's proof theory based on finitist principles, which he and Paul Bernays developed during the 1920s and 1930s.

Reading Hilbert's text, which he republished along with [Hilbert, 1900a] in numerous editions of *Grundlagen der Geometrie*, one might imagine that his lecture in Heidelberg must have stirred up a good deal of excitement and discussion. Hilbert was surely counting on such a reaction, so he must have been disappointed that his audience was preoccupied by what they had just heard. For on this occasion, Hilbert was upstaged by the speaker who preceded him, the the Hungarian mathematician Julius König. His topic, in essence, was Hilbert's first Paris problem, though he took dead aim at Cantor's continuum hypothesis (CH), rather than speaking vaguely about it, somewhat as Hilbert had done four years earlier. Instead, König caused a sensation by answering the problem in the negative: he presented what seemed to be an airtight "proof" that CH was false, indeed, that the cardinality of the continuum was *greater than* any aleph [Moore, 1982, 86–88].[22] This proof was based on technical results that included a formula from Felix Bernstein's recent dissertation. Had König's argument been correct, this would have had a devastating impact on Cantor's whole theory of transfinite numbers, which was predicated on the belief that all (consistent) sets could be well ordered. What Cantor, Hilbert, Felix Hausdorff, and the other experts in the audience witnessed on this occasion was without doubt a dramatic event.

Max Dehn may well have been sitting next to his friend Ernst Zermelo when they heard König speak. Following the Heidelberg Congress, a storm of controversy broke loose over Cantor's theory, partly caused by logical paradoxes as well as problems arising from ill-defined sets. Even more controversial, however, was the argument Zermelo used to prove Cantor's claim that one could always establish a well ordering for any set whatsoever, however large it might be.

Arthur Schoenflies was not only present when König spoke, he was also part of a small group of mathematicians who left for vacation in Switzerland immediately after the Congress ended [Schoenflies, 1922].[23] By chance, this group – Hilbert, Hausdorff, Schoenflies, and Kurt Hensel – ended up staying together in the same hotel in Wengen, a mountain village in the Bernese Oberland. Cantor was also vacationing with two of his daughters in Wengen as well, but at a different hotel. Yet, since the other mathematicans' main topic of breakfast conversation was König's "proof," he was more than eager to come over and join them. Schoenflies mainly recalled how Cantor was entirely convinced that König's proof was somehow faulty, and it seems likely that he and others suspected there was a problem with a result König took from Bernstein's dissertation. Cantor supposedly even joked that he had more faith in the king (König) than in his ministers, presumably meaning Cantor's former student, Felix Bernstein. If that anecdote is true, then he must have already surmised that Bernstein's dissertation – written in Göttingen under Hilbert's supervision – contained an erroneous result.

What happened next is not altogether clear, but it seems that David and Käthe Hilbert returned to Göttingen, whereas Cantor stayed on with his daughters in Wengen. Zermelo went back to Göttingen, as well, though in late September he was staying south of the city in the village of Münden. By the end of the month, Hausdorff had returned to Leipzig, and in a letter to Hilbert from 29 September 1904 he broke the news about the error in Bernstein's dissertation.

---

[22]It should thus be noted that this argument did not pertain to the original form of Cantor's conjecture – as Hilbert had presented it in Paris – but rather to Cantor's claim that $2^{\aleph_0} = \aleph_1$.

[23]Schoenflies appears to have had little sympathy for abstract set theory, whereas Zermelo poured scorn on Schoenflies's competence when it came to such matters [Ebbinghaus, 2007a, 60,76].

> After the continuum problem had tormented me in Wengen nearly like a monomania, my first glance here of course turned to Bernstein's dissertation. The worm is sitting there at exactly the place I suspected, on p. 50: ...Bernstein's consideration leads to a recursion from $\aleph_{\mu+1}$ to $\aleph_\mu$, but it fails for those $\aleph_\mu$ which do not have a predecessor, that is for precisely those alephs which Herr J. König necessarily requires. I had already written in this sense to Herr König on my trip back, so far as I could do so without having Bernstein's work, but I have received no answer. [Purkert, 2015, 16]

Hausdorff soon thereafter clarified the error in a short note published in the DMV's *Jahresbericht* [Hausdorff, 1904] at the end of which he offered these remarks on König's lecture:

> The formula
>
> $$\aleph_\mu^{\aleph_\alpha} = \aleph_\mu \cdot 2^{\aleph_\alpha}$$
>
> obtained by Mr. F. Bernstein ("Untersuchungen aus der Mengenlehre", Diss. Halle 1901, p. 50) by means of unrestricted recursion is thus to be regarded provisionally as unproven. Its correctness appears all the more problematic, as from it, as Mr. J. König has shown, would follow the paradoxical result *that the power of the continuum is not an aleph and that there exist cardinal numbers that are greater than any aleph.* [Hausdorff, 1904, 571]

Zermelo had also been skeptical of König's reasoning, but instead of trying to refute his argument directly, he countered it by coming up with a proof for one of the cornerstones of Cantor's theory: the assertion that *every* consistent set can be well ordered. In conversations with Hilbert's newest star pupil, Erhard Schmidt, Zermelo quickly saw that a simple non-constructive proof for well ordering followed immediately from what some called the choice principle, later known as the Axiom of Choice (AC). This simply asserts that if given a collection of nonempty subsets of a set M, then we can freely pick out one element from each of these subsets. As Zermelo later emphasized, however, this terminology can be misleading because if one has infinitely many subsets, then AC implies the existence of a choice function that "chooses the elements all at once," i.e., independent of time [Ebbinghaus, 2007a, 54–55]. In [Zermelo, 1904], he used the notion of *Belegungen* or coverings, a term Cantor also used in his "Beiträge" [Cantor, 1895].

Retreating to the village of Münden just south of Göttingen, Zermelo found the peace and quiet he needed to write up his argument. One week later, on 24 September, he sent off his proof to Hilbert in the form of a letter. The latter immediately inserted this into the next issue of *Mathematische Annalen*, since the text of [Zermelo, 1904] took up only three pages. Zermelo's proof opened the flood gates of controversy regarding Cantorian set theory, leading to various efforts to secure, modify, or dismiss its main results [Moore, 1982]. Recognizing its profound importance, he sent advance copies of his manuscript to a number of experts and other interested parties. One of those who read it was Max Dehn, who responded positively, evidently unperturbed by Zermelo's use of the choice principle.[24] Dehn's response led Zermelo to send him a postcard on 27 October 1904, which was transcribed and translated in [Ebbinghaus, 2007b]. After thanking him for his friendly assent, Zermelo added these remarks:

> I also received the same day-before-yesterday from the "younger Berlin school," i.e. from [Edmund] Landau and [Isaai] Schur, in fact. But if you only knew how I am bombarded in the meantime by letter with objections from the experts of set theory, from Bernstein, Schoenflies, and König! Some fine polemic will still unfold in the Annalen; but I am not worried about that.

Zermelo's postcard thus reveals that he had been engaged in a flurry of correspondence during the past month, but also that he predicted these private communications would eventually lead to an open debate. This did, indeed, ensue (see [Moore, 1982]), but Zermelo only answered his critics four years later [Zermelo, 1908]. Since Dehn had received no further updates regarding the status of König's claims, Zermelo gave him a short synopsis. This made clear that both he and Julius König had independently found the error in the latter's proof, but could only confirm this after checking Bernstein's dissertation, which was not at hand in Heidelberg. He thus went on to write:

> So you still do not know the fate of König's talk? Innocent soul! Nothing else was really discussed during the entire vacation. K[önig] solemnly informed both Hilbert and me that he revoked the Heidelberg proof. B[ernstein] did the same with his theorem on powers. K[önig] can be happy that the library in Heidelberg had closed so early; otherwise he might have made a fool of himself. Thus I had to wait for checking until my return to G[öttingen]. Then it was instantly obvious. In the meantime, however, K[önig] himself had found that out when elaborating his proof. ... [He points to page 50 of Bernstein's dissertation.] You will see the mistake at once which is the only result K[önig] uses. Blument[h]al says that this is a real "Bernstein ruin." But K[önig] wants to save what can be saved and fantasizes about the set W of *all* order types which *could* be contained in many a set so that my ... [argument] ... would be wrong. Strangely enough, he is backed up in this matter by Bernstein and Schoenflies, by the latter of course in quite a confused and misunderstood manner. The W-believers already deny the def[inition] of subset, perhaps even the principle of the excluded middle. [Ebbinghaus, 2007b, 431]

As noted earlier, Dehn had already learned some of the fundamental ideas of Cantorian set theory as a student. He may well have attended the proseminar Schoenflies offered on this topic, but in any case Hilbert chose this as one of the subjects for his final oral exam. Moreover, Dehn's knowledge of set theory was by no means merely passive. In [Dehn, 1904] he applied point set topology to show the true scope of the Dehn invariant for polyhedra. In his earlier studies related to the third Paris problem, he had merely given examples of pairs of polyhedra with

---

[24]Hilbert had never explicitly appealed to such a general principle in his existence proofs, but Hausdorff later became a strong advocate of AC [Dreben and Kanamori, 1997, 80–81].

the same volume, but which were inequivalent because the values of this invariant differed. Dehn's new results showed that, in fact, there are uncountably many different pairs of such inequivalent polyhedra, and that this is true not only in the Euclidean case but also in non-Euclidean geometries (Chapter 3). As we shall later see, however, Max Dehn never felt drawn to axiomatic set theory, as practiced by Zermelo and Fraenkel, nor did he revel in the abstractions promoted by Cantor and Hausdorff in their works on transfinite arithmetic.

It was probably during the ICM in Heidelberg that Dehn first met the Danish geometer Poul Heegaard, although the latter's name does not appear on the list of participants. Wilhelm Magnus claimed that they got to know one another then and that they left the city on the same train, Dehn going to Hamburg and Heegaard returning to Copenhagen. On the train ride together, according to this story, they supposedly decided to co-author the article on topology for the German Encyclopedia of Mathematical Sciences [Magnus, 1978/79, 134]. In any event, it was around this time that Dehn and Heegaard began their collaboration by adopting a clear division of labor: the older Dane wrote up summaries of the literature, leaving the younger Dehn the task of outlining the foundations of the discipline [Dehn and Heegaard, 1907]. Magnus considered this pioneering effort to systematize combinatorial topology a characteristic example of Dehn's entire mathematical outlook. Although this program failed to provide a lasting framework for research on higher-dimensional manifolds, it nevertheless served to throw new light on deep problems that challenged several of the world's leading mathematicians for decades to come (for further discussion of [Dehn and Heegaard, 1907], see Chapter 4).

Academic Opportunities, 1905–1910

Hilbert was always eager to promote the careers of young mathematicians who studied under him, and he later expressed frustration because of the fact that faculties many times failed to act on his advice. In Dehn's case, Hilbert had already recommended him for an associate professorship in Würzburg in 1903, alongside two others, Felix Bernstein and Gustav Herglotz, though none of the three was chosen. It should be noted, on the other hand, that diplomacy was not Hilbert's strong suit. Unlike Klein, who normally dominated either by persuasion or sheer perseverance, Hilbert was impulsive, opinionated, and arrogant. He did not go out of his way to make enemies, but his authoritative manner often rubbed his fellow mathematicians the wrong way. During the period 1905 to 1910, when Dehn was in Münster, patiently awaiting his first call to a professorship somewhere in Germany, he became entangled in a controversy with Theodor Vahlen, a mathematician whom Hilbert rightfully saw as an usurper in the field of foundations of geometry (see below).

One of the mysteries Hilbert struggled with while writing *Foundations of Geometry* concerned the relationship between the two classical theorems of Pappus and Desargues. For some time it seems he conjectured that the latter was stronger, but it took another five years before Gerhard Hessenberg could clarify this situation. He showed, in fact, that just the opposite is true: Pappus's theorem implies Desargues' theorem, but not conversely [Hessenberg, 1905]. Dehn was vacationing with his family in Hamburg when he received the manuscript of this paper from Hilbert, who was eager to learn his opinion. On returning the paper, Dehn wrote on 12 April 1905 that he had been too busy to study it thoroughly, but expressed confidence in what he had read. He was especially impressed with Hessenberg's elegant proof of Pappus's theorem, and wrote furthermore: "he appears to use in these things essentially the usual tools, your segment arithmetic and the pseudo-geometry from my dissertation. I'm very happy to see this matter now clarified and that the sphere-beings (*Kugelbewohner*) can be blessed without having to believe in continuity."[25]

Furthermore, Dehn reported on his recent results on Euler's formula in $\mathbb{R}^5$ in relation to non-Euclidean spherical excess in $\mathbb{R}^4$ ([Dehn, 1905a] and [Dehn, 1905b] discussed in the chapter following). He also mentioned that he had been asked to fill in for the *Ordinarius* in Kiel during the upcoming summer semester. This was a reference to Paul Stäckel, who had accepted a professorship at the TH Hanover. Hilbert wrote back the very next day, expressing his delight over this turn of events:

> Above all, my hearty congratulations for your [temporary] appointment in Kiel. That makes me happy in the first instance for your sake, but also because it means that finally there are prospects for a Göttingen product to move ahead. But do be very careful around the colleagues in Kiel, as confidentially speaking, faculties can be difficult; one must be ready – I'm thinking of Blumenthal's case – for anything.[26]

Otto Blumenthal, mentioned earlier as the first mathematician in Göttingen to discover Dehn's budding talent before graduating as Hilbert's first doctoral student, was a bit more than two years older than Max Dehn. During the academic year 1904/05 he held a similar temporary professorship in Marburg. Later that same year, Sommerfeld arranged for his appointment at the TH Aachen, where the faculty elevated him to the chair formerly occupied by Lothar Heffter. Although he longed for an appointment at a university, Blumenthal never had the chance to leave Aachen, and in one particularly flagrant case this was due to antisemitism [Rowe, 2018b, 144–150].

Max Dehn was opinionated, too, especially when it came to areas of mathematics in which he had already established his reputation as a leading authority. He was still only 27 years old and patiently waiting for his chance to climb up the academic ladder. Although Hilbert cautioned him to be careful when speaking with the professors in Kiel, he wrote nothing in the very same letter about showing a similar reserve when criticizing the work of a colleague who wrote a mediocre book, one that Hilbert clearly found personally offensive. This colleague was the Greifswald mathematician Theodor Vahlen, whose career trajectory would soon collide with Dehn's in an unfortunate way.

As a backdrop to the section following, it should be mentioned that, only shortly before Dehn wrote his review, he nearly gained his first associate professorship in Greifswald. Although he may not have been aware of this, the Greifswald faculty had placed his name first on their list of candidates for the vacancy created when Gerhard Kowalewski departed for Bonn. Kowalewski's appointment in Bonn was engineered by Eduard Study, who had only recently left Greifswald. Study's successor in Greifswald was his good friend Friedrich Engel, nothing unusual, as

---

[25] Dehn to Hilbert, 12 April 1905, Nachlass Hilbert 67, SUB Göttingen.

[26] Hilbert to Dehn, 13 April 1905, Dehn Papers, Dolph Briscoe Center for American History, University of Texas at Austin.

personal favoritism and compatibility (*Verträglichkeit*) played a large role in academic appointments. In preparing the list of candidates for the empty post of associate professor, both Engel and Greifswald's senior mathematician, Wilhelm Thomé, surely had considerable influence. They noted that Dehn was the youngest of the three nominated. The others named were Vahlen and Felix Hausdorff, an unbaptized Jew, unlike Dehn.[27]

Vahlen was described in the report as a former student of Leopold Kronecker in Berlin, who since 1896 was teaching as a *Privatdozent* in Königsberg. No comparative remarks were made, and the faculty only commented that any of the three would be completely suitable [Brieskorn and Purkert, 2018, 712–713]. Vahlen was chosen by the ministry in 1904 and he remained in Greifswald until 1924. In that year he was forced to resign without pension rights because of political acts directed against the Weimar Republic. These later events as well as Vahlen's singular career as a prominent Nazi figure within the German mathematical community are recounted in [Siegmund-Schultze, 1984]. In 1905, no one had any inkling, of course, of what was to come in the aftermath of the war. Still, in later years the parties involved surely never forgot the acrimonious public exchanges that took place that year.

Dehn's Critique of Vahlen's *Abstrakte Geometrie*

In his letter from April 12 to Hilbert, cited above, Dehn mentioned that he had been asked to review Vahlen's new book on the foundations of geometry. This, too, was welcome news for Hilbert, who had already looked at this work and found it worse than disappointing. As Klaus Volkert has pointed out, he had every reason to feel slighted: Vahlen's *Abstrakte Geometrie* [Vahlen, 1905a] made numerous references to Hilbert's *Grundlagen der Geometrie*, but mainly to criticize small points in it.

Vahlen's book was the first monograph on this subject published in Germany since the appearance of Hilbert's booklet. As we shall see, Vahlen was not without supporters within the German mathematical community, even though he left no lasting mark as a mathematician. Instead, he is solely remembered as a political figure, probably the only professor of mathematics who joined the Nazi Party already in 1924, soon after Hitler's failed *Putsch* in November 1923. Vahlen had the party number 3,961, and after 1933 this early support of the Nazi Party enabled him to become a prominent figure in the regime. In fact, he may have played a role in pushing through Dehn's mandatory retirement in 1935.

In view of these circumstances and the events that would follow, Hilbert's letter is well worth quoting:

> Do indeed take on the review of Vahlen's book. It is altogether important that it receive a decent critique. I received the book yesterday from the book dealer. After paging through it I gained the impression that the whole thing lives from the ideas in my *Grundlagen*, but without saying so. For example, the theorem

---

[27] At this time, Hausdorff had only begun his research on point sets and transfinite arithmetic, which then culminated in 1914 with the publication of his *Grundzüge der Mengenlehre* [Hausdorff, 2002]. Study later brought him to Bonn as Kowalewski's successor in 1910, after which Hausdorff's career swung like a pendulum between Bonn and Greiswald, where he succeeded Engel in 1913, before returning to Bonn in 1921.

stating that the commutative law is not a consequence of the remaining axioms, which cost me a lot of trouble to prove, is simply taken over, and when he does name me, then only to point out some oversight in the first edition. ... Also with respect to other authors, I believe Vahlen acts as though everything stems from him.[28]

Hilbert encouraged Dehn to write a lengthy review for the *Göttingischen Gelehrten Anzeigen*, which he noted paid its reviewers well. Had Max Dehn published his review there, word would have eventually spread, but probably the scandal that ensued would not have occurred. Instead it appeared in the highly visible *Jahresbericht* of the German Mathematical Society (DMV), whose editor, Jena's August Gutzmer, found himself in a difficult position.

Whether Dehn felt prodded by Hilbert's encouragement or not would be hard to say, but he certainly studied Vahlen's book carefully before taking it apart. The scope of this monograph was far broader than Hilbert's, but Dehn hinted that it lacked originality: "In the present work, the author attempts to give a comprehensive and detailed exposition of the foundations of geometry based on methods that have recently been created, especially in Italy and Germany" [Dehn, 1905c, 535]. Aside from this general remark, one finds nothing in the review suggesting that Vahlen dealt with any particular individuals unfairly, as Hilbert insinuated. After a brief paragraph outlining its contents, Dehn simply took up a series of statements, relentlessly exposing them as either obscure, false, or just outright mathematical nonsense. The tone throughout was dispassionate, and of course most of Dehn's criticisms were only intelligible to experts in the field. So, in effect, he was writing for them; others could only draw the obvious conclusion, namely, that this was a worthless book written by an incompetent author.

Vahlen had to respond, of course, and he wasted no time in doing so in [Vahlen, 1905b]. This reply went through all of Dehn's criticisms, point by point, conceding certain shortcomings, but then describing them as the kind of harmless blemishes one finds in any mathematics book. He ended by charging Dehn with unfairness, underscoring that his review failed to mention any of the many positive accomplishments Vahlen took pains to record. Vahlen's long rebuttal closed by accusing Dehn of having merely *thumbed through* his book looking for mistakes instead of actually *reading* it.

The editors of the *Jahresbericht* evidently sent Vahlen's response to Dehn, who submitted a brief reply. This appeared immediately afterward and in it Dehn only bothered to remark that Vahlen's final sentence was totally unfounded: he had, of course, read the book carefully, and since he had expressed what he had to say about it clearly enough, he saw no need to comment further. A number of readers, however, had been puzzled as to why he had not given an overall assessment in his review, and so he offered that here: "Consequently [*Somit*] this work can only be recognized as having limited use: everything of value contained in it, except for some details, can be found presented more clearly and perfectly explained in other works ..." [Dehn, 1905d, 595]. He then named four German authors, one of course being Hilbert. Dehn's mentor had been concerned that Vahlen had appropriated his ideas, whereas this simple remark was far more devastating, since Dehn flatly asserted

---

[28]Hilbert to Dehn, 13 April 1905, Dehn Papers, Dolph Briscoe Center for American History, University of Texas at Austin.

that Vahlen's work was inferior to the ones he had drawn upon, while adding nothing of substance to those works. Vahlen could not very well leave such a charge unanswered, and so he proceeded to spell out eight *new* and *valuable* contributions found in his *Abstrakte Geometrie* and nowhere else in the mathematical literature [Vahlen, 1906a].

The spat between Vahlen and Dehn was observed from afar by Oswald Veblen, whose review of Vahlen's *Abstrakte Geometrie* in the *Bulletin of the American Mathematical Society* surely gave Max Dehn pleasure to read. Much like Dehn, Veblen took brief note of the book's five chapters: I Foundations of arithmetic; II Projective geometry; III Projective geometry; IV Affine geometry; and V Metric geometry. A book on these topics, he wrote, "*would* be very useful for giving a general view of the recent studies on foundations of mathematics," if only it were written with

> that precision of language which is indispensable in any discussion of such a subject. The reader is constantly confronted with statements which are incorrect if taken literally and which, if not taken literally, are open to more than one interpretation. Many of the author's postulates are labeled by him as definitions. Moreover, there are places where it is very difficult to determine which of the previously stated hypotheses are being used and which are not. As a consequence, the reviewer is able to state hardly a single new *result* which is surely established by this book. [Veblen, 1906, 505]

As if these comments were not devastating enough, Veblen let his American audience know that Max Dehn had already taken the time to skewer Vahlen's book for the German Mathematical Society:

> It seems to the reviewer not to be worth while to lengthen this notice with criticisms of details, especially as many of the points that would be mentioned have already been adverted to by Dehn in a review published in the Jahresbericht der Deutschen Mathematiker-Vereinigung, volume 14, page 535. The reader who is interested in such things will find a rejoinder to Dehn by Vahlen on page 591 of the same volume, a retort by Dehn on page 595, and a second "Erwiderung" by Vahlen in volume 15, page 73. With these he may compare a footnote by Schoenflies on page 31, volume 15. [Veblen, 1906]

Turning to that page in Arthur Schoenflies's article, one finds a reference to Dehn's critique of Vahlen's definition of continuity. Echoing Dehn's opinion that this was an illusory concept, Schoenflies claimed it was either nonsense or reduced to Dedekind's notion of continuity. Vahlen denied this, while asserting that his definition was a variant of Veronese's concept. In the decisive footnote mentioned by Veblen, Schoenflies called this an erroneous claim, and then added this stinging remark: "It is very regrettable that a book that strives to treat its subject from all sides comprehensively should suffer from such a fundamental deficiency, which disturbs to a great extent the consistency of the presentation in the arithmetical as well as the geometrical part" [Schoenflies, 1906, 31]. This attack, too, was answered in the *Jahresbericht* in [Vahlen, 1906b].

As noted earlier, Schoenflies had been one of Dehn's teachers when he first came to Göttingen. During the 1890s, he developed a strong interest in early point set topology, about which he wrote reports both for the *Encyklopädie der mathematischen Wissenschaften* (1898) and for the DMV [Schoenflies, 1900]. A protégé of Felix Klein and a personal friend of Hilbert, Schoenflies was a geometer with broad interests. He left Göttingen in 1899 to assume a full professorship in Königsberg, where Vahlen had already been teaching as a *Privatdozent* for the past three years. These circumstances suggest that there was no love lost between Schoenflies and Vahlen, and the former may well have harbored suspicions that his colleague hoped to make a name for himself in geometry at the expense of Hilbert. As for Vahlen and his sympathisers, they were quite astounded by the ferocious tenacity of the attacks on Vahlen's book.

Vahlen himself undoubtedly took note of the fact that his two assailants were both Jews from Göttingen, later characterized as the citadel of "Jewish mathematics" by Nazis in academia. Friedrich Engel shared no such feelings, but he also felt that the attacks against Vahlen had gone too far, and so he complained to August Gutzmer, editor of the *Jahresbericht*. Gutzmer responded on 14 January 1906:

> The unmixed pleasures of life will never be granted to an editor. I have often experienced the truth of that saying, and the case of Vahlen-Dehn is a glowing testimonial as such. But please do not believe that you, with your friendly words, are guilty of contributing to these mixed joys – quite the contrary, I am very grateful to you for speaking out forthrightly! It gives me the opportunity to respond in a similar manner, something I have often done with friends and colleagues whose opinions I value and rely upon.
>
> As for the present matter, I believe that this dispute has not yet come to a close, as I expect to receive a final reply from Vahlen as the party under attack. Once that response has reached my hands, I intend to intervene as editor and declare this to be the last word. If I had wanted to take sides against Vahlen, I would have certainly stated so clearly. But the matter will now take a new turn because the next issue will contain an article by Schoenflies, in which he deals with Veronese's continuity, an issue of decisive importance for this entire dispute.
>
> It was precisely my neutrality that prevented me from writing to Dehn in the manner you suggested. The fact that Dehn's response was extremely sharp could not disqualify it from publication, as it contains no personal attacks. Every reader can decide for themselves whether or not this undermined (*entkräftet*) Vahlen's presentation. Were I in Vahlen's position, I would simply submit a short statement saying that Dehn's review and response are so unusually sharp that I must pass on to others to form their own opinion of the value of the book – one word more than that would be too much. ...[29]

Gutzmer further explained that Dehn's response to Vahlen's defense had arrived very late, which is why Gutzmer had been unable to send it to Vahlen before it

---

[29] Gutzmer to Engel, 14 January 1906, Engel Nachlass, Giessen.

went to press. In fact, Teubner had prepared the whole issue and was ready to send it out, so the publisher had to remove another notice in order to place Dehn's reply immediately after Vahlen's rebuttal. Engel's role here as Vahlen's advocate gave Gutzmer the chance to persuade Vahlen that, as editor, he needed to assume a neutral role, while making clear that he did not condone Dehn's actions. He thus ended by writing:

> I hope very much, dear Engel, that after reading these explanations you will see this matter with somewhat different eyes; presumably Vahlen will see it differently, too, and you are welcome to share these lines with him if you are so inclined. I trust that you know me far too well to imagine that in my capacity as editor of the Jahresbericht I would intervene in a tone like that of Dehn's and take sides in such a difficult problem. Precisely the opposite was my intention.

Engel was evidently satisfied with this response; he sent Gutzmer's letter to Vahlen the following day, adding a note of agreement and urging Vahlen to avoid any sharp formulation in his reply. As planned, Gutzmer then added a footnote stating that this would be the final contribution in this controversy [Vahlen, 1906a, 74]. Schoenflies's critique in [Schoenflies, 1906] was printed in the same issue, however, and so Vahlen had to send yet another response to dispute its claims [Vahlen, 1906a, 74].

For Dehn, however, the consequences of this controversy were yet to come. In 1910, Greifswald's senior mathematician, Wilhelm Thomé, died, and Engel turned to his friend in Bonn, Eduard Study, for advice as to an appropriate successor. Study recommended Dehn, evidently having forgotten the highly visible debate from five years earlier. Engel then reminded him of the "compatibility problem," since Vahlen's presence made Dehn's candidacy unthinkable. Study then replied on 2 October 1910: "When I wrote you, I didn't even think that Dehn and Vahlen don't get along. You will probably have no alternative than to have Vahlen fill the missing *Ordinariat*, and he is certainly not so *much* worse than most of the others." This, indeed, is what happened: Vahlen was thus promoted to a full professorship in Greiswald, whereas Dehn remained in Münster, where he was now in his ninth year as a *Privatdozent*. Finally in 1911, Dehn got his first break when Lothar Heffter left Kiel to assume the professorship in Freiburg previously held by Jacob Lüroth, who died in 1910. This enabled Georg Landsberg, who held an associate professorship in Kiel, to be promoted to full professor, whereupon Dehn was appointed to Landsberg's former position. This appointment became effective on 1 April 1911 and paid Dehn a yearly salary of 2,600 Marks with a supplement of 1,620 Marks for living costs.

Call to Kiel, 1911

Kiel remained a provincial university throughout the nineteenth century with only one professorship in mathematics up until 1874. In that year, Leo Pochhammer, formerly a *Privatdozent* in Berlin, joined the faculty. In 1906, Georg Landsberg from Breslau was appointed as the first occupant of a new *Extraordinariat*, the position Max Dehn would obtain in 1911, when Landsberg succeeded Lothar Heffter as full professor (*Ordinarius*). An expert on algebraic functions, Landsberg is still remembered today as co-author with Kurt Hensel of the monograph *Theorie der algebraischen Funktionen einer Variablen* (1902). He also guided the doctoral work of Jakob Nielsen, who first matriculated at Kiel in 1908. Nielsen eventually took his doctorate under Dehn in 1913, following Landsberg's death a year earlier (see Chapter 6).

Kiel lies only 100 kilometers north of Hamburg, so Max had many opportunities to visit his family. Still, he did not forgo some of the local pleasures provided by living in a port city on the Baltic Sea, one of which was sailing. He surely found this a romantic activity. On a trip back home, he got to meet a young art student from Berlin named Antonie (Toni) Landau, who happened to be boarding with one of Max Dehn's family members. Toni was from Berlin, where her father Isidor was a newspaper editor and theater critic, but also the author of travel books. Her mother Louise worked as a translator, rendering novels by John Galsworthy as well as writings of Theodore Roosevelt into German. Both were originally from Poland. Max invited Toni to go sailing with him, which marked the beginning of a whirlwind romance. She was just nineteen when they married on August 23, 1912.

Only one month later, Georg Landsberg unexpectedly died, a circumstance that seemed to open the way for Dehn to become an *Ordinarius* in Kiel. The discrepancy in salary between associate and full professors was extreme, so that only the independently wealthy could be satisfied with the lower position. His long period of waiting seemed to be over, but in the end this was not to be. We gain a glimpse of what transpired from a letter Dehn wrote to Hilbert on 9 March 1913:

> I owe you my deep gratitude for your efforts on behalf of my promotion. Pochhammer informed me that he read your letter in the faculty meeting. Unfortunately, the success of this proved impossible due to a minority opinion forwarded by two gentlemen, who characterized me as a poor lecturer. Probably the motivation behind this was of a private nature, as objectively this description could only apply to a single course taught two years ago. Since I could easily refute this claim through the success of my present teaching activity, Pochhammer wrote to Elster[30] giving a detailed minority report. This will be of no use at the moment, but it will hopefully help to nullify the unfavorable report in the future.[31]

A few months later, Dehn finally got the call he had long waited to receive – he would succeed Constantin Carathéodory as *Ordinarius* at the Technical College in Breslau. Hilbert may well have had a hand in this, as he had helped orchestrate the negotiations that brought Carathéodory to Göttingen as Klein's successor. In early July, he sent his congratulations to Dehn, who wrote back to say that he had been very angered by the way Hilbert's input had been dismissed by his colleagues in Kiel.[32] He maintained his ties there, however, as he was succeeded by his good friend Otto Toeplitz. Dehn was no doubt greatly relieved to gain a full professorship soon thereafter, though at the newly founded Institute of Technology in Breslau rather than at a university, which would have suited him much better.

---

[30] Ludwig Elster was a ministerial official in Berlin.

[31] Dehn to Hilbert, 9 March 1913, Nachlass Hilbert 67, SUB Göttingen.

[32] Heinrich Wilhelm Jung, who taught at a Gymnasium in Hamburg, was chosen as Landsberg's successor.

Call to Breslau, 1913

The Silesian city of Breslau had a university dating back to 1702. This was an initiative of Jesuits, who were granted this privilege by the Austrian Emperor Leopold I. After Silesia came under Prussian control, this Jesuit university with its philosophical and theological faculties was retained, but it was later combined with the Lutheran university in Frankfurt on the Oder, which was displaced to Breslau in 1811. This gave the Prussian University of Breslau a quite distinctive character. In mathematics, it was strongly represented by two synthetic geometers, Heinrich Schröter and his successor Rudolf Sturm, as well as the invariant theorist Jacob Rosanes. Two of Max Dehn's closest colleagues, Ernst Steinitz and Otto Toeplitz, were students of Rosanes. All three – Steinitz, Toeplitz, and Rosanes – were of Jewish descent. In 1905, Adolf Kneser joined the faculty. He and Dehn shared common mathematical interests, as did Kneser's son, Hellmuth, who profited from their discussions during and after the war (see Chapter 4).

The *Technische Hochschule* in Breslau, located about three miles from the university, was only founded in 1910 with three professorships in mathematics. These were first held by Constantin Carathéodory, Ernst Steinitz, and the geometer Gerhard Hessenberg. When Dehn succeeded Carathéodory in 1913, he found the two remaining colleagues much to his liking; indeed, all three had strong interests in various facets of geometry. As discussed above, in 1905 Hessenberg solved a delicate foundational problem that had baffled Hilbert by proving that the Theorem of Desargues follows from Pascal's Theorem for two lines. Dehn surely already knew several of his earlier papers as well when Hilbert sent him that manuscript.

Hessenberg held the chair in geometry, which meant he taught the standard courses in descriptive geometry. His interests, though, stretched from technology and differential geometry to Cantor's set theory and critical philosophy. Intellectually, he had much in common with Max Dehn, though he exceeded the latter when it came to sarcastic humor. During the first years of the war, he was elected to the office of rector at the TH Breslau, a sign of the high respect his colleagues had for his administrative abilities. As a geometer, Steinitz studied problems involving finite polyhedra, configurations, and groups. These topics were very close to Dehn's interests during the period he taught in Breslau. Indeed, as we will see in Chapter 4, both he and Steinitz stood at the forefront of a movement to promote combinatorial methods in the study of topological problems.

Abraham Fraenkel was studying at the university when Dehn arrived at the TH Breslau. He later recalled how Dehn, Steinitz, and Hessenberg all had to adapt their teaching to the needs of the engineering students, which meant they had little to offer him. Fraenkel nevertheless remembered an incident from that time:

> In the mathematics colloquium I presented my as yet unpublished doctoral dissertation, even though it fell on the Sabbath. It was a complicated matter, but I could not avoid it, and had reasonable success. Only Professor Dehn, whose baptism failed to accelerate his career as he had hoped and expected, commented that young mathematicians should not concern themselves with such abstract subjects, but should instead stick to concrete problems. [Fraenkel, 1967, 101]

Leaving aside the remark about Dehn's baptism – which reflects Fraenkel's opinions as a Zionist – this anecdote surely accords with Dehn's general pedagogical views. As a true disciple of Hilbert, he believed that abstract concepts arose from concrete problems and that no mathematician should neglect the latter in favor of the former when, in fact, they are inextricably connected.

It was during these years in Breslau, interrupted by the Great War, that the Dehns' children were born: Helmut in 1914, his two sisters Maria and Eva in 1915 and 1919, respectively. Their father was conscripted into the army in 1915, working mainly as a coder/decoder near battle sites on the Western front.[33] Helmut confirmed that Max Dehn was less than enthused about military life. In one instance, he was standing by the road when an entourage escorting the Kaiser passed. He failed to notice the other men standing at attention and saluting, but others noticed him with his hands still in his pockets [Yandell, 2002, 122]. Nevertheless, he completed his war service and was discharged without incident as a corporal, in sharp contrast to the experiences of his future colleague Carl Ludwig Siegel (as briefly discussed in Chapter 5).

Unfortunately, Dehn's three years of war service interrupted his work with colleagues in Breslau. Soon afterward, in 1919, Hessenberg left for Tübingen, one of several changes in the mathematical landscape during the immediate postwar period. In some instances, Jewish candidates were among the first to be considered for promotion. In 1920 Toeplitz was elevated to a full professorship in Kiel, and he afterward turned to Klein and Hilbert for advice regarding a vacancy there. There was widespread awareness that many talented Jewish mathematicians had been passed over during the past decade, a circumstance that placed Toeplitz in an awkward position in conducting this search, since he, too, was of Jewish background.[34] Toeplitz also wrote to Eduard Study in Bonn, who advised him to consult the "Old Testament," after which he named several qualified individuals, all of whom happened to be Jewish. The two he mentioned first were Emmy Noether and Arthur Rosenthal, neither of whom were named by Klein or Hilbert.[35] The position in Kiel eventually went to Ernst Steinitz, Dehn's colleague in Breslau, who was high on Hilbert's list.

Since Dehn and Toeplitz were both protégés of Hilbert, they were very likely acquainted with Emmy Noether. She wrote to Dehn on 8 January 1920, answering a no longer extant letter from him. Dehn surely anticipated that Steinitz would soon be leaving Breslau for Kiel, so he wrote to Noether asking her opinion of Werner Schmeidler's abilities. After receiving her positive report, he informed her that it was very helpful, though he did not plan to include Schmeidler's name on the short list of candidates. Noether then replied, in turn:

> I had already heard that Schmeidler will not be on the Breslau list, but I'm pleased that you have a favorable impression of him based on my report. Meanwhile, Toeplitz has offered to let him re-habilitate in Kiel and to take on a large teaching contract

---

[33]In an account of his wartime service at the time he applied for his pension, Dehn listed four different battle sites where he had been stationed: La Bassée, Noyon-Roué, St. Quentin, and Oise.

[34]See the discussion in [Bergmann/Epple/Ungar, 2012, 206–207] and in [Rowe, 2018a, 349–350].

[35]Study to Toeplitz, 1 July 1920, quoted in [Koreuber, 2015, 40]. Eight years later, Noether was also briefly considered for this very same position in Kiel, see [Siegmund-Schultze, 2018].

(with an inflation allowance of around 9,000 Marks). He will no doubt accept, if certain conditions (per diem expenses until he has found a place to live and similar things) are met. You surely know that Toeplitz had originally turned to Nielsen.[36]

Emmy Noether's own earlier attempt to habilitate in Göttingen had led to a famous clash within the philosophical faculty in 1915. It was not until June 1919 that the Berlin Ministry of Education agreed to exempt her from the decree that restricted habilitation to male candidates [Rowe, 2021, 44–61]. Considering that this had transpired only a half-year earlier, her letter to Dehn shows that Noether was not only well connected but her opinions carried real weight as well. Schmeidler and Steinitz both accepted the offers from Kiel, and Dehn then brought Nielsen to Breslau, though he would leave for Copenhagen after just one year (for more on Nielsen, see Chapter 6).

Dehn also decided to leave Breslau in 1921 when he was offered a prestigious chair in Frankfurt. Before departing, he may have remembered Emmy Noether's report when he decided to choose Werner Schmeidler as his successor.[37] Dehn's appointment in Frankfurt was partly due to the support of Arthur Schoenflies, the senior mathematician on the faculty. How these complicated negotiations played out is described in detail in Chapter 5.

References

Bergmann, Birgit; Epple, Moritz; Ungar, Ruti, eds.: *Transcending Tradition: Jewish Mathematicians in German-Speaking Academic Culture*, Heidelberg: Springer.

Blumenthal, Otto: Lebensgeschichte, in [Hilbert, 1935, 388–429].

Born, Max: *Mein Leben: Die Erinnerungen des Nobelpreisträgers*, Munich: Nymphenburger.

Brieskorn, Egbert und Purkert, Walter: *Felix Hausdorff: Gesammelte Werke, Biographie*, Bd. IB, Heidelberg: Springer.

Georg Cantor, Beiträge zur Begründung der transfiniten Mengenlehre, *Mathematische Annalen* 46: 481–512; 49: 207–246.

Dedekind, Richard: *Stetigkeit und irrationale Zahlen*, Braunschweig: Vieweg.

Dehn, Max: Die Legendréschen Sätze über die Winkelsumme im Dreieck, *Mathematische Annalen* 53: 404–439.

Dehn, Max: Ueber raumgleiche Polyeder, *Nachrichten der Königlichen Gesellschaft der Wissenschaften zu Göttingen. Math.-phys. Klasse*, 1900: 345–354.

Dehn, Max: Über den Rauminhalt, *Mathematische Annalen* 55: 465–478.

Dehn, Max: Zwei Anwendungen der Mengenlehre in der elementaren Geometrie, *Mathematische Annalen* 59: 84–88.

Dehn, Max: Die Euler'sche Formel im Zusammenhang mit dem Inhalt in der nichteuklidischen Geometrie, *Mathematische Annalen* 61: 561–586.

Dehn, Max: Über den Inhalt sphärischer Dreiecke, *Mathematische Annalen* 60: 166–174.

---

[36] Noether to Dehn, 8 January 1920, Dehn Papers, Dolph Briscoe Center for American History, University of Texas at Austin.

[37] When another position became vacant in Breslau just one year later, it went to Fritz Noether, whose sister afterward often visited him there during holidays.

Dehn, Max: Besprechung von K.T. Vahlen, Abstrakte Geometrie, *Jahresbericht der Deutschen Mathematiker-Vereinigung* 14(1905): 535–537.

Dehn, Max: Entgegnung, *Jahresbericht der Deutschen Mathematiker-Vereinigung* 14(1905): 595–596.

Dehn, Max and Heegaard, Poul: Analysis situs, *Enzyklopädie der Mathematischen Wissenschaften* III, AB 3, Leipzig: Teubner, 153–220.

Dreben, Burton and Kanamori, Akihiro: Hilbert and Set Theory, *Synthese* 110(1): 77–125.

Ebbinghaus, Heinz-Dieter: *Ernst Zermelo, An Approach to His Life and Work*, Heidelberg: Springer.

Ebbinghaus, Heinz-Dieter: Zermelo and the Heidelberg Congress 1904, *Historia Mathematica* 34(2007): 428–432.

Epple, Moritz: *Die Entstehung der Knotentheorie. Kontexte und Konstruktionen einer modernen mathematischen Theorie*, Wiesbaden: Vieweg.

Fraenkel, Abraham: *Lebenskreise. Aus den Erinnerungen eines jüdischen Mathematikers*, Stuttgart: Deutsche Verlags-Anstalt.

Hausdorff, Felix: Der Potenzbegriff in der Mengenlehre. *Jahresbericht der Deutschen Mathematiker-Vereinigung* 13(1904): 569–571.

Hausdorff, Felix: *Felix Hausdorff: Gesammelte Werke, Grundzüge der Mengenlehre*, Bd. II, E. Brieskorn et al., Hrsg., Heidelberg: Springer.

Hawkins, Thomas: *Emergence of the Theory of Lie Groups: An Essay in the History of Mathematics, 1869–1926*, New York: Springer.

Hessenberg, Gerhard: Beweis des Desarguesschen Satzes aus dem Pascalschen, *Mathematische Annalen* 61: 161–172.

Hilbert, David: Grundlagen der Geometrie, in *Festschrift zur Einweihung des Göttinger Gauss-Weber Denkmals*, Leipzig: Teubner (2. Aufl., 1903); *Grundlagen der Geometrie (Festschrift 1899)*, Klaus Volkert, ed., Heidelberg: Springer.

Hilbert, David: Über den Zahlbegriff, *Jahresbericht der Deutschen Mathematiker-Vereinigung* 8(1900): 180–84.

Hilbert, David: Mathematische Probleme. Vortrag, gehalten auf dem Internationalen Mathematikerkongreß zu Paris, 1900. *Nachrichten der Königlichen Gesellschaft der Wissenschaften zu Göttingen. Math.-phys. Klasse*, 253–297; reprinted in [Hilbert, 1935, 290–329].

Hilbert, David: *Grundlagen der Geometrie*, 2. Auflage, Leipzig: Teubner.

Hilbert, David: Über die Grundlagen der Logik und der Arithmetik, *Verhandlungen des III. Internationalen Mathematiker-Kongresses in Heidelberg 1904*. Leipzig: Teubner, S. 174–185.

Hilbert, David: *Grundlagen der Geometrie*, 7. Auflage, Leipzig: Teubner.

Hilbert, David: *Gesammelte Abhandlungen*, Bd. 3, Berlin: Springer.

Hilbert, David: *Foundations of Geometry*, trans. of tenth ed., LaSalle, Ill: Open Court.

Hilbert, David: *Theory of Algebraic Invariants*, trans. Rienhard C. Laubenbacher, Cambridge: Cambridge University Press.

Hilbert, David: *The Theory of Algebraic Number Fields*, trans. Iain T. Adamson, Berlin, New York: Springer.

Hilbert, David: *David Hilbert's Lectures on the Foundations of Geometry, 1891–1902*, Michael Hallett and Ulrich Majer, eds. Heidelberg: Springer.

Koreuber, Mechthild: *Emmy Noether, die Noether-Schule und die moderne Algebra. Zur Geschichte einer kulturellen Bewegung*, Heidelberg: Springer.

Magnus, Wilhelm (1978/79): Max Dehn, *Mathematical Intelligencer*, 1(3): 132–143.

Meschkowski, Herbert und Nilson, Winfried, eds.: *Georg Cantor Briefe*, Berlin, Springer.

Minkowski, Hermann: *Briefe an David Hilbert*, L. Rüdenberg und H. Zassenhaus, Hrsg., New York: Springer.

Moore, Gregory H.: *Zermelo's Axiom of Choice. Its Origins, Development and Influence*, New York/Heidelberg/Berlin: Springer-Verlag.

Purkert, Walter: On Cantor's Continuum Problem and Well Ordering: What really happened at the 1904 International Congress of Mathematicians in Heidelberg, *A Delicate Balance: Global Perspectives on Innovation and Tradition in the History of Mathematics. A Festschrift in Honor of Joseph W. Dauben,*, David E. Rowe, Wann-Sheng Horng, eds., Basel: Birkhäuser, pp. 3–24.

Rowe, David E.: *A Richer Picture of Mathematics: The Göttingen Tradition and Beyond*, New York: Springer.

Rowe, David E., Hrsg.: *Otto Blumenthal, Ausgewählte Briefe und Schriften I, 1897–1918*, Mathematik im Kontext, Heidelberg: Springer.

Rowe, David E.: *Emmy Noether: Mathematician Extraordinaire*, Cham: Springer Nature Switzerland.

Rowe, David E. und Felsch, Volkmar, Hrsg.: *Otto Blumenthal, Ausgewählte Briefe und Schriften II, 1919–1944*, Mathematik im Kontext, Heidelberg: Springer.

Schoenflies, Arthur: Die Entwickelung der Lehre von den Punktmannigfaltigkeiten, *Jahresbericht der Deutschen Mathematiker-Vereinigung* 8, Heft 2.

Schoenflies, Arthur: Über die Möglichkeit einer projektiven Geometrie bei transfiniter (nicht archimedischer) Maßbestimmung, *Jahresbericht der Deutschen Mathematiker-Vereinigung* 15: 26–41.

Schoenflies, Arthur: Zur Erinnerung an Georg Cantor, *Jahresbericht der Deutschen Mathematiker-Vereinigung* 31: 97–106.

Scott, Charlotte Angas: The International Congress of Mathematicians in Paris, *Bulletin of the American Mathematical Society* 7(2)(1900): 57–79.

Siegmund-Schultze, Reinhard: Theodor Vahlen – zum Schuldanteil eines deutschen Mathematikers am faschistischen Mißbrauch der Wissenschaft, *NTM Schriftenreihe für Geschichte der Naturwissenschaften, Technik und Medizin*, 21(1): 17–32.

Siegmund-Schultze, Reinhard: Emmy Noether: The Experiment to Promote a Woman to a Full Professorship, *Newsletter of the London Mathematical Society* 476(May 2018): 24–29.

Tobies, Renate: *Felix Klein. Visions for Mathematics, Applications, and Education*, Basel: Birkhäuser.

Toepell, Michael-Markus: *Über die Entstehung von David Hilberts "Grundlagen der Geometrie"*, Göttingen: Vandenhoeck & Ruprecht.

Vahlen, Theodor: *Abstrakte Geometrie*, Leipzig: Teubner.

Vahlen, Theodor: Max Dehns Besprechung meiner Abstrakten Geometrie, *Jahresbericht der Deutschen Mathematiker-Vereinigung* 14(1905): 591–595.

Vahlen, Theodor: Erwiderung, *Jahresbericht der Deutschen Mathematiker-Vereinigung* 15(1906): 73–74.

Vahlen, Theodor: Über Stetigkeit und Messbarkeit, *Jahresbericht der Deutschen Mathematiker-Vereinigung* 15(1906): 214–215.

Veblen, Oswald: Review of K.T. Vahlen, *Abstrakte Geometrie*, *Bulletin of the American Mathematical Society* 12(10): 505–506.

Wallis, Ludwig: *Der Göttinger Student. Oder Bemerkungen, Ratschläge und Belehrungen über Göttingen und das Studentenleben auf der Georgia Augusta*, Göttingen: Vandenhoeck & Ruprecht.

Wiener, Hermann: Über Grundlagen und Aufbau der Geometrie, *Jahresbericht der Deutschen Mathematiker-Vereinigung* 1(1891): 45–48.

Wiener, Hermann: Weiteres über Grundlagen und Aufbau der Geometrie, *Jahresbericht der Deutschen Mathematiker-Vereinigung* 3(1893): 70–80.

Yandell, Ben H.: *The Honors Class. Hilbert's Problems and their Solvers*, Natick, Mass.: AK Peters.

Zermelo, Ernst: Hydrodynamische Untersuchungen über die Wirbelbewegungen in einer Kugelfläche, *Zeitschrift für Mathematik und Physik* 47: 201–237.

Zermelo, Ernst: Beweis, daß jede Menge wohlgeordnet werden kann. Aus einem an Herrn Hilbert gerichteten Briefe, *Mathematische Annalen*, 59(1904): 514–516.

Zermelo, Ernst: Neuer Beweis für die Möglichkeit einer Wohlordnung, *Mathematische Annalen*, 65(1908): 107–128.

Zermelo, Ernst: *Ernst Zermelo. Collected Works. Gesammelte Werke*, Heinz-Dieter Ebbinghaus and Akihiro Kanamori, eds., Band 1 (Mengenlehre, Varia), Band 2 (Variationsrechnung, Angewandte Mathematik und Physik), Heidelberg: Springer.

CHAPTER 3

Dehn's Early Mathematics

Jeremy J. Gray and John McCleary

DEHN was a geometer of outstanding fantasy, his problems rooted in the intuitive.[1]

W. MAGNUS AND R. MOUFANG [MAGNUS AND MOUFANG, 1954]

FIGURE 1. Max Dehn aged 24.[2]

1. Introduction

After one semester in Freiburg Max Dehn came to Göttingen for the following Summer semester of 1897, a little before his 19th birthday. As we learned in Chapter 2, he was immediately recognized as a strong student, a *Leitfuchs*. Dehn's later student, Wilhelm Magnus (1907–1990) wrote how "It was a most fortunate coincidence that Dehn met Hilbert during Hilbert's 'geometric period'" [Magnus, 1978/79]. Though we don't have records of precisely which courses he took, we know that Dehn attended several taught by Klein and Hilbert and that he became an eager participant of the mathematical community of Göttingen.

©2024 American Mathematical Society
[1]DEHN war ein Geometer von ausgezeichneter Phantasie, seine Probleme wurzeln in Anschaulichen.
[2]Max Dehn family photos, private collection of Joanna Dehn Beresford

Dehn's earliest works, his doctoral thesis and his work on scissors congruence, owe much to Hilbert's interest in the foundations of geometry. It is in his lectures that we find explicit statements of the problems Dehn tackled. Dehn's strength as a mathematician became apparent in a relatively short time having finished his thesis and the solution to a storied problem around his 22nd birthday. See Figure 1 for his photo around this time.

2. Dehn's thesis

In the standard story of non-Euclidean geometry, the mathematical conclusion is that in two dimensions there is a very limited number of geometries: spherical or elliptic geometry, plane (Euclidean) geometry, and hyperbolic or non-Euclidean geometry. These canonical examples are given by the sphere, the plane, and the non-Euclidean disc with their metrics of constant positive, zero, and negative curvature, respectively.

In more historical terms, as is well known, the story starts with geometry as described by the axioms of Euclid's *Elements*, which is taken to be consistent, and with a concern about the so-called parallel postulate. Numerous investigations into the possibility of a geometry that obeyed all the axioms of the *Elements* except the parallel postulate concluded that there was one other, which was often called non-Euclidean geometry. In addition, if one was prepared to drop another axiom, the indefinite extendability of lines, there was a third, the already well known geometry on the sphere.

The way these results were proved (about which more later) established that there were two trichotomies. In the first, either the angle sum of one (and therefore every) triangle was less than two right angles ($2R = \pi$), or the angle sum equalled $2R$, or it exceeded $2R$. In the second trichotomy, given a line ℓ and a point P not on ℓ, either every line through P met ℓ (there were no parallels to ℓ), or there was a unique line through P that did not meet ℓ, or there were infinitely many lines through P that did not meet ℓ. Happily, these trichotomies fitted exactly: the angle sum of a triangle exceeded $2R$ if and only if there were no parallels; in the Euclidean case, the angle sum is $2R$ and there is a unique parallel to ℓ through P; the angle sum is less than $2R$ and parallels exist but are not unique.

In the various editions of Legendre's *Geometrie* [Legendre, 1794] he explored the status of the parallel postulate, providing some of the last and most sophisticated attempts to prove it from the other axioms of Euclid. What is known as Legendre's first theorem states that there is no geometry in which the angle sum of a triangle is greater than $2R$. The agreement of the trichotomies is his second theorem. In the early 20th century it was discovered that these results had been known well over a century earlier to Gerolamo Saccheri (1667–1733), published in his 1733 book [Saccheri, 1920].

Legendre's proof of the first theorem is a model of simplicity and clarity. Suppose $\triangle A_1 B_1 A_2$ is a triangle whose interior angle sum exceeds $2R$. Extend the base $A_1 A_2$ by n equal segments $A_1 A_2, A_2 A_3, \ldots, A_n A_{n+1}$ and construct n copies of $\triangle A_1 B_1 A_2$ with base $A_i A_{i+1}$ and vertex B_i. Join $B_i B_{i+1}$ giving us n copies of $\triangle B_1 A_2 B_2$ (see Figure 2). Let $\angle A_i B_i A_{i+1}$ be denoted by β and $\angle B_i A_{i+1} B_{i+1}$ by α. Because $\angle B_1 A_2 A_1 + \alpha + \angle B_2 A_2 A_3 = 2R$, we see that $\beta > \alpha$. By Euclid's Proposition I.18, the greater angle subtends the greater side, so $A_i A_{i+1} > B_i B_{i+1}$ for all i.

Euclid's Proposition I.20, two sides of a triangle exceed the third, implies that the path $A_1B_1B_2\cdots B_{n+1}A_{n+1}$ exceeds $A_1A_2\cdots A_{n+1}$. By construction we have $A_1B_1 + nB_1B_2 + B_{n+1}A_{n+1} > nA_1A_2$, and so $2A_1B_1 > n(A_1A_2 - B_1B_2)$. Because $A_1A_2 - B_1B_2 > 0$, the Archimedean Principle implies that there is a value of n for which $n(A_1A_2 - B_1B_2) > 2A_1B_1$. By this contradiction, $\beta \leq \alpha$ and the sum of the interior angles of a triangle is less than or equal to $2R$.

The results of Saccheri that are equivalent to Legendre's first and second theorem also depend on the Archimedean Principle, particularly in his Propositions XI and XII of [Saccheri, 1920]. This result excludes spherical geometry from the discussion of possible Euclidean geometries, tilting Saccheri toward his attack on a geometry where the sum of the angles of a triangle is less than $2R$.

The discovery of non-Euclidean geometry was disturbing. It ended any hope that there was a unique geometry that would therefore be the only possible geometry of physical space. And it challenged the confidence of mathematicians in the quality of their own arguments: There had been a process of rigorously re-examining Euclid's *Elements*, only to discover that the parallel postulate could not be proved after all. To a list of lapses and omissions that Greek and later Arab and Islamic writers had found were added other imperfections, and it seemed best to start again and finish the job that Euclid had begun. (See, for example, the book of Pasch, [Pasch, 1882/1926].)

Hilbert had been thinking about the foundations of geometry since 1891 when he gave a course on projective geometry (to two or three students) in Königsberg. In the notes of his 1894 course on the foundations of geometry we find a list of 38 contemporary books on geometry (See [Hallett and Majer, 2004]).

For the 1891 course, Hilbert divided topics into the geometry of intuition, axiomatic geometry, and analytic geometry. He called attention to a difference between geometry and subjects like number theory and algebra; the latter require only pure thought while geometry depends on physical intuition. Axiomatization of geometry admits the use of the tools of pure thought and axiomatic analysis illuminates the manner in which the axioms support the proofs of fundamental results. Hilbert refers to Hertz [Hertz, 1894] whose 1894 axiomatization of mechanics provided an example of how a physical theory might be framed more formally.

Hilbert attended the 1891 meeting in Halle of the *Deutsche Mathematische Vereinigung* where Hermann Wiener (1857–1939) lectured on the foundations of geometry. Blumenthal remarked in 1935 on Hilbert's excitement following the lecture. Wiener claimed that the major theorems of projective geometry might be proved without continuity assumptions. It was in this context that Hilbert made the well-known remark that it must be possible to replace "point, line, and plane" with "table, chair, and beer mug" without affecting the validity of the theorems of geometry (see the account in [Corry, 2004]).

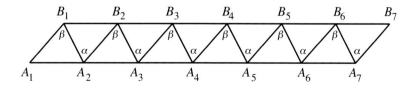

FIGURE 2. Argument of Legendre

In the 1894 Summer semester Hilbert offered a course entitled *Über die Axiome der Geometrie*. He organized the *Kernsätze* (axioms) of Pasch into five groups as he would later in the *Grundlagen der Geometrie*, here with the axioms of continuity and parallels reversed. Of particular importance in these notes (see [Hallett and Majer, 2004]) is the introduction of numerical measures obtained using the cross ratio of projective geometry. The continuity assumption is important here in order to obtain a concept of the real numbers in geometric terms.

In 1895 Hilbert moved to Göttingen, bringing his interest in problems in the foundations of geometry with him. At the time, Göttingen had obtained permission to offer short courses (*Ferienkurse*) in mathematics during Easter break for *Oberlehrer*, teachers at the Gymnasiums. Hilbert's offerings in 1896 and 1898 sought to show that modern research mathematics was not so far removed from the topics taught in the schools. The 1898 *Ferienkurs* was entitled *Über den Begriff des Unendlichen*:

> I wish to show that the *most modern directions* and *the most far flung research* of science stands in the closest relation to the *elements of arithmetic and geometry*, as they are first learned and taught in schools.[3]

The lecture notes focus on the infinite in arithmetic and in Euclidean geometry. Two-thirds of the notes are dedicated to geometry and they suggest a warm-up to his upcoming lecture course in the Winter semester of 1898/99 on the foundations of Euclidean geometry. In Göttingen this topic seemed to come out of the blue. Hilbert entitled his 1898 Winter Semester course *Elemente der Euklidischen Geometrie*. Otto Blumenthal commented how "Surprise and wonder grew however, as the lecture began and a wholely novel content was developed."[4]

The notes of the course led to the famous *Grundlagen der Geometrie*, prepared for the *Festschrift* [Hilbert, 2015] marking the unveiling of the Gauss-Weber memorial in Göttingen. Although it was not the first, Hilbert gave the best framework for the foundations of geometry. He not only provided a clear foundation, he re-invigorated the method of providing axioms in mathematics. To be clear, he did not entirely sort out all of the relations between his axioms and their conclusions; some of his arguments were later improved. On the other hand, Hilbert's axiomatic approach remains the principal way in which new concepts are introduced into mathematics, even in areas far beyond elementary geometry. The *Grundlagen der Geometrie* is also an eloquent example of how a rigorous theory of a subject accessible to beginners can require some very advanced mathematics.

In this book, Hilbert laid down five groups of axioms:

I: connection or incidence (*Axiome der Verknüpfung*);

II: order or betweenness (*Axiome der Anordnung*);

III: parallels (Euclid's axiom, *Axiom der Parallelen*);

IV: congruence (*Axiome der Congruenz*);

V: continuity, the *Archimedean axiom* (later *Axiom der Stetigkeit* after Hilbert added a completeness axiom).

---

[3]Ich möchete zeigen, dass die *modernsten Richtungen* und *die äussersten Fortschritte* der Wissenschaft in engster Beziehung zu den *Elementen der Arithmetik und Geometrie* stehen, wie sie ja zuerst auf den Schulen gelernt und gelehrt werden. [Hallett and Majer, 2004]

[4]Staunen und Bewunderung aber erwachten, als die Vorlesung begann und einen völlig neuartigen Inhalt entwickelte. [Blumenthal, 1935]

He then discussed the compatibility and mutual independence of the axioms, developed what he called an algebra of segments and with it a theory of proportion. He then outlined a theory of plane area that, as was already known, did not depend on the calculus, and discussed the theorems of Desargues and of Pappus (incorrectly called Pascal's theorem in the book).

Each group of axioms is followed by theorems that are consequences of those axioms. In this way it becomes clear how theorems later in the book depend on earlier axioms and their already established consequences. The order of presentation is a little odd – congruence axioms come after the axiom of parallels but do not depend on any assumption about parallels.

The axiom of parallels says that, given a line in a plane and a point in the plane not on that line, there is a unique line in the plane through the given point that does not meet the given line.

The axioms of congruence capture what we mean when we say that two line segments are the same but in different positions, and when two angles are the same but in different positions. They also say (axiom IV, 6) that two triangles are congruent if two sides and their included angle are congruent. Hilbert showed that the axioms of congruence and parallels together imply that the angle sum of any triangle is $2R$.

The theorems in Hilbert's *Grundlagen der Geometrie* follow from the axioms and hence obtain in any situation to which the axioms apply. New concepts, such as the ones introduced by Dehn in his thesis, either follow from Hilbert's examples or they are new but compatible with the *Grundlagen*. In axiomatic analysis such concepts must behave in a manner that allows us to distinguish them from the concepts Hilbert showed follow from the axioms.

Dehn's focus in the 1900 work [Dehn, 1900c] is the so-called axiom of Archimedes. It says:

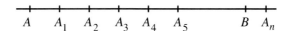

FIGURE 3. The Archimedean Axiom

Let A_1 be any point upon a straight line between the arbitrarily chosen points A and B. Take the points $A_2, A_3, A_4, \ldots$ so that A_1 lies between A and A_2, A_2 lies between A_1 and A_3, etc. Moreover, let the segments

$$AA_1, A_1A_2, A_2A_3, A_3A_4, \ldots$$

be equal to one another. Then, among this series of points, there always exists a certain point A_n such that B lies between A and A_n. (See Figure 3.)

With this axiom any line segment can be used to measure the size of any other line segment. It is so intuitive and indeed inescapable in our experience that, although it had been explicitly used by Archimedes – and, indeed, Euclid – no-one had thought to draw attention to it.

In proving its independence from the other axioms Hilbert introduced non-Archimedean fields. In the planes over such fields the Archimedean axiom fails.

Geometric arguments by continuity, like the proof of Legendre's first theorem, depend on this axiom, and so may fail in these geometries. It was this axiom that Hilbert asked Dehn to investigate. Could there be a geometry in which it was false, and if so what were the theorems in that geometry?

Dehn based his study on that part of the *Grundlagen* which assumes the axioms of incidence (I), of order (II), and of congruence (IV), but not of parallels (III). This defines a space that is known as a Hilbert plane ([Hartshorne, 2000]), in which all but two of the first 28 theorems of Book I of Euclid's *Elements* hold; the exceptions have to do with theorems about when one circle meets another. If the parallel postulate is also assumed, then a Hilbert plane is a model for Euclidean geometry.

Pure incidence theorems also hold in a Hilbert plane, notably Desargues' theorem. However, to prove this theorem, Hilbert assumed the axiom of parallels. He did, however, show that all six of the axioms of congruence are required; his convoluted demonstration of this was soon superseded in 1902 by a much simpler argument by F.R. Moutlon (1872–1952) [Moulton, 1902]. Hilbert also showed that the axioms of congruence can be dropped if an algebra of segments is available instead.

Dehn worked with a Hilbert plane and a new construction that it will be convenient to call a *pseudo-plane* (although Dehn gave it no name). Its defining configuration is the choice of a line and all the lines perpendicular to it. For each line, Dehn added to the Hilbert plane an ideal point where these perpendiculars are understood to meet, which he called the *pole* of the given line. He called the line the *polar* of the point. He then proved [Dehn, 1900c, p. 409] that the collection of all poles of the lines through a given point lie on an (ideal) line, and that all the polar lines of points on that line pass through the given point.

He also proved what we'll call the *perpendicular bisector theorem*: Suppose, as in Figure 4, the line segment AB is bisected at M and ME is the line perpendicular to AB thorough M. If lines a through A and b through B meet on ME, then the base angles at A and B are congruent, and a perpendicular to ME is cut by a and b into congruent segments.

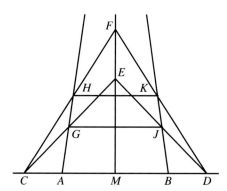

FIGURE 4. Perpendicular bisector theorem.

To define his pseudo-plane, he then chose a point, O, in the enlarged Hilbert plane and removed from that enlarged plane the polar line, t, of the point O. The resulting plane is the pseudo-plane, and his principal theorem is that the axioms defining it generate Euclidean geometry on it, including the axiom of parallels.

On this plane he defined a synthetic concept of *Pseudoparallelen*, pseudo-parallel lines: two lines are pseudo-parallel if they meet at a point on the missing line t. It follows that given a line a and a point A not on a there is a unique line through A that is pseudo-parallel, to a. Furthermore, there will be a pseudo-parallel translation from one segment to another if they can be joined by a sequence of pseudo-parallelograms, quadrilaterals with opposite sides that are pseudo-parallel.

Dehn then defines the synthetic concept of pseudo-congruence: Two line segments are pseudo-congruent if either they are related by a pseudo-parallel translation, or they are related by a pseudo-parallel translation to two segments OE and OF and the line EF passes through the pole of the angle bisector of the angle EOF.

Given a segment AB of a line a and a point A' on a line a' there is a unique point B' on one or other side of the line a' such that AB or BA is pseudo-congruent to $A'B'$ [Dehn, 1900c, p. 413]. In the case of segments on lines that are not pseudo-parallel this requires the perpendicular bisector theorem.

Pseudo-congruence turns out to be an equivalence relation. There is also a concept of pseudo-congruence for angles, which is an equivalence relation, and there is a pseudo-congruence version of the SAS test for congruence in Euclidean geometry: if in a triangle two sides and the included angle are pseudo-congruent then the triangles are pseudo-congruent (that is to say: the remaining sides and angles are pairwise pseudo-congruent). Notice that because pseudo-congruence is a synthetic concept, this is a theorem that must be proved.

Dehn claims that it is easy to see that pseudo-congruence satisfies all the congruence axioms stated by Hilbert, and so the pseudo-plane with the notion of pseudo-parallel obeys all the axioms of Euclid including the axiom of parallels, and the line t plays the role of an 'infinitely distant' line. In particular, the sum of the angles in any triangle is pseudo-congruent to two right angles.

It also follows that just as one may say of any two segments a and a' that either $a < a'$ or $a = a'$ or $a > a'$, one can also say that one segment is less than, equal to, or greater than another with respect to pseudo-congruence: $a \prec a'$ or $a \approx a'$ or $a \succ a'$.

Dehn then observed that Hilbert had introduced his algebra of segments in a way that assumed the axiom of parallels. This algebra had allowed Hilbert to develop a theory of proportion without assuming the Archimedean axiom. It also allowed him to show that the altitudes of a triangle meet in a point as well as Pappus's theorem. But Pappus's theorem is a pure incidence theorem, so it holds automatically in the pseudo-plane. Dehn summed up his principal result this way [Dehn, 1900c, p. 411]: All pure intersection theorems that can be proved on the basis of axiom groups I, II, III, and IV in fact follow from axiom groups I, II, and IV without III (the axiom of parallels).

Dehn next compared congruent and pseudo-congruent segments. Two sequences of points $A_1, A_2, A_3, \ldots$ and $B_1, B_2, B_3, \ldots$ on a line g were congruent if every segment $A_j A_k$ is congruent to the corresponding segment $B_j B_k$. He then proved that if two sequences are congruent then they are obtained one from another by a projection. Thus the cross-ratios of corresponding quadruples of points are equal, and moreover their cross-ratios are pseudo-equal.

This result allowed him to compare congruent and pseudo-congruent segments. He took points O, A, A_1 on a line n and points O, B, B_1 on a line m and such that the segments $OA = a$ and $AA_1 = a_1$ are congruent and the segments $OB = b$ and $BB_1 = b_1$ are congruent, and proved that

$$a \prec, =, \succ a_1 \iff b \prec, =, \succ b_1.$$

Dehn gave two proofs of this claim. The first begins "Since we know that two congruent segments based at O are also pseudo-congruent" (which is not apparent from the paper), this allows him to move all the points onto the same line. He then introduced the cross-ratio of four points $1, 2, 3, 4$ which he defined to be

$$\frac{12.43}{13.42},$$

and used it to give two proofs of the claim. For brevity, we give only the second proof, which, as he said, has its own interest and leads to important results.

It is based on a result about double points in a projective map of a line to itself defined by the map

$$(\bar{A}, O, X, A) \mapsto (O, X, A, A_1),$$

where the points $\bar{A}$ and A define congruent segments on either side of O: $\bar{A}O = OA$. The point A_1 lies on the same side of O as A and the segments OA and AA_1 are also congruent. Therefore, the points $\bar{A}, O, A$ and O, A, A_1 form two congruent sequences.

Let X be a fixed point of the map from one sequence to another, and let the pseudo-lengths of the relevant segments be

$$\bar{A} = -a, \quad OX = x, \quad \text{and} \quad AA_1 = a_1.$$

Then

$$\frac{\bar{A}O.AX}{\bar{A}X.AO} = \frac{OX.A_1A}{OA.A_1X},$$

or

$$\frac{(-a)(x-a)}{(x+a)(-a)} = \frac{xa_1}{a(x+a+a_1)}.$$

This leads to the conclusion that

$$x^2 = \frac{a^2(a+a_1)}{a-a_1}.$$

Applied to the congruent quadruples $(\bar{B}, O, A, B)$ and (O, B, A', B_1) this leads, after a little more algebra, to the conclusion that

$$\frac{a^2(a+a_1)}{a-a_1} = \frac{b^2(b+b_1)}{b-b_1}.$$

If $a \succ a_1$ then the left hand side is positive, and therefore the right hand side must be, and so $b \succ b_1$, as required. The other conclusions fall out similarly.

Finally, in §7, Dehn turned these results to good advantage. He stated and proved the result that if in the pseudo-plane the angle sum of one triangle is less than (resp. equal to, resp. greater than) $2R$, then so it is in every triangle.

To prove it, he observed that it is enough to prove it for right-angled triangles, and took a triangle with right angle at O. As in Figure 5, let the triangle be AOB, let C be the midpoint of AO, and let OD be pseudo-congruent to CA. Let CF be the perpendicular to AO at C, and let it meet AB at F. Let FH be perpendicular to CF at F, meeting OB at H. Let the perpendicular from D to AO meet FH at G. Then $FC \approx GD$, and so triangles FCD and GDO are pseudo-congruent,

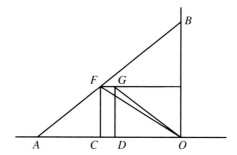

FIGURE 5. Proof of Legendre's theorem.

whereas triangle FCA and FCO are congruent. Therefore the angles FAC and FOC are equal, and the angles FCA and GDO are pseudo-equal.

Suppose now that $OD \prec OC$. Then angle GOD is greater than angle FOC. But the angles of triangle AOB are pseudo-equal to $2R$ and so the angles at O are pseudo-equal to R and the angle sum of triangle AOB is greater than $2R$.

Similarly, if $OD = OC$ or $OD \succ OC$ the angle GOD is equal to or less than angle FOC and the same argument shows that the angle sum of triangle AOB is, respectively, equal or less than $2R$.

To complete the analysis of Legendre's first and second theorem, it remains to connect these triangles to a theory of parallels. In particular, does the angle sum of triangles determine the status of the parallel postulate?

The idea will be to show that it is possible to single out an infinitesimal part of a non-Archimedean field and to use it to define a sort of infinitesimal space within a Hilbert plane. Dehn used a variant of the field Ω, defined in [Hilbert, 2015, §9], which consisted of arbitrary finite extensions of the rationals, $\mathbb{Q}$, that result by adjoining elements of the form $\sqrt{1+\omega^2}$. Dehn took the field $\Omega(t)$ that is defined in the same way but starting with the reals, $\mathbb{R}$.

The allowable transformations of the space Dehn stipulated are the real linear transformations that preserve the null conic $x^2 + y^2 + 1 = 0$; this mimics the usual construction of elliptic geometry. Figures in the space are congruent if one is obtained from the other by an allowable transformation.

Given two elements of this field, $f(t)$ and $g(t)$, Dehn says that $f(t) < g(t)$ if $f(t) - g(t)$ is negative for large enough values of t. This means that every real number is less than t, and more generally that there are non-zero x and y such that

$$-\frac{n}{t} < x, y < \frac{n}{t} \quad (*)$$

for all positive integers n. The domain Dehn defined is the set of all points (x, y) where x and y satisfy the condition $(*)$.

Straight lines in this domain are defined by equations of the form $ux + vy + w = 0$, where the coefficients u, v, w also satisfy condition $(*)$.

Dehn then showed that indefinite repetition of a segment does not leave the domain, nor do rotations of a segment, nor does a segment at right angles to a given segment.

It follows that in this domain axioms of connection, order, and congruence hold, but there are many parallels to a given line through a given point, and the angle sum of every triangle is *greater* than $2R$. Dehn called this non-Archimedean geometry *non-Legendrean*, because the opposite of the theorem Legendre had validly proved in the Archimedean case was true.

Dehn concluded the paper in §9 with a similar argument over a different domain, in which the x and y coordinates of each point satisfy $-n < x, y < n$ for every positive integer n. In this way he exhibited a geometry in which axioms I, II, and IV hold, but there are many parallels to a given line through a given point, and the angle sum of every triangle is equal to $2R$ – a case he called *semi-Euclidean*.

Dehn also investigated the consequences of the axiom groups I, II (suitably modified), and IV together with the assumption that there are no parallels, and showed that in this case the angle sum of every triangle is greater than $2R$. This theorem, he wrote, 'asserts the highly surprising fact that the analogue of Legendre's first theorem in the case of the non-existence of parallels holds without any assumption of continuity.'[5] He was able to remove the need for continuity from the proof.

Dehn summarised his findings in this (translated) table:

Angle sum of triangle	Through a point to a line there are		
	no parallels	one parallel	infinitely many parallels
$> 2R$	Elliptic geometry	(impossible)	Non-Legendrean geometry
$= 2R$	(impossible)	Euclidean geometry	Semi-Euclidean geometry
$< 2R$	(impossible)	(impossible)	Hyperbolic geometry

Let us briefly survey how these conclusions come about, by filling in the gaps in this table:

Angle sum of triangle	Through a point to a line there are		
	no parallels	one parallel	infinitely many parallels
$> 2R$	Elliptic geometry	A	B
$= 2R$	C	Euclidean geometry	D
$< 2R$	E	F	Hyperbolic geometry

Euclidean geometry is traditional, and freshly re-established by Hilbert in the *Grundlagen der Geometrie*. Elliptic geometry is a close relative of the geometry on a sphere. Hyperbolic geometry is the discovery of Bolyai and Lobachevskii in the 1830s. By Legendre's second theorem, any remaining geometries must require the negation of the Archimedean axiom.

---

[5]Dieser Satz spricht die höchst überraschende Thatsache aus, dass das Analogon der ersten Legendre'schen Satzes Für den Fall der Nichtexistenz von Parallelen (ohne Voraussetzung der Stetigkeit) gilt.

Hilbert had shown, without any reference to the Archimedean axiom, that if the parallel axiom holds then the angle sum of every triangle is $2R$ – this rules out there being geometries that would occupy boxes A and F.

That leaves boxes C and E, which are ruled out by Dehn's highly surprising fact.

3. Hilbert's Third Problem

Hilbert's celebrated collection of 23 problems [Hilbert, 1900], announced at the 1900 International Congress of Mathematicians in Paris, was intended to shape mathematical research in the 20th century. Many of the problems can be approached through the axiomatic method that Hilbert developed in his exploration of the foundations of geometry. Briefly put, the axiomatic method strives to provide a description of a mathematical phenomenon by the axioms that characterize it. By examining the relations among the axioms one can reveal the underlying principles supporting the phenomenon (see [Corry, 2004]). The independence of the axioms acts as a critical tool by suggesting possible alternative constructions by assuming some axioms but not others. The Parallel Postulate of Euclid presents a motivating example. Hilbert's axiomatic approach also enabled him to develop a new theory of planar area independent of the Archimedean axiom.

Area of polygons in the plane

In Chapter IV of the *Grundlagen* Hilbert developed the theory of area for polygons (*Flächeninhalt*). His approach is based on Euclid's – two polygonal figures have the same content (*Inhalt*) if you can decompose one of them into finitely many polygonal pieces that can be rearranged to obtain the other polygon. In this case, we say that the polygons are *scissors congruent* (SC). The popular puzzle *Tangrams* produces scissors congruent polygons in the plane, for example, as in Figure 6.

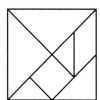 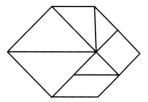

FIGURE 6. Tangrams

For Euclid two polygons are (of) equal (content) if by adding or subtracting congruent polygons to each given polygon, we arrive at a pair of scissors congruent polygons. We say that the polygons are *stably scissors congruent* (SSC) in this case. For example, Figure 7 illustrates the best known proof of the Pythagorean theorem where four copies of a right triangle are added to the square on the hypotenuse and to the pair of squares on the legs to give congruent squares. Subtracting the four copies of the triangle leaves two figures of the same area, which is the Pythagorean theorem. Euclid's "equal" polygons in Book I are stably scissors congruent – Propositions 34 to 45 from Book I constitute Euclid's theory of area for polygons.

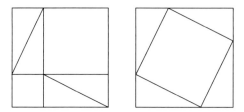

FIGURE 7. Proof of the Pythagorean theorem.

Stably scissors congruent and scissors congruent pairs of polygons have the same area. (If equals be subtracted from equals, the remainders are equal.) This fact leaves open whether the converse holds: if two polygons have the same area are they scissors congruent? William Wallace (1768–1843), Farkas Bolyai (1775–1856), who first posed it, and K. L. Gerwien (1799–1858) independently answered the question in the early nineteenth century. The proof may be illustrated as in Figure 8. Given a polygon P, say a pentagon $ABCDE$, first triangulate it, and then decompose each triangle into a rectangle by halving an altitude and forming the rectangle of that height on the base.

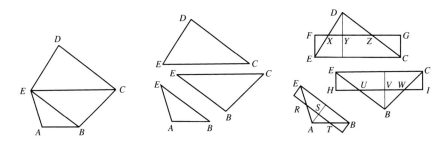

FIGURE 8. Pentagon to rectangles.

Next choose a unit length, OP. It would be nice to swap each rectangle for a rectangle with one side of unit length. Such a swap can be done by applying Euclid's Proposition I.43, another example of a stable scissors congruence. Given a rectangle $EFGC$, extend the side FG to FX with $GX \cong OP$ and EF to EQ so that the diagonal XC meets the line EF at Q. Then complete the rectangle $XFQP$ as shown in Figure 9.

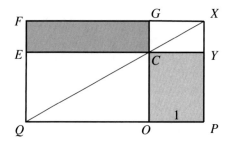

FIGURE 9. Equal area rectangles.

The triangles △CGX and △QEC have been added to the rectangle EFCG, and the triangles △XYC and △COQ have been added to OPYC. But △XQP and △QXF are congruent, as are the pairs △QEC and △QOC, and △XYC and △CGX (I.34). Subtracting the triangles △CGX and △QEC, and △XYC and △COQ, we are left with figures of the same area.

But is the rectangle OPYC scissors congruent to the original rectangle? This question is answered by the Wallace-Bolyai-Gerwien theorem.

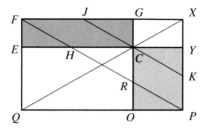

Figure 10. Wallace-Bolyai-Gerwien theorem.

Joining the opposite diagonal of the figure from I.43 (Figure 10), we obtain triangles △FEH ≅ △XYC by Side-Angle-Side. But △XYC ≅ △ROP by ASA. Hence, we can decompose part of each rectangle with congruent right triangles. (See Figure 10.)

Let JK be the line through C parallel to FP. The parallelograms $FHCJ$ and $PRCK$ have the same area by viewing them on congruent bases FH and RP. Euclid's Proposition I.36 gives a stable scissors congruence between $FHCJ$ and $PRCK$. Thus the rectangles $OPYC$ and $EFCG$ are stably scissors congruent.

With such decompositions we can break up each rectangle from the polygon into parts that can be reassembled into a rectangle on a side of unit length. Stacking these rectangles together we obtain a rectangle on a side of unit length equal in area to our original pentagon. For any other polygon of the same area as the pentagon, the same process produces a congruent rectangle on a side of unit length. The two polygonal decompositions may be superimposed to obtain a common polygonal decomposition, and so the two figures are scissors congruent. For polygons, equal area is equivalent to scissors congruence.

Hilbert showed (Kapitel 4 of the *Grundlagen*) that stable scissors congruence of parallelograms does not imply scissors congruence without continuity using his construction of a non-Archimedean field.

Hilbert introduced a measure of content for triangles and showed how it behaved as a version of area for polygons and hence is proportional to area. The measure, half the height times the base, may be taken for area and can be shown to have the property that two polygons have the same area if and only if they have the same measure. Hilbert also showed that this statement does not require the Archimedean axiom.

Volume of polyhedra in space

In a letter to Christian Ludwig Gerling (1788–1864), Gauss drew attention to the apparent lack of a theory of volume based on scissors congruence in three dimensions. He suggested that it was a shame that no such theory existed, even for polyhedra that are symmetric. He mentioned an argument of Legendre in his *Éléments de Geometrie* [Legendre, 1794] proving that two tetrahedra on bases of equal area and of equal height have the same volume. Legendre's argument is shown in Figure 11. If the tetrahedron $ABCS$ is greater than the other tetrahedron $abcs$, let x be a height for which the plane parallel to the plane of the bases cuts $ABCS$ in a triangle for which the prism $ABCX$ has volume the amount by which $ABCS$ exceeds $abcs$.

Choose an integer value k for which $AT/k < Ax$ and subdivide the height AT into k equal pieces. Form the prisms exterior to $ABCS$ and interior to $abcs$ on the triangles made by the intersection of the planes parallel to the base at each multiple of AT/k. Legendre had proved that prisms of the same height on bases of the same area have the same volume, so prism $DEFG$ and $defa$ have the same volume, $GHIK$ and $ghid$, etc. Summing the volumes of the exterior prisms in

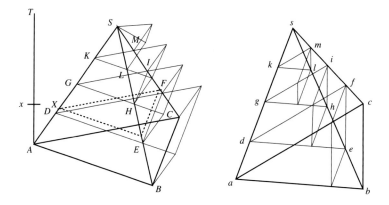

FIGURE 11. Volume argument of Legendre.

$ABCS$ is greater than the volume of $ABCS$, and the sum of the volumes of the interior prisms in $abcs$ is less than the volume of $abcs$. The difference between these sums is greater than the difference between the volumes of the tetrahedra. But the correspondence between volumes of the construction shows that the volume of $ABCD$ is the difference of the sum of the volumes of the exterior prisms and the sum of the volumes of the interior prisms. Hence the volume of $ABCD$ exceeds the volume of $ABCX$. However, these prisms are on the same base so the height of $ABCD$ must be greater than the height Ax of $ABCX$, a contradiction. By a symmetric argument, the volumes of the two tetrahedra are equal. This argument is what Gauss identifies as "... *nicht unabhängig von der Exhaustionsmethode*...". Legendre's proof establishes the equality of the volumes of polyhedra that are symmetric but not congruent by orientation preserving motions.

What Gauss desired was an argument by scissors congruence where one polyhedron is decomposed into polyhedral pieces and then reassembled with exactly these pieces to obtain the second polyhedron. No reflections are required. Gerling responded to show that if two polyhedra were symmetric across some plane, then they are, in fact, scissors congruent in this sense. Gauss praised this result in a letter of April 14, 1844 to Gerling, in which he was critical of Legendre's proof above.

Focusing on Legendre's Proposition VI.14 [Legendre, 1794], we can ask if two tetrahedra on bases of equal area and equal heights are scissors congruent. This particular question was stated explicitly in a book [Stolz, 1885] of Otto Stolz (1842–1905) that Hilbert referred to in his 1898 *Ferienkurs*. There he paraphrased Stolz to report:

> Following Stolz, it is to date not known whether 2 pyramids of equal base area and height are provably scissors congruent (*inhaltsgleich*) in the narrow sense.[6]

This statement became the third problem on Hilbert's list of millennial problems. In the 1902 translation of Mary Winston [Newson, 1902], it states in its entirety:

3. THE EQUALITY OF TWO VOLUMES OF TWO TETRAHEDRA OF EQUAL BASES AND EQUAL ALTITUDES

> In two letters to Gerling, Gauss* expresses his regret that certain theorems of solid geometry depend upon the method of exhaustion, *i. e.*, in modern phraseology, upon the axiom of continuity (or upon the axiom of Archimedes). Gauss mentions in particular the theorem of Euclid, that triangular pyramids of equal altitudes are to each other as their bases. Now the analogous problem in the plane has been solved.† Gerling also succeeded in proving the equality of volume of symmetrical polyhedra by dividing them into congruent parts. Nevertheless, it seems to me probable that a general proof of this kind for the theorem of Euclid just mentioned is impossible, and it should be our task to give a rigorous proof of its impossibility. This would be obtained, as soon as we succeeded in *specifying two tetrahedra of equal bases and equal altitudes which can in no way be split up into congruent tetrahedra, and which cannot be combined with congruent tetrahedra to form two polyhedra which themselves could be split up into congruent tetrahedra.*‡

---

* Werke, vol. 8, pp. 241 and 244.

†Cf. beside earlier literature, Hilbert, Grundlagen der Geometrie, Leipzig, 1899. ch. 4. [Translated by Townsend, Chicago, 1902.]

‡Since this was written M. Dehn has succeeded in proving this impossibility. See his note: "Ueber raumgleiche Polyeder," in Nachrichten d. K. Gesellsch. d. Wiss. zu Göttingen, 1900, and a paper soon to appear in Math. Annalen [vol. 55, 465-478].

As the footnote ‡ to the English translation of Hilbert's problems states, Dehn had provided a solution to the third problem shortly after the Paris International Congress. The actual lecture in Paris did not feature all of Hilbert's 23 problems [Gray, 2000]. We can speculate that Hilbert discussed the problem with Dehn before Paris, and that Dehn may have read Stolz's book [Stolz, 1885] in the year

---

[6]Nach Stolz [Stolz, 1885, p. 78] is es bisher nicht gelungen, 2 Pyramiden von gleichen Grundlinie und Höhe als inhaltsgleich im engeren Sinne nachzuweisen.

that he was writing his dissertation. Dehn's experience with the role of continuity in geometry in his doctoral work made it natural for him to tackle such a problem.

In what follows, we first discuss work on scissors congruence carried out before Hilbert's *Grundlagen*. We then present Dehn's solution to the third problem in the context of Hilbert's investigations of the foundations of geometry. Finally we consider the subsequent work motivated by Dehn's achievement.

Before Dehn

The Wallace-Bolyai-Gerwein theorem provides a foundation for the computation of the area of polygons in the plane via finite processes. A concrete example of the analogous argument in three dimensions is the volume of a pyramid on a square base with dihedral angles $\pi/4$ between the faces and the base. Six copies of this symmetric polyhedron fill a cube and so the volume of the pyramid is a sixth the volume of the cube (see Figure 12). Euclid's arguments in Book XI permit comparison between the volumes of parallelepipeds and rectangular solids on a unit square base. Hence the cube on a unit side may be taken as a standard of volume.

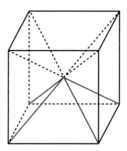

FIGURE 12. Volume of a certain pyramid.

This argument was extended considerably by M.J.M. Hill (1856–1929) in an 1895 paper where he introduces three species of tetrahedra whose volume can be determined by finite means. For example, in §4 of [Hill, 1896] he shows how to "construct a tetrahedron whose volume can be determined without using the proposition that tetrahedra on equal bases and having equal altitudes are equal ...", which is Legendre's main tool. Hill's arguments enlarge on Gerling's observation that polyhedra that are symmetric across a plane have the same volume. His families of tetrahedra enjoy symmetries that allow them to be embedded in prisms whose volumes are known from Euclid.

The limitations of scissors congruence as a theory of volume of polyhedra was considered by others before the publication of the *Grundlagen* and Hilbert's third problem. Two efforts stand out, having been cited by Dehn in his *Annalen* paper. The first [Bricard, 1896] is by Raoul Bricard (1870–1944) appearing in 1896, and the second [Sforza, 1897] by Giuseppe Sforza (1858–1927?) in 1897. The papers present arguments that are underdeveloped, leaving room for Dehn to fill in the missing details.

Bricard was a professor of geometry in Paris in positions for which he taught engineering mathematics. He is well known for his work on flexible polyhedra and on kinematic geometry. He was the president of the Société Mathèmatique de France in 1910.

In *Sur une question de géométrie relative aux polyèdres* [Bricard, 1896] Bricard calls attention to the key issue of the need for "infinitesimal calculus" to compute volumes of polyhedra. Apparently inspired by conversations with a M. Hoffbauer, a lieutenant of the artillery, he considers a polyhedron P decomposed into subpolyhedra p_1, p_2, ... which are then reassembled into another polyhedron P'. Bricard considers an edge of one of the subpolyhedra and the sum of the dihedral angles around that edge in the decompositions of P and P'. Such a sum must be $4R = 2\pi$, if the edge appears in the interior of the polyhedron; $2R$, if the edge appears on a face of P or P'; or a dihedral angle of P or P' if the edge lies along an edge of one of the polyhedra. If we designate the dihedral angles of P by A, B, ... and those of P' by A', B', ..., then Bricard claims that *"on arrivera facilement"* at the relation

$$mA + nB + \cdots - m'A' - n'B' - \cdots \equiv 0 (\bmod 2R),$$

where the coefficients m, n, ..., m', n', ... are integers, not all zero.

Bricard concludes correctly that it is necessary that a certain integral linear function of the dihedral angles of the polyhedra gives a multiple of $2R$. He also considers the case of stable scissors congruence and concludes the same condition. The main example is the pair of a regular tetrahedron and a cube of the same volume. The dihedral angles θ of a regular tetrahedron satisfy $\cos\theta = 1/3$. In a footnote, Bricard shows that θ is not a rational multiple of π. The dihedral angles of a cube, on the other hand, are all R and even in number. It follows that no integral combination of θ can give a rational multiple of $2R$, while even integral combinations of R will. Thus a regular tetrahedron cannot be scissors congruent to a cube.

Bricard goes on to suggest that

> The transformation in question will therefore be impossible in the general case. It follows that one cannot establish the equivalence of two tetrahedra having the same base and the same height, but not satisfying the other particular condition, without having recourse to the decomposition into infinitely small elements.[7]

This essentially addresses Hilbert's statement of the third problem before it was stated. Bricard finished the paper with a discussion of the case where two tetrahedra are symmetric, the result of Gerling, for which he gives the same proof.

The second reference of Dehn is to a paper by Sforza of the Instituto Tecnico Reggio Emilia in Northern Italy. Sforza wrote papers on topics in algebra and geometry, notably on non-Euclidean volumes. He was a translator of the book of Julius Petersen, *Theorie der algebraischen Gleichungen* [Petersen, 1878], bringing an important source on Galois theory to Italy.

Sforza's approach to analyzing scissor congruence focused on the vertices of a polyhedral decomposition of a polyhedron P. He calls a vertex of a subpolyhedron of the decomposition a *knot* and says that P is divided into *knotted* parts. His main theorem states:

---

[7]La transformation en question sera donc impossible dans le cas général. Il en résulte que l'on ne peut établir l'équivalence de deux tétraèrres ayant même base et même hauteur, et ne satisfaisant pas à d'autre condition particulière, sans avoir recours à la décomposition en éléments infiniment petits.

Theorem. *If a convex polyhedron P is divided into knotted convex polyhedral pieces, the sum of the solid angles of the parts differs by a multiple of π from the sum of the associated multiples of the dihedral angles of P by a number not less than 2.*

The use of solid angles is confusing, but the analysis by knots explains it. He considers four cases: (1) the knot is internal to P; (2) the knot is in the interior of a face of P; (3) the knot is on an edge of P; and (4) the knot is a vertex of P. Case (1) leads to a copy of 4π while case (2) adds a copy of 2π. Case (3) contributes copies of the dihedral angles of P and case (4) contributes a value of the sum of the dihedral angles in which the vertex appears minus $(n-2)\pi$, which is the sum of the interior angles of an n-gon, where n is the number of edges meeting at the vertex.

Summing all of these contributions leads to a sum S which satisfies

$$S \pm K\pi = (r_1 + 2)D_1 + (r_2 + 2)D_2 + \cdots.$$

The values r_i count the number of internal knots on an edge of P whose dihedral angle is D_i.

Sforza next proves that the dihedral angle θ of a regular tetrahedron and the dihedral angle δ of an isosceles trirectangular tetrahedron are related by $2\delta = \pi - \theta$. He also gives a complete proof that θ is incommensurable with π. Going further he proved that the dihedral angles of the octahedron, dodecahedron, and the icosahedron are all incommensurable with π.

As a corollary of these lemmas Sforza can prove that the regular tetrahedron and the isosceles trirectangular tetrahedron are not scissor congruent. Without reference to Bricard's work, Sforza has answered Bricard's question, and Hilbert's later Paris question as well. Furthermore, the only regular polyhedron scissors congruent to a cube is the cube itself. While some authors [Bartocci, 2012] have written that Sforza plagiarized Bricard's work, he did develop the ideas further, and filled in details of the key arithmetic facts. However, his construction lacks details that leave gaps. For example, his *knots* must be shared with adjoining subpolyhedra in order to use the (possibly erroneous) value of the spherical angle.

These two flawed attempts, together with the critical importance placed on the question by his doctoral advisor, gave Dehn the opportunity to turn his geometric intuition to the problem.

Dehn's argument

In two papers [Dehn, 1900d, 1900] and [Dehn, 1902a, 1902], Dehn established necessary conditions for two arbitrary polyhedra to be scissors congruent. Dehn's first paper appeared in the *Nachrichten von der Gesellschaft der Wissenschaten zu Göttingen*, having been presented at the 27 October 1900 meeting by Hilbert, less than three months after his 7 August lecture in Paris. In fact, the first complete list of Hilbert's 23 problems appeared in the same issue as Dehn's paper. In 1902 Dehn gave a more thorough exposition of his results in the *Mathematische Annalen*.

Scissors congruence and stable scissors congruence are foundational concepts to describe the relation of having equal area between polygons. In the *Grundlagen* Hilbert uses *flächengleich* for scissors congruent polygons, and *inhaltsgleich* for stably scissors congruent polygons. In Dehn's Göttingen report [Dehn, 1900d] he generalizes *flächengleich* to *raumgleich* for scissors congruent polyhedra, and retains *inhaltsgleich* for stably scissors congruent polyhedra. Dehn sought more precise terms in the *Annalen* paper. He termed scissors congruent polygons and polyhedra as *zerlegungsgleich*, and *endlichgleich* or *ergänzungsgleich* for the relation of stable scissors congruence – he acknowledges Heinrich Liebmann (1874–1939) for the suggested terminology. During the 1950s revival of interest in these relations among polyhedra, Hugo Hadwiger (1908–1981) adopted Dehn's terms [Hadwiger, 1957]. Later, V.G. Boltyanskii (1925–2019) introduced the terms *equidecomposable* for scissors congruent and *equicomplementable* for stably scissors congruent in his book [Boltyanski, 1978]. We will use the more colorful terms of (stably) scissors congruent [Sah, 1979]

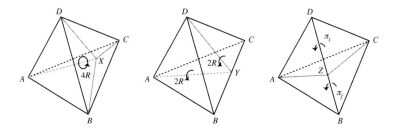

FIGURE 13. Dihedral angles.

To motivate his approach to the third problem, Dehn first considered the two-dimensional case. Suppose P is a polygon in the plane, decomposed into subpolygons. At each vertex of the polygons in the subdivision, the sum of the interior angles at that vertex is either $4R$ if the vertex is in the interior of the polygon, $2R$ if it lies on a side of the polygon, or an interior angle of P if it coincides with a vertex of P. It follows that the sum W of all the interior angles of the subpolygons is equal to $S + 2nR$, where S is the sum of the interior angles of P (see Figure 13).

When P is scissors congruent to P' we get $W = S' + 2n'R$ where S' is the sum of the interior angles of P'. Since the sum W of all the interior angles of the subpolygons is shared by P and P', $S+2nR = S'+2n'R$, or $S \equiv S' \pmod{2R}$. This observation is trivial in Euclidean geometry, however, it remains true in spherical and non-Euclidean geometry where it is more meaningful.

To extend this argument to three dimensions, suppose Π is a polyhedron. Denote the dihedral angles of Π by $\pi_1, \pi_2, \ldots,$ and the dihedral angles of the subpolyhedra by $\tau_1, \tau_2, \ldots$. When we sum the dihedral angles of the subpolyhedra of Π we can get $4R$ from an interior edge, $2R$ from an edge in a face of Π, and we can get multiples of a dihedral angle π_j of Π.

For example, consider Figure 14 which is Figure 3 in [Dehn, 1902a] (we have added color for clarity):

The subpolyhedra have their dihedral angles labelled in Figure 15: In this case, the sum of the dihedral angles leads to the following relation:

$$\tau_1 + \tau_2 + \tau_3 + \tau_4 + \tau_5 + 2\tau_6 + \cdots + \tau_{18} = 2\pi_1 + 2\pi_2 + \pi_3 + \cdots + \pi_6 + 8R.$$

The summand $2\tau_6$ comes from the division of CD into CF and FD; the summand $2\pi_1$ comes from the division of AB into AE and EB. In general, we get a relation

$$\Lambda(\tau_1, \tau_2, \ldots) = L(\pi_1, \pi_2, \ldots, 2R)$$

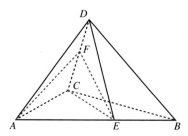

FIGURE 14. Dehn's figure.

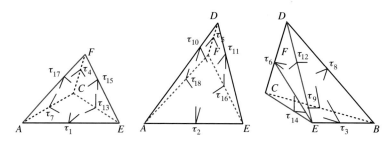

FIGURE 15. The subpolyhedra.

between the sum of all of the dihedral angles of the subpolyhedra in the decomposition and the associated integral linear combination of the dihedral angles of Π and the angle $2R$.

Suppose Π and Π' denote two polyhedra that are scissors congruent to one another. Then there is a decomposition of each polyhedron into a finite collection of subpolyhedra $\{P_1, P_2, \ldots, P_k\}$ and $\{P'_1, P'_2, \ldots, P'_k\}$ with $\Pi_1 = P_1 \cup P_2 \cup \cdots \cup P_k$ and $\Pi_2 = P'_1 \cup P'_2 \cup \cdots \cup P'_k$ for which the subpolyhedra P_i (resp., P'_j) meet one another at most along faces, and P_i is congruent to P'_i for all i.

With these data, the associated integral linear combinations of dihedral angles satisfy
$$L(\pi_1, \pi_2, \ldots, 2R) = \Lambda(\tau_1, \tau_2, \ldots) = L'(\pi'_1, \pi'_2, \ldots, 2R).$$
Ignoring the contributions of multiples of $2R$, we get
$$L(\pi_1, \pi_2, \ldots, 0) \equiv L(\pi'_1, \pi'_2, \ldots, 0) \pmod{2R}.$$
This relation is a precise statement of the key argument of Bricard [Bricard, 1896].

Dehn's analysis reaches further with the introduction of a geometric representation of the data from a polyhedral decomposition. In [Dehn, 1900d] the representation appears in a rectangle, and in [Dehn, 1902a] it is part of a cylinder of radius one. To construct the representation (see Figure 16), let p_i denote the length of the edge in Π with dihedral angle π_i, and, for subpolyhedra, and let t_i denote the length of the edge associated with the dihedral angle τ_i. When we interpret the products $t_i \tau_i$, as the areas of rectangles, their sum can be pictured: Adding in the unshaded portion we obtain the relation
$$t_1 \tau_1 + t_2 \tau_2 + \cdots + p_1(4R - \pi_1) + p_2(4R - \pi_2) + \cdots + (h_1 + h_2 + \cdots)2R = h \cdot 4R.$$
Here h_i is a length of an edge whose sum of dihedral angles is $2R$ (an edge in a face of Π), and h is the sum of all the lengths of the edges of the subpolyhedra.

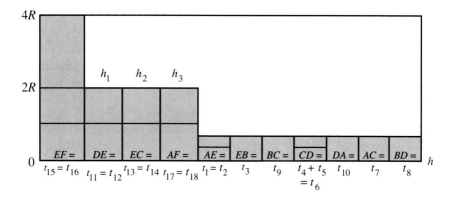

FIGURE 16. Dehn's rectangular subdivision.

Taken modulo $2R$ we get the relation
$$t_1\tau_1 + t_2\tau_2 + \cdots = p_1\pi_1 + p_2\pi_2 + \cdots \pmod{2R}.$$
When two polyhedra Π, Π' are scissors congruent, we have
$$p_1\pi_1 + p_2\pi_2 + \cdots \equiv p'_1\pi'_1 + p'_2\pi'_2 + \cdots \pmod{2R}.$$
This sum of the products of edge lengths and dihedral angles has become known as the *Dehn invariant* of a polyhedron. Notice that is is obtained from the unshaded parts of the rectangle constructed from the decomposition and it is intrinsic to each polyhedron, independent of the decomposition. A necessary condition for scissors congruence of two polyhedra is the equivalence mod $2R$ of these sums.

Dehn went much further, however, to observe that the data on lengths, $t_1, t_2, \ldots$, $p_1, p_2, \ldots, h_1, h_2, \ldots, h$ satisfy a set of linear relations with rational coefficients. For example, in Dehn's figure 3, $t_1 + t_3 - p_1 = 0$, $t_1 - t_2 = 0$, $t_4 + t_5 - p_2 = 0$, as well as, $t_{11} + t_{12} = h_1$, etc. Taking a complete and independent system of such relations determines the data $t_1, t_2, \ldots, p_1, p_2, \ldots, h_1, h_2, \ldots, h$, and furthermore, another solution to the system of linear relations, $\bar{t}_1, \bar{t}_2, \ldots, \bar{p}_1, \bar{p}_2, \ldots, \bar{h}_1, \bar{h}_2, \ldots, \bar{h}$ also implies
$$\bar{t}_1\tau_1 + \bar{t}_2\tau_2 + \cdots + \bar{p}_1(4R - \pi_1) + \bar{p}_2(4R - \pi_2) + \cdots + (\bar{h}_1 + \bar{h}_2 + \cdots)2R = \bar{h} \cdot 4R.$$
This follows from the geometric origin of the quantities involved. Having established this connection, Dehn observes that there are *rational solutions* to the complete system of linear relations and so there is a choice of rational numbers $r_{t_1}, r_{t_2}, \ldots$, $r_{p_1}, r_{p_2}, \ldots, r_{h_1}, r_{h_2}, \ldots, r_h$ satisfying
$$r_{t_1}\tau_1 + r_{t_2}\tau_2 + \cdots + r_{p_1}(4R - \pi_1) + r_{p_2}(4R - \pi_2) + \cdots + (r_{h_1} + r_{h_2} + \cdots)2R = r_h \cdot 4R.$$
Thus
$$t_1\tau_1 + t_2\tau_2 + \cdots = p_1\pi_1 + p_2\pi_2 + \cdots + r2R,$$
with all coefficients rational. Clearing denominators, we can take the coefficients to be integers. When two polyhedra are scissors congruent, these data can be satisfied with the same r_{t_i} which leads to the relation
$$r_{p_1}\pi_1 + r_{p_2}\pi_2 + \cdots = r_{p'_1}\pi'_1 + r_{p'_2}\pi'_2 + \cdots + r \cdot R, \quad \spadesuit$$
with all coefficients integers.

From this relation Dehn establishes the results of Bricard and Sforza. In the case of a regular tetrahedron and a prism [Bricard, 1896], the relation ♠ implies that the dihedral angle of every edge of the regular tetrahedron $\tau = \arccos(1/3)$ is a rational multiple of π. By a thorough and beautiful number-theoretic argument, Dehn [Dehn, 1900d] shows this is impossible. For the regular tetrahedron and the isosceles trirectangular tetrahedron of the same volume [Dehn, 1900d], the relevant dihedral angle for the isosceles trirectangular tetrahedron is ρ which satisfies $2\rho = 2R - \tau$ and so, to be scissors congruent, $\tau \equiv -2\rho \pmod{R}$ from which it follows that τ is a rational multiple of π, again a contradiction. This case solves Hilbert's third problem.

In the *Annalen* paper [Dehn, 1902a] Dehn considers the case of a regular tetrahedron being scissors congruent to a pair of regular tetrahedra. In this case, from the volumes we deduce that $a^3 = b^3 + c^3$ where a, b, and c are the sides of the tetrahedra. Dehn's reduction to sides of rational length leads him to a case of Fermat's Last Theorem and an impossibility. In the final section of the paper, Dehn extends his analysis to stable scissors congruence to show that the same relations over the rational numbers must hold as in the case of scissors congruent polyhedra.

The flow of ideas from number theory and elementary geometry to scissors congruence results mark Dehn's papers on Hilbert's third problem as original and deep. One is reminded of the manner in which Hilbert used non-Archimedean fields, an idea from his work on number theory, to establish independence results in the *Grundlagen*. The transcendence of π was proved in 1882 by Ferdinand von Lindemann (1852–1939), who was the Hilbert's doctoral advisor. This fact played a key role in the papers of Bricard, Sforza, and Dehn on scissors congruence.

4. Further geometry

The heart of the argument in Dehn's solution to Hilbert's third problem is the use of linear systems over the integers that organize the relations among the edge lengths of the constituent subpolyhedra. This system determines the value of a planar area made up of rectangles with sides an edge length and associated dihedral angle. In his 1903 paper *Über Zerlegung von Rechtecken in Rechtecke* [Dehn, 1903], Dehn studied the general question: is there a one-parameter (*eingliedrige*) family of polygons for which every polygon in the plane has a decomposition into pieces from this family? As a first step, Dehn considers the case of rectangles in the plane with the family of polygons in some collection of rectangles whose edges are parallel to a given rectangle.

Generally, if a given rectangle has sides x_0 and y_0 and the subrectangles have sides x_i and y_i for $i = 1, 2, \ldots, n$, then there are integral linear systems

$$S_x : \begin{cases} l_1^x(x_0, x_1, \ldots, x_n) = 0 \\ l_2^x(x_0, x_1, \ldots, x_n) = 0 \\ \vdots \end{cases} \text{ and } S_y : \begin{cases} l_1^y(y_0, y_1, \ldots, y_n) = 0 \\ l_2^y(y_0, y_1, \ldots, y_n) = 0 \\ \vdots \end{cases}$$

that are independent equations, and which represent the relations among the sides in each direction. By assumption the subrectangles fill the given rectangle and so we have the nonlinear relation given by planar area:

$$x_0 y_0 = x_1 y_1 + x_2 y_2 + \cdots + x_n y_n.$$

For example, suppose $y_1 = r_1 x_1, \ldots, y_n = r_n x_n$ for a fixed set of numbers r_1, ..., r_n. In this case, Dehn proves that if a given rectangle can be composed of a number of subrectangles, each subrectangle satisfying one of the relations $y_i = r_i x_i$, then the sides of the given rectangle are in a rational relationship to one another. An immediate corollary is that if a square is composed of a number of squares, then all of the sides of the subsquares are commensurable with the side of the given square.

Thus conditions on the subrectangles restrict the rectangles that can be tiled with members of the family. Dehn ends the paper showing that a one-parameter family of rectangles can tile at most a one-parameter family of rectangles. (See [Wagon, 1987] and [Aigner and Ziegler, 2018] for proofs of related results.) The framework and proofs in [Dehn, 1903] are clearly parallel to his work on scissors congruence.

In 1904 Dehn added to the picture of scissors congruence of polyhedra. In *Zwei Anwendungen der Mengenlehre in der elementaren Geometrie* [Dehn, 1904a] he shows that there are uncountably many pairs of polyhedra with the same volume that are not scissors congruent. Furthermore, this holds in both Euclidean space and in non-Euclidean space. The proofs are applications of the uncountability of the real numbers in any interval.

The notion of scissors congruence for polygons applies to spherical polygons. In *Über den Inhalt sphärischer Dreiecke* [Dehn, 1905] Dehn proved the spherical analogue of the Wallace-Bolyai-Gerwein theorem. The argument establishes that spherical excess is invariant under scissors congruence, that is, if a spherical triangle is decomposed into subtriangles, then the angular excess of the given triangle is the sum of the excesses of the subtriangles. He goes on to prove that a right triangle on the sphere is scissors congruent to a triangle with two right angles. Then that every triangle is scissors congruent to two equal triangles with two right angles. It follows that scissors congruence is an adequate notion of area for polygons on the sphere, and in this way, a theory of area can be developed without the axiom of continuity. Dehn ended the paper with an outline of the hyperbolic case where triangles with two vanishing angles play the role of the double right angle triangles on the sphere.

The last paper from this period in Dehn's career marks a transition from geometric to topological questions. In *Die Eulersche Formel im Zusammenhang mit dem Inhalt in der Nicht-Euklidischen Geometrie* [Dehn, 1905e] of 1905, Dehn returned to Legendre and Legendre's proof of the Euler relation $v - e + f = 2$. In particular, for a convex polyhedron, central projection onto a sphere from an interior point gives a decomposition of the sphere into spherical polygons. Spherical excess, angle sums, and the total surface area of the sphere lead to a proof of Euler's relation. Dehn, with his command of decompositions, spherical excess, and relations among dihedral and spherical angles, set out to obtain a generalization of Euler's relation in dimensions four and five by similar arguments. He acknowledges Poincaré's discovery [Poincare, 1895] of the same relation: $M_0 + M_2 = M_1 + M_3$ where M_i is the number of faces of dimension i and the object is a convex four-dimensional polytope, bounded by tetrahedra. By a careful analysis of boundary relations, Dehn arrived at several relations, aided by the Euler relation in lower dimensions. In five dimensions the simplest boundary pieces are called *Pentatetrahedra*, each consisting of 5 tetrahedra, 10 triangles, 10 edges, and 5 vertices, that is, the boundary of a four-simplex. When all boundary pieces are *Pentatetrahedra* Dehn deduces three linear relations among the combinatorial data including $M_0 + M_2 + M_4 = M_1 + M_3 + 2$, the Poincaré-Euler relation.

Having established these relations, Dehn considered the case of four-dimensional non-Euclidean spaces, where angle excess and defect can be applied directly. He ends the paper with speculation about higher-dimensional cases. He conjectured that in $\mathbb{R}^n$ there are $\left[\frac{n}{2}\right] + 1$ relations between the numbers of the various dimensional faces of a figure built out of simplices. These relations were established by D.M.Y. Sommerville (1879–1934) in 1927 [Sommerville, n.d.] and they have become known as the *Dehn-Sommerville relations*.

For an alternate account of Dehn's early work, see the paper of Stillwell, [Stillwell, 2002].

5. Reception of Dehn's work

Dehn's solution to Hilbert's third problem generated a flurry of interest, especially in the volumes of the *Mathematische Annalen* that immediately followed the publication of his paper. The following year, in volume 56, K.Th. Vahlen (1869–1945) obtained an extension of Dehn's results by further decomposition of the subpolyhedra and the construction of a net from the associated data. He arrived at a relation of Bricard type which allowed him to show that Platonic solids of equal volume fail to be pairwise scissors congruent.

In volume 57, 1903, two papers appeared by authors at the Novorossysky University in Odessa. The first [Kagan, 1903] by Benjamin F. Kagan (1869–1953) offered a simplification of Dehn's argument by focusing on the assignment of integral *masses* to each edge in a decomposition subject to some compatibility conditions. Dehn's papers were difficult to read and Kagan's simplification was welcome, even to Dehn, who refers to Kagan's work in a review of a related paper (JFM 35.0498.01). Kagan led the geometry group in Odessa and edited the collected works of Lobachevsky. He also published a two-part book entitled *Foundations of Geometry*.

The second paper was by S.O. Schatunovsky [Schatunovsky, 1903], translated from the Russian for the *Annalen* and titled *Über den Rauminhalt der Polyeder*. The point of this paper is different from the goal of Hilbert's third problem. Schatunovsky asks if there is a notion of *value* assignable to polyhedra that shares the basic properties of volume. In Hilbert's *Grundlagen*, Chapter IV is dedicated to the study of the area of polygons in the plane. His *Flächenmass* associates a value to each polygon that may be used as a measure of area. Its properties coincide with stable scissors congruence for polygons. Schatunovsky provides axioms for a valuation on polyhedra that lead to a proof that scissors congruent polyhedra have the same valuation. In particular, these axioms do not involve the Archimedean Axiom. Schatanovsky points out that Dehn's results describe how continuity affects the outcome of the construction of valuations.

On 6 February 1903 the Royal Danish Academy of Sciences posted a gold medal problem:

> "Determination of the necessary and sufficient conditions for two polyhedra to be composed of a finite number of pairs of congruent parts ..."

The prize submissions closed 31 October 1904 without a medal winner. As we will see below, the problem was not forgotten.

A member of the Danish Academy at the time and probably a member of prize committee, Christian Juel (1855–1935), published a paper in 1903 in Danish, *Om endelig ligestore Polyedre* (On finitely equivalent polyhedra) [Juel, 1903], in which he gives a decomposition of one of Hill's polyhedra into parts that reassemble a rectangular parallelepiped (Figure 17). He also discusses the problem of when a polyhedron is scissors congruent to a cube. Juel proposed an example of a prism

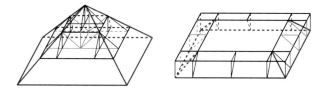

FIGURE 17. The Juel decomposition.

that incorrectly has nonzero Dehn invariant. Later results showed Juel to be mistaken.

The place of Hilbert's third problem in work on the foundations of geometry generated some discussion on the teaching of geometry. In a 1904 article, Hans Keferstein [Keferstein, 1904] reviewed Dehn's work in considerable detail in the *Kleinere Mitteilungen* section of the *Zeitschrift für mathematischen und naturwissenschaftlichen Unterricht*, a pedagogical journal.

In Italy Dehn's work was surveyed in the chapter *Sulla teoria della equivalenza* by Ugo Amaldi of the second edition of the collection of essays *Questioni riguardanti le matematiche elementari* edited by Federigo Enriques that appeared in 1912 [Enriques, 1912]. The ideas inspired Onorato Nicoletti (1872–1929) who undertook a penetrating analysis of Dehn's work and the related papers. In three papers, all titled *Sulla equivalenza dei poliedri* [Nicoletti, 1914] appearing in 1913, -14, -15, respectively, he recast Dehn's results in a more general framework that anticipated the current formulation and presented derivations of previous results in a clear and applicable formulation. He was able to discern further conditions for scissors congruence by combining Dehn's work on the decomposition of rectangles by rectangles with the work on scissors congruence. He also considered the problem of the converse.

Dehn's work turned to topology after 1910. The notion of scissors congruence as a foundation for volume of polyhedra in space became more subtle thanks to Dehn's results. Interest in axiomatic geometry continued and in 1921 Wilhelm Süss (1895–1958) published his first paper [Süss, 1921] treating a version of scissors congruence based on *Cavalieri equivalent tetrahedra*, that is, tetrahedra on bases of the same area and of the same height. Süss shows that over non-Archimedean fields, there are tetrahedra of equal volume that are not Cavalieri scissors congruent, that is, there is no decomposition of each tetrahedron into sub-tetrahedra that are pairwise Cavalieri equivalent. Finally, Süss showed that stable Cavalieri scissors congruence gives a satisfactory theory of polyhedral volume. Dehn reviewed the paper [Süss, 1921] for the *Journal für Mathematik* (JFM 48.0640.04). Süss spent the years 1923–28 in Japan where he shared his ideas of scissors congruence, leading to a paper [Nakajima, 1926] by Soji Nakajima who proved that any polyhedron of volume one shared the same Dehn invariant value with a tetrahedron with one face an equilateral triangle of side one.

Early in the twentieth century, integration, the general method of computing areas and volumes, received focused axiomatic development that led to the satisfactory formulation of Henri Lebesgue (1875–1941). In 1938 Lebesgue gave a lecture [Lebesgue, 1939] in Krakow taking up the notion of scissors congruence, which he called *équivalent de façon finie*, in the context of a foundation for the volumes of polyhedra. He presented a clear and careful exposition of Dehn's work and related developments, leading to the idea of *équivalence par multiplication* when a polyhedron is decomposed into pieces that are all similar to one another. Lebesgue's paper stimulated renewed interest in the notion of scissors congruence.

In a 1939 paper *Om Polyedres Rumfang* [Jessen, 1939] based on a 1936 lecture, the Danish analyst Børge Jessen (1907–1993) presented a complete proof of Dehn's results, citing Keferstein, Lebesgue, Nicoletti, and Enriques. In 1939, Dehn spent three months in Copenhagen en route to Trondheim and eventually the U.S. (see Chapter 9). Jessen undoubtedly met Dehn and spoke with him about scissors congruence. In a subsequent paper [Jessen, 1941] *En Bemærkning om Polyedres Volumen*, Jessen takes volume to satisfy "1) congruent polyhedra have the same volume, 2) the volume of a polyhedron composed of several polyhedra is the sum of their volumes, 3) the volume of a prism is the usual." He concludes that when two polyhedra have the same volume, they must be scissors congruent. Dehn's results imply that these assumptions do not characterize volume. To make the argument of this theorem, Jessen introduced a group of polyhedra $\mathfrak{G}$ in which a decomposition of a polyhedron P into subpolyhedra $P_1, P_2, \ldots, P_n$ corresponds to the sum in $\mathfrak{G}$ given by $P = P_1 + \cdots + P_n$. Volume is an instance of a *volume function* f defined on $\mathfrak{G}$ satisfying 1) $f(P) = f(Q)$ when P is congruent to Q; 2) $f(P+Q) = f(P) + f(Q)$ for all P and Q in $\mathfrak{G}$; and 3) $f(P)$ is the volume of P when P is a prism. Jessen proves that if $f(P) = f(Q)$ for all volume functions, then P is scissors congruent to Q. In his analysis, he also proves that $\mathfrak{G}$ has a vector space structure over the field of rational numbers. Jessen ends the paper with: "For various remarks in connection with the present study, I am grateful to Professor Dehn."

The Swiss mathematician Hugo Hadwiger (1908–1981), independently of Jessen, considered *polyhedron-functionals* satisfying motion-invariance, additivity, and normalized to give 1 on the unit cube. Like Jessen, Hadwiger [Hadwiger, 1949a] proved that these conditions lead to the usual volume on tetrahedra. He also established the existence of a basis for the family of polyhedra in 3-space for which any polyhedron A has a finite decomposition $A \sim \xi E + \sum a_n A_n$ with E the unit cube, $\xi, a_n \in \mathbb{R}$, and A_n taken from the basis. With this basis, found by transfinite induction, further study of polyhedron-functionals could proceed [Hadwiger, 1949b]. He next introduced functionals Φ defined on polytopes in $\mathbb{R}^k$ that satisfy $\Phi(A + B) = \Phi(A) + \Phi(B)$ when A and B are polytopes for which the interior of $A \cap B$ is empty, and $\Phi(A) = \Phi(B)$ when A is G-congruent to B for a fixed group G of transformations on $\mathbb{R}^k$. For example, $\Phi(A) = c \operatorname{Vol}(A)$ satisfies these conditions when G is a group of isometries.

Hadwiger proved Jessen's theorem for higher dimensions [Hadwiger, 1950a], and determined how the translation subgroup of the full group of symmetries controlled scissors congruence of k-dimensional parallelotopes [Hadwiger, 1950b]. These investigations opened up a new research area dependent on the choice of group of motions and dimension [Hadwiger, 1957]. In 1965 V.B. Zylev proved a general result for polyhedra in Riemannian manifolds: stable scissors congruence implies scissors congruence in this general setting [Zylev, 1965].

The Swiss mathematician Jean-Pierre Sydler (1921–1988) published *Sur la décomposition des polyèdres* [Sydler, 1944] in 1944 where he introduced a geometric construction for a polyhedron P and a set of positive real numbers $a_1, \ldots, a_n$ satisfying $a_1 + \cdots + a_n = 1$, for polyhedra $p_1, \ldots, p_n$ with p_i similar to P with ratio a_i, together with a cube R for which P is scissors congruent to $p_1 + \cdots + p_n + R$. Sydler obtained as a consequence that a regular tetrahedron is not scissors congruent to a set of $k > 1$ many regular tetrahedra, a generalization of Dehn's result for $k = 2$. His decomposition also implies a condition for a polyhedron to be scissors congruent to a cube – P must be scissors congruent to k polyhedra all similar to P.

Sydler obtained a doctorate in 1946 at the *Eidgenössische Technische Hochschule (ETH)* in Zürich under the direction of Louis Kollros and Ferdinand Gonseth in algebraic geometry. He became Kollros's assistant for a few years before he took a position in 1950 in the library at the *ETH*. In a long series of incremental advances, he incorporated the work of Hadwiger during the 1950s (Sydler 1952, 1959). In 1965 [Sydler, 1965] he succeeded in giving a complete proof of the sufficiency of Dehn's theorem, that is, if two polyhedra have the same volume and the same Dehn invariant then they are scissors congruent.

Sydler's result was enthusiastically received. The Royal Danish Academy of Sciences gold medal of 1903 was awarded to Sydler, as described in [Hadwiger, 1967], "On December 2, 1966, Sydler was able to accept the award ... in the Danish Embassy in Bern as part of a small celebration." By 1968 [Jessen, Karpf, and Thorup, 1968] Jessen and his collaborators, Jørgen Krapf and Anders Thorup, had reformulated Sydler's proof in an algebraic framework to give a simpler proof of the Dehn-Sydler theorem [Jessen, 1968].

The algebraic structures consider by Jessen, Karpf, and Thorup were soon understood in terms of homological algebra. This direction has been richly developed by Chih-Han Sah (1934–1997) [Sah, 1979] and Johan Dupont (1944–2015) [Dupont, 2001]. Most recently scissors congruence has been greatly enriched through the addition of the methods of algebraic K-theory [Zakharevich, 2016].

Although Dehn's solution of Hilbert's third problem was the earliest success of the proposed 23 problems, its development shows Hilbert's foresight. He was able to choose problems of depth and ongoing fascination. Dehn's work stands out as a powerful insight into the geometry of polyhedra that has led in many varied directions.

Acknowledgments

The authors thank Jesper Lützen for help with details of the Danish gold medal, and to Anders Thorup for recollections on the writing of [Jessen, Karpf, and Thorup, 1968]. Thanks to Elliot Schreiber for help with some subtle German passages. And thanks to Marjorie Senechal, David Rowe, and Jemma Lorenat for encouragement of this project.

References

Aigner, M. and Ziegler, G. M., *Proofs from The Book*, Sixth ed., Springer, Berlin, 2018.

Bartocci, C., *Una piramide di problemi. Storie di geometrie da Gauss a Hilbert*, Milano: Raffaello Cortina Editore, 2012.

Blumenthal, O., Lebensgeschichte [David Hilbert], in Hilbert, D., *Gesammelte Abhandlungen*. Dritter Band. Berlin: Springer 1935, 388–429.

Boltyanskiĭ, V. G., *Hilbert's Third Problem* (in Russian), Izdat. "Nauka", Moscow, (1977), Translated from the Russian by Richard A. Silverman, With a foreword by Albert B.J. Novikoff, Scripta Series in Mathematics, V.H. Winston & Sons, Washington, D.C.; Halsted Press [John Wiley & Sons], New York-Toronto-London, 1978.

Bricard, R., Sur une question de géométrie relative aux polyèdres, Nouv. Ann., **15**(1896), 331–334.

Corry, L., *David Hilbert and the Axiomatization of Physics (1898–1918)*, Kluwer Academic Publishers, Dordrecht, 2004.

Dehn, M., Die Legendréschen Sätze über die Winkelsumme im Dreieck, Math. Ann., **53**(1900) 404–439.

Dehn, M., Ueber raumgleiche Polyeder, Nachr. Ges. Wiss. Göttingen, Math.-Phys. Kl., **1900**, 345–354.

Dehn, M., Über den Rauminhalt, Math. Ann., **55**(1902), 465–478.

Dehn, M., Über Zerlegung von Rechtecken in Rechtecke, Math. Ann., **57**(1903), 314–332.

Dehn, M., Zwei Anwendungen der Mengenlehre in der elementaren Geometrie, Math. Ann., **59**(1904), 84–88.

Dehn, M., Über den Inhalt sphärischer Dreiecke, Math. Ann., **60**(1905), 166–174.

Dehn, M., Die *Euler*sche Formel im Zusammenhang mit dem Inhalt in der nichteuklidischen Geometrie, Math. Ann., **61**(1905), 561–586.

Dupont, J.L, *Scissors congruences, group homology and characteristic classes*, Nankai Tracts in Mathematics, **1**, World Scientific Publishing Co., Inc., River Edge, NJ, (2001).

Enriques, F., Questioni riguardanti le matematiche elementari. Vol. I. Critica dei principii, Bologna: Zanichelli. First edition 1900, (1912).

Gauss, C.-F., *Werke, vol. 8*, Teubner, Leipzig, Germany, (1900).

Gray, J.J., *The Hilbert challenge*, Oxford University Press, Oxford, (2000).

Hadwiger, H., Zerlegungsgleichheit und additive Polyederfunktionale, Arch. Math., **1**(1949), 468–472.

Hadwiger, H., Zum Problem der Zerlegungsgleichheit der Polyeder, Arch. Math. (Basel), **2**(1949/50), 441–444 (1951).

Hadwiger, H., Zur Inhaltstheorie der Polyeder, Collectanea Math., **3**(1950), 137–158.

Hadwiger, H., Translative Zerlegungsgleichheit k-dimensionaler Parallelotope, Collectanea Math., **3**(1950),11–23.

Hadwiger, H., *Vorlesungen über Inhalt, Oberfläche und Isoperimetrie*, Springer-Verlag, Berlin-Göttingen-Heidelberg, (1957).

Hadwiger, H., Neuere Ergebnisse innerhalb der Zerlegungstheorie euklidischer Polyeder, Jber. Deutsch. Math.-Verein., **70**(1967/68), 167–176.

Hallett, M., Majer, U., *David Hilbert's Lectures on the Foundations of Geometry 1891–1902*, Springer-Verlag Berlin Heidelberg, Germany, (2004).

Hartshorne, R., *Geometry: Euclid and beyond*, Undergraduate Texts in Mathematics, Springer-Verlag, New York, (2000).

Hertz, H., *Die Prinzipien der Mechanik*, Leipzig: Barth (Meiner) 1894.

Hilbert, David: *Grundlagen der Geometrie*, in Festschrift zur Einweihung des Göttinger Gauss-Weber Denkmals, Leipzig: Teubner (2. Aufl., 1903); *Grundlagen der Geometrie* (Festschrift 1899), Klaus Volkert, ed., Heidelberg: Springer 2015.

Hilbert, D., Mathematische Probleme. Vortrag, gehalten auf dem internationalen Mathematiker-Congress zu Paris 1900, Nachr. Ges. Wiss. Göttingen, Math.-Phys. Kl., **1900**, 253–297, translated in [44], and in [17], pp. 240-282.

Hill, M.J.M., Determination of the volumes of certain species of tetrahedra without employment of the method of the limits, Proc. Lond. Math. Soc., **27**(1896), 39–53.

Jessen, B., Om Polyedres Rumfang (On the volume of polyhedra), Mat. Tidsskr. A, **1939**, 35–44.

Jessen, B., En Bemærkning om Polyedres Volumen (A remark on the volume of polyhedra), Mat. Tidsskr. B, **1941**, 59–65.

Jessen, B., The algebra of polyhedra and the Dehn-Sydler theorem, Math. Scand., **22**(1968), 241–256.

Jessen, B., Karpf, J., and Thorup, A., Some functional equations in groups and rings, Math. Scand., **22**(1968), 257–265.

Jessen, B. and Thorup, A., The algebra of polytopes in affine spaces, Math. Scand., **43**(1978), 211–240.

Juel, C., Om endelig ligestore Polyedre (On finitely equal polyhedra), Nyt Tidss. for Math., **14**(1903), 53–63.

Kagan, B., Über die Transformation der Polyeder, Math. Ann., **57**(1903), 421–424.

Keferstein, H., Die Bedingung für die Zerlegbarkeit von inhaltsgleichen Polyedern. Bericht über zwei Aufsätze von M. Dehn-Münster i. W, Zs. f. math. u. naturw. Unterr. **35**(1904), 111–123.

Lebesgue, H., Sur l'équivalence des polyèdres, en particulier des polyèdres réguliers, et sur la dissection des polyèdres réguliers en polyèdres réguliers, Ann. Soc. Polon. Math., **17**(1939), 193–226.

Legendre, A.-M., *Eléments de géométrie*, Chez Firmin Didot, Paris, (1794).

Magnus, W., Max Dehn, Mathematical Intelligencer, **1**(1978), 132–143.

Magnus, W. and Moufang, R., Max Dehn zum Gedächtnis, Math. Ann., **127**(1954), 215–227.

Moulton, F.R., A simple non-Desarguesian plane geometry, Trans. Am. Math. Soc., **3**(1902), 192–195.

Nakajima, S., Zur Endlichgliechheit der Polyeder, Tohoku Mathematical Journal, First Series, **27**(1926), 293–296.

Newson, M.W.: David Hilbert, *Mathematical problems*, Bull. A.M.S. **8**(1902), 437–479.

Nicoletti, O., Sulla equivalenza dei poliedri (1, 2, 3), Rom. Acc. L. Rend. (5), **22**(1913), 767–770; Rend. Circ. Mat. Palermo **37**(1914), 47–75; **40**(1915), 194–210.

Pasch, M., *Vorlesungen über neuere Geometrie*, Teubner, Leipzig, (1882), Second edition (1926) Springer-Verlag, Berlin. Mit einem Anhang: M. Dehn. *Die Grundlegung der Geometrie in historischer Entwicklung.*

Petersen, J., *Theorie der algebraischen Gleichungen*, Copenhagen: A.F. Host & Sohn, (1878), Italian translation: Julius Petersen, *Teoria delle equazioni algebriche*, trans. Gerolamo Rizzolino e Giuseppe Sforza, Naples: Libreria B. Pellerano, 1891.

Poincaré, H., Analysis situs, Jour. École Poly., (2)**1**(1895) 1–123.

Saccheri, G., *Euclides Vindicatus*, edited and translated by George Bruce Halsted., Chicago, The Open Court Publishing Company (1920).

Sah, C.H., *Hilbert's third problem: scissors congruence*, Research Notes in Mathematics, **33**, Pitman (Advanced Publishing Program), Boston, Mass.-London, (1979).

Schatunovsky, S.O., Über den Rauminhalt der Polyeder, Math. Ann., **57**(1903), 496–508.

Sforza, G., Un'osservazione sull'equivalenza dei poliedri per congruenza delle parti, Periodico di Mat., **12**(1897), 105–109.

Sommerville, D.M.Y., The relations connecting the angle-sums and volume of a polytope in space of n dimensions, Proc. Royal Soc. London (A)**115** 91927), 103–119.

Stolz, O., *Vorlesungen über allgemeine Arithmetic. Nach den neureren Ansichten. Erster Theil: Allgemeines und Arithmetik der reelen Zahlen.* Teubner 1885, Leipzig.

Stillwell, J., Max Dehn and geometry, Math. Semesterber. **49**(2002), 145–152.

Süss, W., Begründung der Lehre vom Polyederinhalt, Math. Ann., **82**(1921), 297–305.

Süss, W., Über Endlichgleichheit im Raum, Jahresber. Dtsch. Math.-Ver., **31**(1922), 19–21.

Sydler, J.-P., Sur la décomposition des polyèdres, Comment. Math. Helv., **16**(1944), 266–273.

Sydler, J.-P., Sur les conditions nécessaires pour l'équivalence des polyèdres euclidiens, Elem. Math., **7**(1952), 49–53.

Sydler, J.-P., Sur quelques polyèdres équivalents obtenus par un procédé en chaînes, Elem. Math., **14**(1959), 100–109.

Sydler, J.-P., Conditions nécessaires et suffisantes pour l'équivalence des polyèdres de l'espace euclidien à trois dimensions, Comment. Math. Helv., **40**(1965), 43–80.

Vahlen, K.Th., Über endlichgleiche Polyeder, Math. Ann., **56**(1903), 507–508.

Wagon, S., Fourteen proofs of a result about tiling a rectangle, Amer. Math. Monthly, **94**(1987), 601–617.

Zakharevich, I., Perspectives on scissors congruence, Bull. Amer. Math. Soc. (N.S.), **53**(2016), 269–294.

Zylev, V. B., On the equi-dissectability of two equi-completable polytopes, Dokl. Akad. Nauk SSSR, **161**(1965), 515–516.

CHAPTER 4

Dehn's Early Work in Topology

Cameron McA. Gordon and David E. Rowe

This chapter describes Dehn's best-known papers in topology, but also some of his less well-known works in cases where these shed light on his development as a topologist. In only a few cases will we enter into details, as the larger aim is to shed light on the intellectual journey that began around 1904 when Dehn shifted his research interests away from foundations of geometry in order to pursue very different problems in the field of topology. Before picking up that story, though, we first need to say a few words about the general development of topology, a field that was only beginning to find its bearings during Dehn's prime years as a research mathematician.

In fact, calling him (or any mathematician) from that time a topologist can easily create a false impression. Prior to the First World War, for example, the number of mathematicians who had published important work on 3-manifolds could be counted on one hand. Aside from the papers by Poincaré and Dehn, there was Poul Heegaard's dissertation [Heegaard, 1898] and Heinrich Tietze's *Habilitationschrift* [Tietze, 1908], and little more. Even during the 1920s, there were very few works of interest beyond some short papers by James W. Alexander and Hellmuth Kneser's paper [Kneser, 1929], discussed below in connection with the difficulties that arose in proving Dehn's lemma. Dehn's contemporaries thought of him as a geometer who worked on problems in analysis situs, the traditional term for topology. By the 1930s, when the first major textbooks began to appear, this older name largely disappeared; only around then did one talk about mathematicians who were topologists.

The intellectual roots of topology were, of course, far older. They dated back to the nineteenth century or even further, if one focuses on isolated findings, such as Euler's polyhedral formula, that were later identified as topological invariants. Within the realm of analytic geometry, the study of the algebraic and differential aspects of curves and surfaces in space led to the problem of classifying these objects and their properties. The notion of invariance was particularly well suited to projective geometry, which emerged side-by-side with algebraic invariant theory. Felix Klein's "Erlangen Program" (1872) extended that framework by exploiting the nascent idea of a group of transformations, another idea borrowed from algebra. Geometrical research could then be undertaken more systematically by fixing a group and studying properties that remain invariant under its transformations. One then has metric geometries (Euclidean and non-Euclidean) as special cases of projective geometry, since the latter group contains the former ones. The still

---

©2024 American Mathematical Society

larger group of topological transformations studies invariants that are inherited by these other geometries.

Although Klein couched this framework in terms of a theory of manifolds of arbitrary dimension, his approach was largely inspired by projective geometry and the vision of groups that act on a manifold globally. Riemann's notion of a manifold was local and led to the general theory of differentiable manifolds, even though differential geometry was not his main motivation.[1] In connection with topology, Riemann's pioneering work in complex analysis had a profound influence. It was in the context of studying the connectivity properties of Riemann surfaces that he laid the groundwork for homology theory, which then came fully into view with Poincaré's work on higher-dimensional manifolds. Like Riemann, Poincaré invented new tools for studying the topological properties of the various constructs that arose from his work in analysis, including differential equations and celestial mechanics.[2]

Today, mathematicians take for granted that one must carefully distinguish between topological, piecewise-linear, and differentiable manifolds, but in fact these distinctions were unclear for a very long time. Indeed, Dehn's earliest topological work was motivated by a general program that aimed to provide a common foundation for these theories. He thus hoped to place the theory of topological manifolds on a rigorous basis by exploiting combinatorial methods first introduced by Poincaré, though in an informal and highly non-rigorous fashion. Dehn was one of the very first mathematicians to seize on and develop parts of Poincaré's intellectual legacy in this new corner of mathematics. Part of his interest related to basic homological properties of n-manifolds as set forth in [Poincaré, 1895], including the generalized Euler characteristic and Poincaré duality. Dehn's most important contribution toward rigorizing this part of Poincaré's work came when he advanced a formal framework for combinatorial topology in the report he co-authored with Poul Heegaard on "Analysis Situs" [Dehn and Heegaard, 1907]. This theoretical interest was entirely akin to the Hilbertian program that aimed to place all of mathematics on rigorous axiomatic foundations, a vision certainly antithetical to Poincaré's understanding of mathematical knowledge.

Interestingly enough, Dehn's own research style hardly reflected a dogmatic rejection of informal arguments. After pointing the way to a fully rigorous approach to manifold theory, he became fascinated by the problems Poincaré raised in [Poincaré, 1904], his fifth and final supplement to his 1895 paper on "Analysis Situs." Therein Max Dehn came face-to-face with Poincaré's example of a homology sphere, a 3-manifold M^3 with the homological properties of a 3-sphere, but which was topologically distinct from S^3 because Poincaré proved that its fundamental group $\pi_1(M^3)$ was nontrivial. This example also led him to ask a very natural question: if a closed manifold M^3 is simply connected, must it then be homeomorphic to S^3? Dehn thought the answer was surely: yes. His instincts were, in fact, seldom wrong, but like Hilbert he sometimes underestimated the difficulties that can arise when trying to prove a plausible conjecture.

Back when he was studying under Hilbert in Göttingen, both thought they knew the answer to the latter's third Paris problem. In that case, though, Dehn answered

---

[1] For an insightful survey of the early history of the manifold concept, see [Scholz, 1980]; later developments up to ca. 1950 are described in [Scholz, 1999].

[2] For an overview of Poincaré's famous papers, see [Sarkaria, 1999] and [Gray, 2013, 427–466]; on their place within the larger history of 3-manifolds, see [Gordon, 1999, 451–462] and [Volkert, 1994].

the question at hand by finding an invariant that enabled him to distinguish between different polyhedra of equal volume (see Chapter 3). To answer Poincaré's question in the positive sense, on the other hand, was a different sort of problem altogether: this meant, he somehow would have to prove that a closed 3-manifold with the homotopic properties of an S^3 was, in fact, an S^3. As described in [Gordon, 1999] and [O'Shea, 2007], the theory of 3-manifolds began with Poincaré's fundamental contributions from this period, culminating with the famous conjecture that the three-sphere S^3 was the only closed 3-manifold with trivial fundamental group. Max Dehn soon set his eyes on proving this, and the last section of [Dehn, 1910] shows that this goal was still very much on his mind at that time.

Little could Dehn have imagined that he had stumbled upon an incredibly difficult problem, although Poincaré seems to have sensed its difficulty when he wrote, at the end of [Poincaré, 1904]: "But this question would lead us too far afield." By the 1920s, it would be dubbed "Poincaré's conjecture," and by mid-century, after some leading experts in topology failed to crack it, this plausible-sounding guess came to be recognized as *the* outstanding open problem in topology. That status was confirmed in the year 2000, when the Clay Mathematics Institute selected it as one of its seven Millennium Problems. Max Dehn continued to ponder the Poincaré conjecture well after 1910, whereas before that time it accompanied him throughout a new and formative period in his research career. The years 1907 to 1910 marked a major turning point in Dehn's topological interests, during which he gained a deep appreciation for some of the major difficulties that arise in dimension three. When considering Dehn's place in the history of modern topology, we can easily appreciate why he never felt any incentive to go beyond low-dimensional topology.[3] Dehn's tastes ran toward concrete mathematics anchored in geometry; general topology and set theory in the spirit of Felix Hausdorff had no appeal for him, nor was he inclined to study new homology theories developed for more abstract topological spaces. He found plenty to do just working on problems that were nearer to hand, and so he chose to attack these using geometric groups, hyperbolic geometry, and a vivid imagination.[4]

The authors would like to thank John McCleary for his comments on the text and Cynthia Hog-Angeloni for helpful conversations on the Dehn-Kneser correspondence.

Dehn's First Topological Studies

Although Max Dehn's first published works in topology appeared some years after his dissertation [Dehn, 1900a] and habilitation thesis [Dehn, 1902], he already had a keen interest in fundamental topological problems dating back to his days as a student [Guggenheimer, 1977]. This is evident from [Dehn, 1899], an unpublished manuscript in Dehn's estate that gives the first axiomatic proof of the Jordan curve theorem for polygons. The precise period during which he wrote this paper

---

[3]For an overview of the dramatic breakthroughs made in low-dimensional topology during the 1950s and 60s, see [Milnor, 2009].

[4]Beyond John Stillwell's several studies dealing directly with Dehn's mathematics, one can also gain a clear impression of the larger context of that work from his very readable book, *Classical Topology and Combinatorial Group theory* [Stillwell, 1980]. This was written just during the time when Bill Thurston began promoting his famous geometrization program, which eventually opened the way to Perelman's proof of the Poincaré conjecture along with a complete classification of 3-manifolds.

is unknown, but it can be dated provisionally based on the fact that Dehn made use of Hilbert's plane axioms for incidence and order, which he first unveiled in his lecture course from 1898/99. The main novelty Dehn introduced in his manuscript was to show that the Jordan theorem for polygons required nothing more than these basic axioms. His proof was thus entirely elementary and required nothing like the segment arithmetics Hilbert had derived based on Pappus and Desargues [Hilbert, 1899/2015, 108–111, 131–133]. Dehn's argument also follows a general line of attack that Arthur Schoenflies introduced in [Schoenflies, 1896]. As one of Dehn's professors in Göttingen, Schoenflies was not only an important influence on his early education. He was also one of the key figures pursuing the line of ideas on foundations of geometry that Hermann Wiener first set forth in [Wiener, 1891].

When Dehn's paper re-emerged in the 1970s, it immediately aroused considerable interest. Dehn's estate was then still in the hands of Wilhelm Magnus, who made several interesting comments about this paper of some 2,000 words in [Magnus, 1978/79]. First, he pointed out that the Jordan curve theorem appears without proof in Section 4 of Chapter 1 of Hilbert's "Grundlagen der Geometrie" [Hilbert, 1899/2015], which made it likely "that Hilbert had given this problem to Dehn ...as a mere exercise, and that this fact kept Dehn from publishing it" [Magnus, 1978/79, 133]. Yet even if this were the case, the mystery as to why this text never found its way into print remains. Indeed, as Magnus noted, Dehn also proved the analogous theorem in three dimensions, namely, that a simple closed polyhedron divides three dimensional space into two parts, using only Hilbert's axioms of incidence and order for solid geometry. Moreover, Dehn's arguments, even though entirely elementary, were in several respects novel, and some of his results filled lacunae in Hilbert's famous text.

In his lecture course on "Grundlagen der Euklidischen Geometrie," Hilbert developed a theory of area for plane rectilinear figures on the basis of Pappus's theorem. He then presented the final version of this theory as Chapter 4 of [Hilbert, 1899/2015], though he neglected a key step (which Euclid also ignored in his theory of area): to prove that a non-convex polygon can be triangulated by diagonals. The first published proof of this theorem appeared in [Lennes, 1911], but Guggenheimer emphasized that [Dehn, 1899] already contains a rigorous proof, which makes it all the more remarkable that Hilbert passed over this critical step. The problem of finding efficient algorithms for triangulating arbitrary polygonal figures has, in fact, spawned a large and still growing literature.

In the 1970s, Heinrich Guggenheimer studied Dehn's manuscript and published two noteworthy papers related to it. The first, [Guggenheimer, 1977], is an historical study that discusses Dehn's results in the context of the mathematical literature related to Jordan's curve theorem. The second paper, [Guggenheimer, 1978], elaborated on Dehn's work by showing its close connection with a special case of Schoenflies' Theorem. This case asserts that every homeomorphism between a triangle and a Jordan polygon can be extended to a homeomorphism between their respective interiors. Guggenheimer showed how this followed directly from Jordan's theorem for polygons and a result he called Dehn's theorem, from which one can find a homeomorphism of the plane that maps a given Jordan polygon onto a triangle. He emphasized that this aspect of Dehn's papers was especially remarkable not only in view of the fact that it antedated Schoenflies' paper [Schoenflies, 1906] by seven years but also because the proof only involved elementary axiomatic geometry, whereas Schoenflies' proof involved convoluted arguments that relied on point set topology.

The Jordan curve theorem became a topic of intense interest soon after Camille Jordan presented a partial proof in [Jordan, 1893]. This "intuitively obvious" theorem states that a simple closed curve C in the plane $\mathbb{R}^2$ (i.e., the homeomorphic image of a circle) is the common boundary of two connected disjoint plane subsets G_1, G_2, where $G_1 \cup G_2 = \mathbb{R}^2 - C$. Jordan's proof was grounded in analytic geometry, and he simply assumed the validity of the theorem for the case of a polygonal curve. A few years later, Schoenflies filled this gap to give a complete proof for a certain class of curves [Schoenflies, 1896]. Guggenheimer notes that his argument is valid for piecewise C^1 curves of finite order, meaning that an arbitrary line meets the curve in only finitely many points. Schoenflies used the latter property to define interior points P as those for which any ray passing through P necessarily meets the curve in an odd number of points. He also used this property to prove that any two interior points can be joined by a polygonal arc that lies in the interior.

Dehn's proof essentially follows Schoenflies by focusing on the interior points in a convex region of a polygon. The simplest case is that of a triangle, which corresponds to Pasch's axiom, or Axiom II.4 in Hilbert's axioms of order. If Δ is a triangle with vertices ABC, then the interior points P have the property that every line through P intersects one of the intervals $\overline{AB}, \overline{BC}, \overline{CA}$. In the course of his proof, Dehn shows that for any polygon that is not a triangle, at least one diagonal lies in the interior, thereby leading to a decomposition into two polygons with fewer vertices. Repeating this process then produces the desired triangulation. Dehn proceeds by examining convex vertices along the way.

If Ψ is the initial polygon, then a vertex P_i of Ψ is a *convex corner* if there exists an A on $P_{i-1}P_i$ and a B on P_iP_{i+1} such that the open segment AB is in the interior of Ψ. The convex corner P_i is then called an *ear* if no point of Ψ lies in the open triangle $P_{i-1}P_iP_{i+1}$ or on the open segment $P_{i-1}P_{i+1}$. Dehn proves that every Jordan polygon has at least two ears, and so he can proceed by lopping these off in the course of the decomposition. He argues as follows: suppose P_i is a convex corner that is *not* an ear, then there exists a vertex P_j in the open triangle $P_{i-1}P_iP_{i+1}$ such that the open halfplane of $P_{i+1}P_j$ contains only the vertex P_i. This means that P_iP_j is a diagonal contained in the interior of the polygon, which can thus be divided into two Jordan polygons. Repeating these steps for each convex corner, one obtains a triangulation of Ψ after a finite number of steps.

Magnus noted that Dehn's manuscript was found in a folder labeled "Beitraege zur Analysis Situs. Prinzipien der Geometry" (Contributions to Analysis Situs. Principles of Geometry). Based on his published works, one might well imagine that Dehn's interests in topology arose only after his immersion in Hilbert's axiomatic approach to geometry. In fact, both interests were closely linked from the beginning, as can be seen from the arguments in [Dehn, 1899], which predate Dehn's studies on piecewise-linear topology. Beyond the strong accent on axiomatics, this manuscript also reflects Dehn's general approach to topology, which sought to avoid the need for continuity principles. This was, in fact, one of the central motivations behind Hilbert's "Grundlagen der Geometrie." Wilhelm Magnus emphasized, however, that Dehn was less inspired by detailed axiomatic studies than in providing "a solid and, as far as possible, a simple foundation to a theory" [Magnus, 1978/79, 134]. A striking example of this can be found in the article "Analysis Situs" [Dehn and Heegaard, 1907], which he co-authored with Poul Heegaard for the German *Encyclopedia of the Mathematical Sciences*. Although this article was mainly known for its thorough treatment of surfaces, including the classification of compact surfaces, it also laid the groundwork for a combinatorial approach to the study of manifolds of higher dimension. Henri Poincaré had already opened that door, but Max Dehn was one of the first to walk through it and explore these new possibilities.

Dehn meets Heegaard meets Poincaré

In 1904, topology was a many-splendored thing, but it was far from being an independent field of mathematical research. Dehn's work on foundations of geometry clearly went hand-in-hand with an interest in related aspects of analysis situs. But foundational studies were only one of the avenues leading to topological problems. In the case of Poul Heegaard, his interests, like those of Henri Poincaré, arose from geometrical problems in the theory of complex functions (in Poincaré's case, automorphic functions and differential equations).

Heegaard was seven years older than Dehn and had studied briefly in Göttingen under Felix Klein, who in 1894 gave him the topic for his dissertation [Heegaard, 1898]. This contains the main ideas behind what came to be known as Heegaard diagrams and Heegaard splittings, which became central techniques for the theory of 3-manifolds. Appreciation was not long in coming, as Heegaard later recalled:

> I had sent my dissertation to Picard and Poincaré. The latter asked me about different things that he had not understood in the Danish text. Thus I wrote a summary of the dissertation in French for him. This led him to write a paper supplementing his original treatise, and thus my dissertation became known abroad even if it was written in Danish. In Denmark public opinion held it worthless and completely ridiculous. One of my foreign friends noted this in a conversation with one of the older mathematicians who had to admit at the same time that he had not read it. [Munkholm and Munkholm, 1999, 932]

Heegaard's approach to 3-manifolds was based on decompositions into handlebodies B_1 and B_2, each of genus n. He took a system of n canonical curves on B_1 and their homeomorphic images on B_2 and showed that these determined the topology of a 3-manifold M completely. One of the first important uses of these so-called Heegaard diagrams was the construction of a homology sphere in [Poincaré, 1904]. Beyond this, however, Heegaard's work had an immediate relevance for Poincaré's definition of the Betti numbers, as the latter had failed to take account of torsion. Heegaard showed that one can detect the difference between S^3 and real projective 3-space homologically by means of torsion, since S^3 has no torsion, whereas $\mathbb{P}_3(\mathbb{R})$ has torsion number 2. Poincaré then reworked his homology theory in two *Compléments* to [Poincaré, 1895], which he published in [Poincaré, 1899] and [Poincaré, 1900]. It would appear likely that Dehn's first encounters with these ideas came about through his collaboration with Heegaard, which began soon after they met each other at the Heidelberg Congress in 1904.

Although little is known about their relationship, Dehn and Heegaard have long been linked as co-authors of the survey article, "Analysis Situs" [Dehn and Heegaard, 1907]. This collaboration came about largely through happenstance, as initially Heegaard planned to write this report on his own. Klein almost surely had a hand in this, since Heegaard had worked closely with him during the mid-1890s. The fact that Poincaré's "First Complement" was written in response to [Heegaard, 1898] also must have left a deep impression on Göttingen's senior mathematician. Since he oversaw the entire Encyclopedia project and knew Heegaard's interests well, Klein was in an ideal position to recommend the otherwise little-known Dane to Franz Meyer, who edited the volumes on geometry.

This choice would be otherwise difficult to explain, as at this stage in his career Heegaard had no publication record in mathematics at all! Already before completing his dissertation in 1898, he had married and begun having children, circumstances that forced him to take on a very heavy teaching load at various institutions in Copenhagen, especially the military and naval academies. Heegaard agreed, nevertheless, to take on this project, in part no doubt simply to broaden his own knowledge. He later described this as a pleasant diversion, as he

> started the work with great pleasure and – lack of time notwithstanding – finished an outline and a bibliography. However, it was difficult to get the time and quiet needed to work out the theoretical introduction. Moreover, quite senselessly, I let myself be influenced by a variety of malicious comments on my work in topology. Therefore, I asked Franz Meyer for an assistant. It was then arranged that I should write the article with the young German mathematician Max Dehn, Dr. from Göttingen. ...
>
> In the meantime, Dehn had become Privatdozent in Kiel, and in the summer of 1905 I went down there to work with him. I now initiated him into my viewpoints and he began to work on the general introduction, which he finished beautifully during the next winter. [Munkholm and Munkholm, 1999, 932–933]

Heegaard's recollection of having worked with Dehn during the summer of 1905 was obviously correct, as it was during that semester that Dehn substituted for Paul Stäckel in Kiel. Otherwise, however, his text leaves the impression that Dehn merely carried out ideas Heegaard already had in mind. Although we have no way of knowing, this seems highly unlikely, as the "theoretical introduction" has a very different character than many parts of what follow it. In many places, the text merely provides dense descriptions of and references to the relevant mathematical literature, much like what one finds in other articles written for the German Encyclopedia project. Still, whatever the case may have been, this clearly turned out to be a successful collaboration. A footnote on page one indicates that Heegaard had been responsible for the literature search, whereas Dehn took responsibility for the final form of the article. This being so, it comes as no great surprise to see that he structured this field of research around principles that strongly echo his own earlier work and that of his former mentor.

Heegaard's account also omitted an interesting anecdote that makes clear his own role in gaining Dehn's cooperation. He surely did ask Meyer to appoint a co-author (not an assistant), but he already had someone in mind for this job. After the Heidelberg Congress, Dehn and Heegaard left together on a train bound for Hamburg (see the section "ICM in Heidelberg 1904" in Chapter 2). Heegaard probably told his student Ingebrigt Johannsson about the conversation he and Dehn had that day, although Johannsson may well have heard this from Dehn during the early 1930s, when he studied with him in Frankfurt. In 1948, Johannsson mentioned this in his obituary for Heegaard, which puts the matter in a very different light:

> Dehn believed that one should postulate just enough axioms to let the topological essence stand out clearly, something that had never been done before. Here, in the railroad compartment, combinatorial topology was created. Heegaard was enthusiastic and proposed that they write the article jointly [Munkholm and Munkholm, 1999, 933]

Putting this scant evidence together, we can confidently assert that Dehn wrote the introductory section and that he and Heegaard probably discussed the overall arrangement of the material already drafted when Dehn joined the project. Two topics in the report required a more probing treatment: the topological classification of surfaces and the purely combinatorial development of homology based on Poincaré's two papers [Poincaré, 1899] and [Poincaré, 1900]. Since the first of these papers was written in direct reaction to a criticism Heegaard had raised in his dissertation [Heegaard, 1898], it would seem likely that he had already written something about Poincaré's new proof of duality. It was in this "First Complement" that Poincaré introduced the fundamental concepts for a new combinatorial approach to homology. This was precisely the aspect of most interest to Dehn, so it would seem very likely that he was responsible for putting Poincaré's results into a purely combinatorial form in [Dehn and Heegaard, 1907, 178–186]. This discussion in the report led over to a consideration of an homology sphere constructed by means of a Heegaard diagram, which will be taken up below. As for the quite detailed treatment of surfaces in [Dehn and Heegaard, 1907, 189–205], it seems likely that Dehn wrote up its middle section (pp. 191–196), which contains hardly any references. Taking Heegaard's personal situation into account, he probably at that time had no opportunity to pursue original work of this type.

Moritz Epple has emphasized the modernist style that Dehn brought to this project, even going so far as to call [Dehn and Heegaard, 1907] a manifesto for modern topology [Epple, 1999a, 229–235]. Whether or not that characterization appears apt, this text surely does reflect a strong desire to identify the fundamental features of topological knowledge and to articulate these as key organizational principles. Dehn's modernist outlook was apparent not only in view of the axiomatic framework he introduced. Even more striking were his informal remarks in a short section on methodology. There he underscored the promise of the approach previously outlined, namely to establish firm foundations for topology by using methods that enable the field to be seen as a branch of combinatorics. The payoff in doing so came from the security provided – a topological proof no longer had to depend on *Anschauung* or what came to be called "hand-waving arguments." Dehn made clear that *Anschauung*, analogies, and heuristic arguments had an important role to play in the course of finding new results, but he drew the line when it came to arguments that purported to be fully rigorous.

Only a year afterward, Heinrich Tietze and Ernst Steinitz published closely allied works in analysis situs that did much to promote the same cause (see [Scholz, 1999, 48–50] and [Volkert, 1994, 162–182]). The lengthy study [Tietze, 1908] was especially noteworthy for having introduced lens spaces, which played an important role in the early history of 3-manifolds. Tietze conjectured that the two lens spaces $L(5,1)$ and $L(5,2)$ were not homeomorphic, even though their fundamental groups were identical, and in 1919 James W. Alexander proved this was indeed the case. That finding clearly raised the stakes for the Poincaré conjecture. Steinitz was particularly emphatic in supporting the approach taken by Dehn and Heegaard, as he shared their view that topology should shed all ties with continuity assumptions:

> It is therefore certainly a welcome sign of progress that the article on analysis situs in the *Encyclopedia of the Mathematical Sciences* introduces an axiom system for the discipline that is at least sufficient for 3-dimensional space. This will of course neither solve nor remove the important and difficult problems offered by analysis situs in the space of ordinary geometry, but it opens the possibility of further developing the field of analysis situs with complete rigor without having to wait until those questions have been resolved. [Steinitz, 1908, 30]

The modern field of topology would not emerge in fulsome form until the 1930s, so describing its principles, methods, and goals at this early stage was clearly a pioneering effort. This may seem very surprising, in view of the dearth of major results then in print. In fact, the lion's share of the literature compiled and described in [Dehn and Heegaard, 1907] dealt with space curves or systems of curves on surfaces. Knot theory in the British tabular tradition was presented in great detail, though mainly as a compendium of interesting results and speculations [Epple, 1999a, 125–160]. This corner of topology was still very much a descriptive, rather than a deductive science based on clear first principles [Epple, 1999b]. Given the importance of knots in Dehn's early topological work, it also comes as a surprise to see that the report makes no mention of their most important invariant: the fundamental group of the complementary space. Thus, the old problem of classifying knots remained at a standstill, quite untouched by the programmatic framework Dehn and Heegaard had in mind. Far more influential were those parts of their report that dealt with surfaces and higher-dimensional manifolds.

Was the time really ripe for such an abstract combinatorial approach using a system of axioms that (ideally) served as the last resort for rigorous proof? On the one hand, Dehn stressed the importance of intuition, both for finding new results as well as justifying them. On the other, his goal was to outline a formal framework independent of any appeal to concrete visual experience. That meant, in his view, constructing abstract combinatorial/geometric foundations for topology. Naturally, such a theory drew on results obtained earlier by those who worked with mixed methods. Poincaré had utilized simplicial structures to compute Betti numbers, torsion numbers, and the fundamental group, but these were additive features of the manifold under study, which had differentiable properties as well. Dehn, on the other hand, took the simplicial structure to be the essential defining feature of a manifold. This approach was well illustrated by the careful discussion of canonical forms for surfaces in [Dehn and Heegaard, 1907]. Their proof of the classification theorem for 2-manifolds was the first to meet Hilbertian standards of rigor, but as Stillwell points out they lacked the tools needed to make significant progress in higher dimensions [Stillwell, 2012]. Rather surprisingly, groups are nearly absent in [Dehn and Heegaard, 1907]. They briefly mention the fundamental group, whereas group theory only entered the terrain of homology in the mid-1920s at the initiative of Emmy Noether [Rowe, 2021, 117–118].

Poincaré's conjecture emerged only after some missteps that led to his discovery of a homology 3-sphere – i.e., a closed 3-manifold with the same homological structure as S^3 – that is not homeomorphic to S^3. In modern terms, a homology n-sphere is an n-manifold X with homology groups identical to those of an n-sphere, thus $H_0(X,\mathbb{Z}) = H_n(X,\mathbb{Z}) = \mathbb{Z}$ and otherwise $H_i(X,\mathbb{Z}) = \{0\}$. Before the advent of homology groups in the late 1920s, this was expressed by saying that X is connected and its only higher Betti number is $b_n = 1$. Initially Poincaré conjectured that these conditions characterized a topological n-sphere, but then he discovered an example of a homology 3-sphere, presented in [Poincaré, 1904] and that we will call M_{Poin}, which was not homeomorphic to S^3 because $\pi_1(M_{Poin}) \neq \{0\}$. (Poincaré was the first to introduce the fundamental group π_1, which Kneser preferred to call the path group, a name others briefly used, but which did not catch on). Later, it was shown that M_{Poin} can be constructed by identifying opposite faces of a dodecahedron while twisting these through a certain angle so that they align, and so this is now often referred to as Poincaré's dodecahedral space. Its fundamental group $\pi_1(M_{Poin})$ has order 120 and is called the binary icosahedral group; it is now known that M_{Poin} is, in fact, the only homology 3-sphere with a nontrivial *finite* fundamental group.

In their *Enzyklopädie* article, Dehn and Heegaard discussed Poincaré's counterexample, which refuted the Frenchman's original conjecture that S^3 could be characterized by its homology alone [Dehn and Heegaard, 1907, 186–188]. Although the authorship of this part of their report is uncertain, subsequent events suggest that Dehn was responsible for presenting what he thought was a simpler version of Poincaré's example. Using a slightly modified construction of an M^3, Dehn gave a straightforward geometric argument showing that this 3-manifold was not homeomorphic to S^3, thereby avoiding Poincaré's algebraic argument involving the fundamental group. Dehn's M^3 was very similar to Poincaré's, as both are described by a pair of curves on the boundary of a standard genus 2 handlebody. Dehn's argument turned out to be faulty, however, as he himself realized soon afterward: the diagram shown in the report (Figure 1) corresponds, in fact, to an S^3.

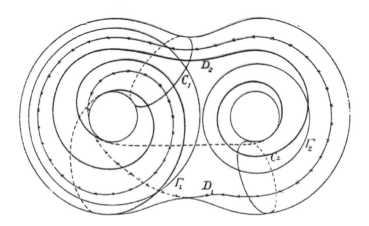

FIGURE 1. A Heegaard surface for S^3

Dehn's construction uses two copies R and R' of a genus 2 handlebody, with boundaries r and r', respectively. By gluing R and R' together along r and r' so as to identify a set of curves in the first surface with corresponding curves in the second, one can ensure that the resulting manifold M^3 satisfies $H_1(M^3, \mathbb{Z}) = \{0\}$, which implies (by Poincaré duality) that M^3 is a homology 3-sphere. In Figure 1 the two circles C_1, C_2 bound in the handlebody R, whereas the closed curves Γ_1, Γ_2 are so chosen that $\Gamma_i \cap C_j$ is a single point when $i = j$ and otherwise they do not meet. Finally, we see that the closed curves D_1, D_2 satisfy (with the appropriate orientations)

$$D_1 + 2\Gamma_1 + \Gamma_2 + C_1 \sim 0, \ D_2 + 3\Gamma_1 + 2\Gamma_2 + C_1 + C_2 \sim 0,$$

which is equivalent to

$$\Gamma_1 + 2D_1 - D_2 + C_1 - C_2 \sim 0, \ \Gamma_2 - 3D_1 - 2D_2 - C_1 + 2C_2 \sim 0.$$

Citing results proved elsewhere in their article, Dehn and Heegaard note that C_1, C_2 and D_1, D_2 are equivalent curve systems. This means that one has a homeomorphism $f : r \longrightarrow r'$ such that $f(D_1) = C_1'$, $f(D_2) = C_2'$. They then define M^3 as $R \cup_f R'$, and note that not only C_1, C_2 but also D_1, D_2 bound in M^3, and because of the above relations, so do Γ_1, Γ_2. Furthermore, every curve in M^3 is homologous to a linear combination of $C_1, C_2, \Gamma_1, \Gamma_2$, which means $H_1(M^3, \mathbb{Z}) = \{0\}$.

This part of the argument given in [Dehn and Heegaard, 1907, 187–188] was sound, and the construction was completely analogous to Poincaré's construction of M_{Poin} in [Poincaré, 1904]. As mentioned already, it seems very likely that Dehn wrote up this part of the article, even though the general method was first presented in Heegaard's dissertation [Heegaard, 1898]. Poincaré knew at least parts of this work through him, although Poincaré never claimed that closed 3-manifolds always admit a Heegaard splitting. Regarding M_{Poin}, Poincaré's remark that "we will restrict ourselves to giving one example" makes it seem likely he realized that one could construct other examples of homology 3-spheres in a similar fashion, but that there was no systematic way of showing that their fundamental groups were nontrivial. He gave an explicit presentation of $\pi_1(M_{Poin})$, namely:

$$\langle a, b : a^4 b a^{-1} b = 1, b^{-2} a^{-1} b a^{-1} = 1 \rangle.$$

This becomes trivial when abelianized, whereas adding the relation $(a^{-1}b)^2 = 1$ yields a presentation for the icosahedral group

$$\langle a, b : (a^{-1}b)^2 = a^5 = b^3 = 1 \rangle.$$

So $H_1(M_{Poin})$ is trivial, whereas $\pi_1(M_{Poin})$ is not.[5]

This was the specific context in which Poincaré formulated the famous conjecture that later bore his name, until Perelman proved it some twenty years ago [O'Shea, 2007]. Dehn seems to have thought that his example did not require calculating the fundamental group, since the argument presented in [Dehn and Heegaard, 1907, 187–188] depends only on properties of the curves C_1, C_2, D_1, D_2 and Γ_1, Γ_2. Dehn still assumed the validity of his geometric argument that his example was not S^3 when he submitted the manuscript in January 1907. Teubner must

---

[5] Jeremy Gray has pointed out that Poincaré may well have arrived at his example of a homology sphere M_{Poin} by first solving a group-theoretic problem, namely, to find a presentation with the same number of generators and relations, which trivializes when abelianized. The answer he found was the binary icosahedral group, which can also be defined as $SL(2, \mathbb{Z}/5\mathbb{Z})$; see [Gray, 2013, 465–466].

have sent back proofs in the early part of that year, as the report appeared before Dehn noticed that the matter was more subtle than he had realized. Probably he recognized the error in the fall of 1907, after which he dashed off [Dehn, 1907] for the *Jahresbericht* of the German Mathematical Society, alerting its readers to the mistake in the report on "Analysis Situs." Dehn states that in fact the homology 3-sphere described in the *Encyklopädie* article is S^3. To obtain Poincaré's example M_{Poin} one replaces the curves D_1, D_2 by two other curves E_1, E_2 equivalent to those shown in Figure 4 of [Poincaré, 1904, 46].[6] The curves Γ_1, Γ_2 are then not null-homotopic in M_{Poin}, and so this 3-manifold is not homeomorphic to S^3.

Dehn's correction in the *Jahresbericht* was presumably seen by many readers, as by this time Franz Meyer was publishing a regular column informing the DMV's members about recent additions to the ongoing *Encyklopädie* project. On the other hand, they would have had to read carefully in order to appreciate Dehn's new construction of a homology sphere, which he briefly described directly afterward. His idea was to take two copies of the complement of a knotted open solid torus in S^3 and glue these together along their boundaries in such a way that a meridian of each is identified with a latitude of the other. The resulting 3-manifold will be a homology sphere that cannot be homeomorphic to S^3, since it contains a torus (namely the common boundary of the two knot complements) which does not bound a solid torus on either side. This simple and evidently very general method completely evades the problem of examining the fundamental group, and it may have anticipated a famous result that Dehn only proved three years later by making use of Dehn's Lemma [Stillwell, 1979]. In [Dehn, 1910] he showed that a closed space curve without singularities is unknotted if and only if its associated fundamental group is abelian.

Dehn's new construction, to be more precise, takes two copies K, K' of the 3-sphere S^3 minus the tubular neighborhood of a knot k and glues them together via a homeomorphism f of the boundaries $\partial K, \partial K'$ to produce a 3-manifold D. The mapping f matches a curve l on ∂K which bounds in K with a curve m' on $\partial K'$ which is non-bounding in K', and likewise matches an m on ∂K which is non-bounding in K with an l' on $\partial K'$ which bounds in K'. The idea is to choose a "meridian" m on ∂K, thus a generator of $H_1(K)$, and let l be a simple curve on ∂K which meets m exactly once and equals $0 \in H_1(K)$. Then let m', l' be corresponding curves on $\partial K'$. The resulting 3-manifold D is then a homology sphere, and Dehn argues further that $D \neq S^3$ because a torus cannot separate S^3 into two pieces, neither of which is a solid torus.

Dehn's final conclusion was correct, but this was only later proved in (Alexander 1924). John Stillwell pointed out that this construction may have put him on the trail that led to his famous lemma [Stillwell, 1979]. He considered it likely that Dehn soon recognized the difficulties involved in proving the final claim, which may have motivated him to introduce the new methods in [Dehn, 1910]. Indeed, Stillwell pointed out that with the help of "Dehn's Lemma," one can give a group-theoretic justification for the above construction. First, he noted that the fundamental group of the torus $\pi_1(\partial K) = \mathbb{Z} \times \mathbb{Z}$ either injects in $\pi_1(K)$ or collapses to $\mathbb{Z} = H_1(K)$. (This follows from the Loop Theorem, which was proved by Papakyriakopoulos in 1957.) The latter possibility, however, is ruled out by Dehn's lemma. This means

---

[6]The picture drawn in Poincaré's paper can also be found in [Sarkaria, 1999, 164] and in [Gray, 2013, 463].

that the inclusion homomorphisms $\pi_1(\partial K) \longrightarrow \pi_1(K)$ and $\pi_1(\partial K') \longrightarrow \pi_1(K')$ are injective and $\pi_1(D)$ is then the amalgamated free product

$$\pi_1(K) *_{\mathbb{Z} \times \mathbb{Z}} \pi_1(K'),$$

obtained from the isomorphism of $\mathbb{Z} \times \mathbb{Z}$ which exchanges m and l. This shows that $\pi_1(D)$ is infinite, which proves that D cannot be an S^3. Since the same construction with an unknotted curve k clearly does yield an S^3, we have an effective procedure for converting a curve k to a 3-manifold $D(k)$ such that k is knotted if and only if $D(k) \neq S^3$. So any algorithm capable of recognizing S^3 among 3-manifolds would need to be strong enough to recognize when a curve is unknotted.

By 1908, Dehn was deeply immersed in Poincaré's latest work on the fundamental group and homology spheres, and especially the later famous Poincaré conjecture. In fact, as Klaus Volkert aptly noted, he nearly became its first victim when he briefly thought he had found a proof [Volkert, 1996]. Writing to Hilbert in great excitement, he urged his former mentor to submit his manuscript at the next meeting of the Göttingen Scientific Society, as he otherwise feared that a certain rival might claim this prize; Dehn obviously speculated that Poincaré himself was close to cracking this problem!

Luckily, he also sent a copy to the Austrian topologist Heinrich Tietze, whom he met later that year at the International Congress of Mathematicians held in Rome [O'Shea, 2007, 140–141]. Dehn had hoped to unveil the key ideas in his proof on that very occasion, but Tietze persuaded him that there was a fly in the ointment he had prepared. The manuscript had already gone to the printer, but Dehn had not yet returned corrected page proofs, so he wrote Hilbert in haste, informing him that he had to withdraw his paper, which seems to have since disappeared without a trace:

> To my great regret, I must inform you that my paper, which you submitted to the society, is incorrect. Herr Tietze, to whom I had sent the proofs, pointed out to me in Rome that my axiom for ordinary 3-space, "separation by arbitrary closed 2-complexes," is not correct. With it, collapse nearly all of the results in my paper. I have little hope of mastering this problem in a short time, and in any case it appears to me impossible to maintain a good part of my work in its present form. I must therefore withdraw the paper. ...At the same time, I was informed early enough by Herr Tietze to hinder me from holding my lecture.[7]

Dehn might well have remembered the commotion caused by König's sensational lecture at the Heidelberg ICM four years earlier (see Chapter 2), though in that case it took a few weeks for the experts on Cantorian set theory to uncover the error. He surely felt deeply relieved that Tietze had informed him of his mistake ahead of time.

Dehn's Papers on 3-Manifolds and Knots, 1910–1914

In Stillwell's opinion, Dehn's earliest topological investigations were hampered by his lack of appreciation for group theory [Stillwell, 2012]. After 1908, though, Dehn's awareness of the importance of groups became more and more evident. This transition in the orientation of his research can be seen from notes for unpublished

---

[7]Dehn to Hilbert, 16 April 1908, Nachlass Hilbert 67, SUB Göttingen.

lectures on group theory, which he presented in Münster during the period 1909–1910. The first two chapters from these appear in English translation in [Dehn, 1987].

Max Dehn's pioneering work on three-dimensional topology is best known from his seminal paper [Dehn, 1910], which was translated by John Stillwell in Max Dehn, *Papers on Group Theory and Topology* [Dehn, 1987, 92–126]. In his introduction to this paper, Stillwell notes the remarkable set of topics that Dehn here introduced for the first time in print:

> ...statement of the word and conjugacy problems for finitely presented groups, realisation of such groups as fundamental groups of 2-complexes, Dehn's lemma, the diagram of the trefoil knot group, equivalence between triviality of a knot and commutativity of its group, construction of homology spheres by surgery, and proof that one of them (actually Poincaré's homology sphere) has finite fundamental group [Dehn, 1987, 86].

One key tool for this undertaking was Dehn's *Gruppenbilder* (see Chapter 7), which he introduced at the beginning of his paper. These are similar to Arthur Cayley's group diagrams for finite groups. Dehn also discusses finite examples, such as the icosahedral group, but his diagrams broke new ground by including presentations of infinite groups, an approach that led to modern geometric group theory. Another important tool was his lemma, about which more below. A third new instrument in Dehn's toolkit was Poincaré's fundamental group, which Dehn employed as a topological invariant for distinguishing 3-manifolds constructed by means of knots. Using these techniques, Dehn was able to construct a whole infinite family of homology spheres, using their nontrivial fundamental groups to show that these were not 3-spheres.

The method he employed – an early variant of what came to be called Dehn surgery[8] – was simply to remove a knotted solid torus from an S^3 and then sew it back in differently. Since there are infinitely many ways to identify a pair of canonical curves on a solid torus with a pair of curves on the boundary of the knotted hole in S^3, this technique leads to infinitely many possible 3-manifolds and as many homology spheres. Dehn also regarded his construction as a method for showing that a knot K is nontrivial. For if the fundamental group of one of the homology spheres obtained by Dehn surgery on K is nontrivial, then it follows that $\pi_1(S^3 - K) \neq \mathbb{Z}$ and so K is nontrivial. The converse – if K is nontrivial, then any manifold obtained by nontrivial Dehn surgery on K has nontrivial fundamental group – was the longstanding Property P conjecture, which was finally proved in 2004.

As discussed above, one can easily identify a strong Hilbertian current behind Dehn's transition from foundations of geometry to topology, but for [Dehn, 1910] the key motivation clearly came from Poincaré's famous topological papers, in particular [Poincaré, 1895] and [Poincaré, 1904]. In fact, Dehn was still very much in the hunt for a way to use Heegaard splittings to crack open the Poincaré conjec-

---

[8] For an overview of modern Dehn surgery, see [Gordon, 2009].

ture.[9] In his introduction, he spelled out this motivation very clearly:

> [§2 of Chapter 3] deals with the important problem of the topological characterization of ordinary space, without, however, resolving the problem. It treats the question of how ordinary space may be topologically defined through the properties of its closed curves, and how to make it possible to decide whether or not a given space is homeomorphic to ordinary space. The history of this problem began when first Heegaard ... and then Poincaré ... pointed out that in order to characterize ordinary space it does not suffice to assume that each curve bounds, possibly when multiply traversed. Indeed the manifolds with torsion show this. Then Poincaré proved in [Poincaré, 1904], by construction of a "Poincaré space," that it is even insufficient for each curve to bound when traversed once. It now is natural to investigate whether it suffices to suppose that each curve in the space bounds a disk. This is also suggested at the end of Poincaré's work. However, the reduction of the problem given in the present work does not appear to lead directly to a solution. A deeper investigation of the fundamental groups of two-sided closed surfaces seems to be unavoidable [Dehn, 1987, 93–94].

Dehn's introduction described the overall structure of [Dehn, 1910], which consisted of three chapters. In the first, he presented the three main tools: the *Gruppenbild*, fundamental groups of 2-complexes, and "Dehn's Lemma." He shows that a finitely presented group can be described by a 1-complex (*Gruppenbild*), and it can also be realized as the fundamental group of a 2-complex. Chapter 2 deals with knots and the fundamental group of their complements, but also groups of 3-manifolds obtained from knots by what is now known as Dehn surgery. This chapter has five sections and is entitled knots and groups; we describe its main results in what follows. Chapter 3 on 3-dimensional manifolds gives a brief account of Dehn's reflections in the direction of the Poincaré conjecture.

At the outset of Chapter 2, Dehn recalled two definitions he and Heegaard introduced in [Dehn and Heegaard, 1907], while noting the distinction between homeomorphic and isotopic objects in ordinary space (here S^3). Any two closed space curves C and C' are homeomorphic, in particular equivalent to an unknotted curve S^1. But for C to be isotopic to S^1 – thus mappable to it via transformations of the ambient space S^3 – requires that C bound a disk $D^2 \subset S^3$. Dehn adds that he will later give a criterion for deciding whether or not a space curve is knotted (namely, precisely when its fundamental group is nonabelian). As Epple points out [Epple, 1999a, 238–239], Dehn never mentions how Tietze showed by this means that the trefoil knot and the unknot are not isotopic [Tietze, 1908, 82]. This result was, of course, long considered intuitively obvious, so Tietze's explicit proof provides an indication of the great value he placed on rigorous methods. Dehn would later go a great step further by proving that left- and right-handed trefoil knots – although they have isomorphic fundamental groups – are *not* isotopically equivalent [Dehn, 1914].

---

[9]For a discussion of Dehn's strategy in his 1910 paper, see [Gordon, 1999, 476–477].

Before presenting knot groups, however, Dehn first laid the groundwork for his new method for constructing Poincaré spaces, i.e. homology 3-spheres.[10] As noted earlier, the method described in [Dehn, 1907] was, in fact, sound, although it used a fact that was not proved until later. In the meantime, Dehn found a related construction that made direct use of the fundamental group, by extending the complement space A of a knotted solid torus to obtain a 3-manifold with trivial first homology, but nontrivial fundamental group. Before presenting the full construction, he first described the homological properties of special curves on the surface of a torus.

He thus considered a knotted space curve K surrounded by a tube T, the surface of a twisted solid torus $E_3^1 \cup E_3^2$, where E_3^1, E_3^2 are 3-cells that meet along two 2-cells E_2^1 and E_2^2. Let β be the boundary of the 2-cell E_2^1. Dehn points out that any non-self-intersecting closed curve C, which does not separate T and meets β just once, will together with β form a basis for the homology of T. He claims that among such curves, there is a $\lambda \subset T$ that bounds in A. This is easy to see since homologically C is equivalent to a multiple of β, so $C \sim n \cdot \beta$, and therefore $C - n \cdot \beta \sim 0$. But one can always find a nonsingular curve λ that meets β once for which $C - n \cdot \beta \sim \lambda$, and hence $\lambda \sim 0$ in A. Furthermore, λ is isotopic to K, since together they bound a non-singular strip in the interior of T.

It was in this context that Dehn first indicated the role of his lemma, which he attempted to prove in Chapter 1. Dehn's Lemma asserts that a singular disc $D' \subset M^3$ without singularities on $\partial D'$ can be replaced by a nonsingular disc D with the same boundary. If λ bounds a singular disc in A, then extending A to the strip joining λ and K yields a singular disc D' with $\partial D' = K$, and since there are no singularities on $\partial D'$, the lemma asserts that K bounds a nonsingular disc D. But this means K is unknotted, so λ is null-homotopic in A if and only if K is unknotted.

Up to this point, Dehn had merely sketched the geometrical constructions required without drawing on group theory. He now interrupted his presentation to pick up his earlier discussion of fundamental groups of 2-complexes, where here these are built from knot projections. Knot groups had been introduced earlier by Wilhelm Wirtinger and Heinrich Tietze, but Dehn's method for giving a presentation of the group of a knot K was different. He considered a diagram of K, with $(n-1)$ crossings, say. Regarding the crossings as vertices, the diagram defines a 1-complex, and the complement of a neighborhood of this 1-complex is a genus n handlebody. Dehn observed that the complement of K may be obtained from this handlebody by attaching $(n-1)$ thickened 2-cells, one at each crossing of the diagram. In this way he arrived at a presentation of the group of K with n generators and $(n-1)$ relations. In preparation for his construction of Poincaré spaces, he focused on the simplest of all cases, namely, a trefoil knot, which has just three crossings. The corresponding knot diagram then has four bounded regions (Figure 2).

To each bounded region, Dehn associated a loop C_i, $i = 1, \ldots, 4$, which passes through a base point P outside the plane of the diagram and pierces the region once. These are the generators, whereas the relations are read off at each crossing by

---

[10]Dehn's paper contains several drawings, but none for this key construction. For a clear explanation of both this and the earlier construction sketched in [Dehn, 1907], both illustrated with pictures, see [Epple, 1999a, 267–271].

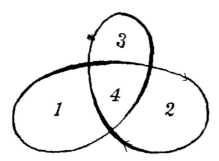

FIGURE 2. Diagram for a trefoil knot

passing around the adjacent regions taking note of the orientations. For the trefoil knot, these are then $C_1 C_4^{-1} C_2 = C_2 C_4^{-1} C_3 = C_3 C_4^{-1} C_1 = 1$, which determine the knot group; Dehn denotes it by G_K (later by $G_{K\ell}$). As mentioned above, Poincaré gave a presentation for $\pi_1(M_{Poin})$ based on just two generators. Dehn could have easily eliminated two of the generators to obtain $C_4^{-2} C_1 C_4 C_1 = 1$ or $C_4^{-3}(C_4 C_1)^2 = 1$. Then letting $A = C_4^{-1}$, $B = C_4 C_1$ we get the standard presentation for a trefoil knot: $\langle A, B : A^3 B^2 = 1 \rangle$. This makes it the simplest example of a torus knot, the family whose group presentation is given by $\langle A, B : A^p B^q = 1 \rangle$, where p, q are coprime.

Dehn's interest in knot groups, though, was mainly to use them to study the fundamental groups of the associated 3-manifolds, not to study their presentations systematically. Returning to the discussion of the space curves K and λ, he noted that the latter (which bounds in A) spans a disc precisely when K is unknotted, which is equivalent to the condition that G_K is abelian. This condition is then satisfied when the substitution s_λ in G_K that corresponds to the closed curve λ is the identity. So here $s_\lambda = 1$ means that K and the torus T are unknotted, and all the curves on T are equivalent in A to some multiple of a curve that loops once around β. Thus, in the unknotted case the group G_K will be isomorphic to $\mathbb{Z}$.

Dehn next considered a curve $\rho \subset T$ such that $\rho \sim \beta + n \cdot \lambda$. He noted that $s_\beta = C_i \in G_K$, where C_i is associated with a bounded region adjacent to the unbounded region, and therefore if G_K is abelian then $s_\rho = C_i$, since $s_\lambda = 1$. Dehn then extended the complement space A by attaching a thickened 2-cell to A along a strip neighborhood of β on T, and noted that for any knot K this gives a 3-cell. Therefore adding the relation $C_i = 1$ to G_K results in the trivial group. Consequently, if K is unknotted then G_K also becomes trivial on adding the relation $s_\rho = 1$. On the other hand, if K is knotted this will not be the case in general.

Dehn took up the simplest case: a trefoil knot with the group presentation given above for G_K, an infinite nonabelian group. The closed curve λ corresponds to $C_1 C_2 C_3 C_2^{-1} C_2^{-1} C_2^{-1} \in G_K$. To construct a Poincaré space Φ, Dehn attaches a thickened 2-cell to A along a neighborhood of a curve ρ on T that crosses each of λ and C_1 once. This gives a 3-manifold Φ with boundary a sphere such that $H_1(\Phi) = \{0\}$. It can thus be obtained from a Poincaré space with the same fundamental group by removing an open 3-cell. The fundamental group G_Φ of Φ is the quotient of G_K defined by adding the relation $s_\rho = 1$. Since $s_\rho = C_1 C_2 C_3 C_2^{-1} C_2^{-1}$, G_Φ has

the presentation with four generators C_1, C_2, C_3, C_4 and four relations
$$C_1 C_4^{-1} C_2 = C_2 C_4^{-1} C_3 = C_3 C_4^{-1} C_1 = 1, \ C_1 C_2 C_3 C_2^{-1} C_2^{-1} = 1.$$
Dehn next constructs the *Gruppenbild* for G_K before building the corresponding structure for G_Φ, which turns out to be a group of order 120 that has the icosahedral group as a quotient group. Since Poincaré's homology sphere had precisely the same fundamental group – as Dehn well knew – he surely assumed that the closed M^3 obtained from Φ by filling its boundary surface with a ball was homeomorphic with M_{Poin}. Evidently, though, he saw no easy way to prove this, and so he made no mention of the possibility that his new construction was very likely a new way of obtaining M_{Poin}.[11]

For the *Gruppenbild* of G_K, Dehn took a strip bordered by a composition of C_4 edges (Figure 3). Starting with a vertex on the left, one easily sees that the triangles correspond to the three relations. He notes further that there are three types of vertices: those where C_1 and C_2^{-1} go out, another for C_2 and C_3^{-1}, and a third with C_3 and C_1^{-1}. Since the relations in G_K are contained already in this single strip, the *Gruppenbild* is constructed by attaching two more copies of it to each boundary so that each vertex has segments with these types, so that altogether all eight possibilities are represented:
$$C_1, \ C_2, \ C_3, \ C_4, \ C_1^{-1}, \ C_2^{-1}, \ C_3^{-1}, \ C_4^{-1}.$$
After attaching these pairs of strips to each boundary of the original one, we obtain four free boundaries, so repeating the procedure *ad infinitum* leads to the group diagram.

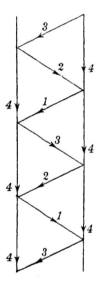

FIGURE 3. Part of the *Gruppenbild* for a Trefoil Knot

Using this, Dehn confirms two properties of G_K. First, it is nonabelian, as for example the polygonal path $C_1 C_4 C_1^{-1} C_4^{-1}$ is open in the graph. Second, the curve λ corresponding to $C_1 C_2 C_3 C_2^{-1} C_2^{-1} C_2^{-1} \in G_K$ also does not close. Thus λ does

---
[11]Seifert and Weber later proved that this was indeed the case.

not bound a disc in the complement space A, only a surface of higher connectivity, proving that K is indeed knotted.

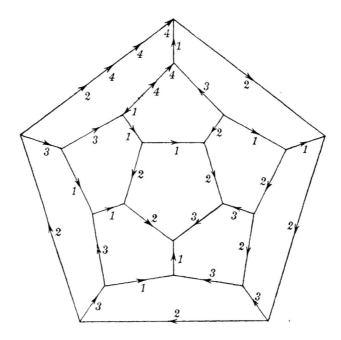

FIGURE 4. Dehn's modified dodecahedron

To obtain the *Gruppenbild* of G_Φ, Dehn introduced the modified dodecahedron shown in Figure 4. He had earlier introduced the "genuine" dodecahedral figure as the *Gruppenbild* of the icosahedral group of order 60 (by duality, this is the same as the rotation group of the dodecahedron). He explained how this figure arises by manipulating the group diagram of G_K to take account of the added relation $C_1C_2C_3C_2^{-1}C_2^{-1} = 1$. This modified figure has twelve faces (counting the boundary face), but only ten are pentagons. The upper left face is a decagon that shares the "triple edge" $C_1^{-1}C_4C_4$ with the heptagon below it, whereas its topmost segment is a "quadruple edge" composed of $C_2C_4C_4C_4$. Among the twelve relations shown in these polygons, three derive from the additional relation $C_1C_2C_3C_2^{-1}C_2^{-1} = 1$ corresponding to the curve ρ. The others arise from a combination of this relation and the three from G_K. One easily finds:

$$C_2^5 C_4^3 = 1, \qquad C_2^5 C_4^{-3} = 1,$$

where the first is shown by the boundary of Figure 4. From these relations, it follows that

$$C_1^{10} = C_2^{10} = C_3^{10} = C_4^6 = 1.$$

Dehn now uses these derived relations to describe the *Gruppenbild* of G_Φ in terms of that of G_K. For this purpose, he takes part of the latter, consisting of five neighboring strips of the type shown in Figure 3, where the diagonals are formed from C_4 edges (see Figure 5). These two boundaries are identified at corresponding vertices, as shown by the darkened segment joining two of them, which corresponds to the relation $C_1C_2C_3C_2^{-1}C_2^{-1} = 1$. The various polygons in the dodecahedral

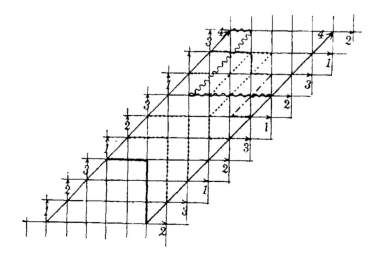

FIGURE 5. Construction of the *Gruppenbild* for G_Φ

figure are indicated by dotted lines as well, whereas the relation $C_2^5 C_4^{-3} = 1$ is shown by the wavy lines. After identifying the two boundaries, we see from $C_4^6 = 1$ that any six edges along that chain also close, which means that 5×6 vertices lie on a torus. Furthermore, none of the polygons from the dodecahedral figure separate this surface, so we can imagine passing this torus through Figure 4. From $C_1^{10} = C_2^{10} = C_3^{10} = 1$ we see that there are altogether 120 vertices, showing that G_Φ is a group of order 120. Dehn imagined that G_Φ was isomorphic with the icosahedral group extended by reflections, but this turned out to be mistaken. Still, his virtuosity in manipulating these *Gruppenbilder* was on full display here.

As if this were not enough, Dehn went on to indicate that the same techniques can be used to construct any number of Poincaré spaces starting with a trefoil knot and a suitable attaching curve ρ. If the curve so chosen intersects β n times and λ once where $n > 1$, then for the associated space Φ the *Gruppenbild* of G_Φ can be derived from a tessellation of the hyperbolic plane by $(6n-1)$-gons, which meet in threes at each vertex. The trefoil can also be described as the $(2,3)$-torus knot. Finally, Dehn made some brief remarks about the Poincaré spaces that can be obtained in the same way from the $(2, 2m+1)$-torus knots with $m > 1$; these, too, lead to tessellations of the hyperbolic plane. Regarding the general utility of group diagrams, Dehn had this to say:

> In order to present the meaning of the group diagram more strongly, we remark that each closed curve on the diagram corresponds to a substitution which is $= 1$, and to a curve on the manifold contractible to a point; i.e., one which bounds a singular disc. The above arrangement of the group diagram not only helps to prove that space curves are knotted, or that the spaces are not homeomorphic to ordinary space, but *it also solves the problem of deciding, for a given curve of the complement space or Poincaré space respectively, whether it is contractible to a point or not.* [Dehn, 1987, 121]

As noted already, an important tool for Dehn was his lemma, which he formulated as follows:

> Let C_2 be a 2-complex in a homogeneous manifold M_3. Let K be a [simple closed] curve in C_2 which bounds a (singular) disc E_2' in M_3. If E_2' has no singularities on its boundary, then K bounds a completely nonsingular disc in M_3.

Dehn's appeal to a 2-complex in his statement of the lemma suggests he had in mind the seam surface N_2 of a Heegaard splitting and that his principal motivation was his attack on the Poincaré conjecture in Section 2 of Chapter 3. A modernized statement of Dehn's Lemma would make explicit reference to mappings, and so might read:

> Let f be a piecewise-linear (PL) map of a disk $D^2 = int(D^2) \cup \partial D^2$ into a 3-manifold M^3, where the singularity set $S(f) \subset int(D^2)$. Then there exists another PL map f^* which is an embedding of D^2 into M^3 and such that $f^* = f$ on ∂D^2.

Dehn conceived of this as analogous to a lower-dimensional result for surfaces: If C is a connected 1-complex in a 2-manifold M_2, then any two points $\{P,Q\} \in C$ can be connected by an arc in C, i.e. a polygonal path free of singularities. Beyond its role in Dehn's approach to the Poincaré conjecture, he also applied his lemma to prove the important theorem that a knot is trivial if and only if its fundamental group is isomorphic to $\mathbb{Z}$, as indicated earlier.

In the final two sections of his paper, Dehn first shows that every closed 3-manifold M can be obtained by sewing a 3-ball along its boundary onto a seam surface N_2 in M. (This 2-complex N_2 is what Heegaard called the "nucleus"). Letting N_1 be the 1-skeleton of N_2, Dehn notes that a neighborhood of N_1 in M is a (possibly nonorientable) solid handlebody, whose complement is also a solid handlebody. He thus concludes that every closed 3-manifold has a Heegaard splitting, though he does not mention Heegaard here at all. He also notes how this implies that every closed 3-manifold is the union of four 3-balls.

Dehn then turns to his reduction of the Poincaré conjecture, beginning with the observation that one may assume that N_1 consists of a wedge of circles. Since the manifold M is simply connected by hypothesis, each of these circles bounds a singular disk. *If* it were possible to choose these disks to have no singularities on their boundaries, then Dehn's Lemma would give a system of embedded disks, with the same boundaries, meeting only at a single point. A neighborhood of the union of these disks would be a 3-ball, whose boundary S is contained in the solid handlebody A that is the complement of a neighborhood of N_1. Dehn now asserts that the 2-sphere S bounds a 3-ball in A, which would mean that M is the union of two 3-balls along their boundaries, and therefore homeomorphic to S^3. Whether he continued to pursue this *Ansatz* or some other line of attack on the Poincaré conjecture after 1910 is unclear, but we know from a letter he wrote to Hellmuth Kneser in 1929 that Dehn still continued to think about this problem some twenty years later.

Dehn's argument in his "proof" of the lemma involved removing the singularities in the immersed disc in a systematic way. Such singularities typically arose as double lines where two sheets crossed one another. Removing a double line by means of a local move was a procedure closely analogous to a method for removing singularities in the case of algebraic curves, a technique popularized during the late

1800s by Felix Klein in his lecture courses, illustrated here by the simple case of quartic curves (Figure 6).

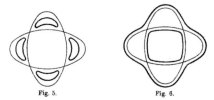

FIGURE 6. Four cases of quartic curves, Felix Klein, *Gesammelte Mathematische Abhandlungen* Band 2, S. 111.

A nonsingular quartic curve C_4 has genus $p = 3$, which means that a *singular* quartic can have at most 3 double points. If a curve given by a fourth-degree polynomial $F_4(x,y) = 0$ should have 4 double points, then the polynomial is reducible, e.g., $F_4(x,y) = f_2(x,y)g_2(x,y) = 0$. The two conics corresponding to $f_2(x,y) = 0$ and $g_2(x,y) = 0$ may have up to 4 real points of intersection. Assuming this to be the case, these curves can be regarded as the contour for a nonsingular quartic given by $F_4(x,y) = f_2(x,y)g_2(x,y) = \epsilon$, for a small nonzero number ϵ. Clearly, this has the effect of moving the curve a small distance from the two intersecting conics, thereby desingularizing the curve. Note, further, that this can be done in two distinct ways, depending on whether $\epsilon > 0$ or $\epsilon < 0$.

One can liken Dehn's technique for removing pieces of double lines on a singular disc in M^3 with the even simpler case of removing the intersection point of two lines in a plane. If the product of the two linear equations is given by $f_1(x,y)g_1(x,y) = 0$, then setting $f_1(x,y)g_1(x,y) = \epsilon$ produces a hyperbola with the given lines as asymptotes. Furthermore, this desingularization can be done in two ways, depending on whether ϵ is positive or negative. Max Dehn called his method for removing pieces of double lines "Umschaltung" ("switch over"). Figure 7 shows that this topological procedure can be seen as the higher-dimensional analogue corresponding to the process that transforms two intersecting lines into a hyperbola. Each such switch over can therefore be carried out in two different ways. The trickiness then becomes a matter of deciding which of these two possible choices to use in removing *all* the double lines in a disc immersed in a 3-manifold.

In his paper, Dehn noted that one must also take into account that as many as 3 sheets may cross one another, giving rise to a finite number of triple points. Higher-order singularities than this could be removed, however, so these triple points presented the worst-case scenario for the topologist. As described below, nearly twenty years later Hellmuth Kneser stumbled over this problem, which he and Dehn struggled with, but were unable to resolve. Subsequent efforts to prove

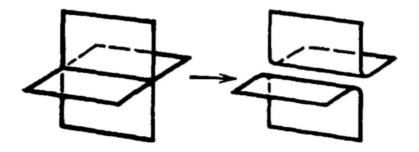

FIGURE 7. Dehn's method for removing double lines by a "switch over" [Johansson, 1935, 314]

Dehn's Lemma in the 1930s led to new insights, though none that pointed the way to a final proof. Thus, much like the Poincaré conjecture, the central lemma in [Dehn, 1910] became one of the major open problems of modern topology.

In the years immediately afterward, Max Dehn focused mainly on central problems in combinatorial group theory (about which, see Chapters 6–8). This interest was strongly connected with knot groups and their diagrams, whereas Dehn's interest in knot theory per se appears to have been quite limited. Heegaard presumably had already done the research on that topic, which fills several pages in [Dehn and Heegaard, 1907]. As mentioned earlier, their report contained nothing at all about knot groups. Probably at some point Dehn acquired considerable knowledge of the earlier British tradition, in particular the work of Peter Guthrie Tait, which relied entirely on heuristic methods and intensive empirical investigations [Epple, 1999a, 125–160]. This "tabular tradition" aimed to classify the prodigious number of different types that could arise, an undertaking that lacked firm foundations. That alone would no doubt have discouraged Dehn from throwing himself into this fray. His own interests in topology all aimed to make progress by adopting straightforward combinatorial methods.

Nevertheless, Dehn made one particularly noteworthy contribution to the classification problem by giving the first proof of chirality in knots. Fittingly, in [Dehn, 1914] he took up the simplest case: the trefoil knot. The phenomenon of achiral (or as Tait called them amphicheiral) knots was long known, having been pointed out by J.B. Listing in 1847. An achiral knot is one that is isotopically equivalent to its reflection, or mirror image. The simplest example is the figure-eight knot, which has a knot diagram with four crossings and five regions; Dehn gave a group presentation for it in [Dehn, 1914, 222]. One of Tait's famous claims was that all amphicheiral knots had even crossing numbers, but this turned out not to be true, as a counterexample was found in 1998. The Tait conjecture was proven correct, however, if restricted to alternating knots. Clearly, it took many decades to develop the kinds of techniques needed to prove such theorems. As a pioneer in this respect, Dehn emphasized that even long obvious facts about knots stood in need of proof (here he alluded to a similar remark in [Tietze, 1908, 97]).

Dehn had already come up with a new way of viewing the *Gruppenbild* for the group $G_{K\ell}$[12] of the trefoil knot in [Dehn, 1912], which he used to solve the conjugacy problem for G_{Kl}. The construction uses the generators C_1, C_2, C_3 to build a net of triads in the hyperbolic plane, over which Dehn erected a cylindrical structure using the C_4 vertically (recall Figure 3) [Dehn, 1987, 173–176]. Since none of the segments built from the C_1, C_2, C_3 close, the graph in the hyperbolic plane is an infinite tree Γ. Making use of this visualization, Dehn was able to analyze the automorphisms of G_{Kl}, and by this means he proved that there was no orientation-reversing homeomorphism of S^3 taking the trefoil knot to itself. He did this by observing that such a homeomorphism would give rise to an automorphism of G_{Kl} that either took β to β and λ to λ^{-1} or β to β^{-1} and λ to λ, where β and λ are the oriented curves on a torus surrounding Kl described earlier. By determining the automorphism group $\mathrm{Aut}(G_{Kl})$ of G_{Kl} he showed that this was impossible.

In the final part of [Dehn, 1914], he also briefly discussed the case of the figure-eight as the simplest achiral knot. Bill Thurston later highlighted its key importance within his geometrization program for 3-manifolds. Its *Gruppenbild* was constructed by one of Dehn's students in Kiel, Fritz Klein, but this was never published. Another Dehn student, Hugo Gieseking, who took his Ph.D. in Münster in 1912 after Dehn had already left, also found representations of *Gruppenbilder* by means of hyperbolic geometry. Thus, before he arrived in Breslau in 1913, Dehn had already guided two promising students into this new area of research. Both of them, however, like so many others, never survived the war. World War I also represented a major break in Dehn's career, after he had planted so many seeds for the future of topology. We touch on some of the later developments in the final section that follows.

Topology in the 1920s

As described in the final section of Chapter 2, Dehn had two congenial colleagues at the *Technische Hochschule* in Breslau, namely, the aforementioned Ernst Steinitz and the geometer Gerhard Hessenberg. All three, however, found their lives disrupted by the war, and soon after it ended they went their separate ways. Hessenberg received an attractive offer from the TH Charlottenburg, but owing to the political chaos in Berlin he decided to accept a call from Tübingen instead. In Chapter 2, we described the situation in Breslau in 1920, when Steinitz left to accept a professorship in Kiel. One year later, Dehn was bound for Frankfurt.

Not many records have survived that shed light on Dehn's years in Breslau, but a few sources point to important contacts he made at that time. At the nearby university, he befriended Erhard Schmidt, another distinguished student of Hilbert, as well as the somewhat older mathematician Adolf Kneser. During the war years, Schmidt taught a course on point set topology that was attended by two young students, Heinz Hopf and Hellmuth Kneser, who also took courses with his father. Hopf probably never studied under Dehn, but Kneser definitely did, even if only informally. Hopf broke off his studies after the outbreak of the Great War to join the army. During the height of the fighting on the West front, he was seriously wounded at Verdun and sent home to recuperate. It was during this period of convalescence that he heard Schmidt's lectures and caught fire. Later he took his

---

[12] $K\ell$ is short for *Kleeblattschlinge* or clover-leaf knot.

doctorate under Erhard Schmidt, who in 1917 was called to a chair at the University of Berlin. In his famous dissertation, Hopf proved that any simply connected complete Riemannian 3-manifold of constant sectional curvature is globally isometric to Euclidean, spherical, or hyperbolic space. Soon thereafter, he proved what today is called the Poincaré-Hopf theorem, which connects the zeros of a vector field on a manifold to its Euler-Poincaré characteristic.

Hellmuth Kneser spent five semesters in Breslau before going on to Göttingen, where Richard Courant appointed him as his *Assistent*, soon to be joined by a young number-theorist by the name of Carl Ludwig Siegel (see Chapter 5). Kneser's interests were exceedingly broad, even by the standards of those days. In 1921, he took his doctorate under Hilbert with a dissertation on mathematical methods in quantum theory. This was published the following year in *Mathematische Annalen* around the time Kneser became a *Privatdozent* in Göttingen. Then, at age 27, he succeeded Johann Radon in 1925 as full professor in Greifswald. Certain details regarding Kneser's relationship with Max Dehn can be gleaned from the letters they exchanged, some of which are cited below. Kneser's correspondence with other mathematicians, including Heinz Hopf, has yet to be explored, however. Here we can only touch on a few key aspects.

In a broad sense, Dehn's program for higher-dimensional topology continued to guide an important line of research throughout the 1920s. One of its key advocates during that decade was Hellmuth Kneser, who offered a survey of the field in a lecture delivered at the September 1924 meeting of the German Mathematical Society held in Innsbruck. His paper, "The Topology of Manifolds" [Kneser, 1926], amplified on that lecture, though Kneser omitted the part in which he gave a short overview of the main results on the topology of manifolds. Instead, his accent fell on combinatorial methods, which he claimed were the most promising for making progress on the purely topological properties of manifolds. At the same time, he stressed that the concept of manifold had to be formulated broadly enough to embrace important examples from other fields. These included Riemann surfaces and related geometrical constructs associated with functions in one or more complex variables, as well as regular surfaces in 3-space and phase spaces in dynamical systems. As pointed out earlier, Poul Heegaard's dissertation [Heegaard, 1898] had been motivated by similar concerns.

Kneser began by citing Hausdorff's axioms for a topological space in *Grundzüge der Mengenlehre* [Hausdorff, 2002]. To define an n-manifold, he introduced two additional axioms: 5) each point has a neighborhood that is an open topological n-ball; 6) the closed n-manifolds satisfy the Heine-Borel theorem (compactness criterion). Rather surprisingly, he considered 5) to be problematic since it amounts to importing a special construct into the picture, a blemish he would have liked to remove [Kneser, 1926, 3]. Condition 6) was mainly a way to eliminate certain more exotic examples with uncountable bases, etc.; here he cited an example recently given by Pavel Alexandrov.

Kneser next indicated how one could use combinatorial methods to obtain invariants like the Euler characteristic without having to appeal to the properties of $\mathbb{R}^n$. In an appendix he gave a short proof that if two manifolds have isomorphic decompositions as cell complexes, then they are topologically homeomorphic. The converse does not necessarily hold, however, and he described the types of difficulties that can arise. Kneser saw [Dehn and Heegaard, 1907] as providing a fundamental starting point for a combinatorial approach to topological manifolds. He briefly described how they made use of elementary transformations (e.T.) to go from one cell decomposition to another.

This leads to what Kneser called the *Hauptvermutung*, which Dehn and Heegaard took to be essential for a purely combinatorial theory of manifolds:

> If Z_1 and Z_2 are two decompositions of a manifold into elementary cells, then one can through (finitely many) e.T. of Z_1 obtain Z_2 or an isomorphic decomposition of it. [Kneser, 1926, 6]

Steinitz had already raised the question of whether this was true in all dimensions [Steinitz, 1908]. Kneser noted that the *Hauptvermutung* had been proved for dimensions 2 and 3, citing work by Bela Kerékjártó and Robert Furch. He himself regarded it as plausible enough that he assumed its validity for arbitrary dimensions in the remainder of his report.[13]

Kneser elaborated on this combinatorial program for topology at some length. He raised the possibility of developing a complete system of combinatorial invariants, which would then hold for all subdivisions, thereby yielding a system of topological invariants for n-manifolds. This, he suggested, would open the way to creating a new discipline grounded on the still unproven *Hauptvermutung*, and he proceeded to sketch what could be done along these lines, restricting himself to the case of compact manifolds. Following this approach, one can define combinatorial invariants for a great number of topological concepts: connectivity, orientability, homology, Betti and torsion numbers, the Kronecker characteristic, and fundamental groups. There remained a number of technical difficulties, however, and Kneser listed four unproven theorems that were indispensable in order to apply the theory fruitfully. Finally, he briefly took note of two approaches that skirted the *Hauptvermutung* altogether. The first was the older *méthode mixte* long associated with Poincaré's work, but which Oswald Veblen had recently refined and developed further. The second, set forward by Hermann Weyl, was a more complicated axiomatic approach than the one offered by Kneser. In the years that followed, the former method gained momentum from L.E.J. Brouwer's work along with the advent of homology groups, a breakthrough that took place in Amsterdam in 1925. A key figure in this connection was the Austrian topologist Leopold Vietoris (of Mayer-Vietoris fame), who was a member of Brouwer's group at that time.

It was also during that year that Heinrich Tietze came to Munich from Erlangen. Although the precise circumstances remain unclear, Tietze was already preparing a new report on topology for the *Encyklopädie*. In all likelihood, he discussed this project with Kneser at the September 1924 meeting in Innsbruck, as Tietze thanked him for his help in acquiring access to foreign literature during the time when Kneser was still in Göttingen. Brouwer soon got wind of this project and took a deep interest in promoting this work, enlisting Vietoris as a co-author for the report [Tietze/Vietoris, 1929, 141]. Three others from Brouwer's circle contributed to it as well: Pavel Alexandrov, Witold Hurewicz, and Karl Menger. When Dehn and Heegaard published their report 22 years earlier, analysis situs was little more

---

[13]The related question as to whether and when a manifold can be triangulated drew the attention of several researchers after this time. Tibor Radó proved this for general surfaces in 1925, but it was not until 1952 that Edwin Moise proved that topological 3-manifolds were triangulable.

than an assortment of results closely tied with classical geometry and complex analysis. By the year 1929, it was a full-blown international research field still exploring new directions.

Hellmuth Kneser was still in Göttingen in October 1924, when he received a letter from Brouwer alerting him that a bright young man from Amsterdam was about to arrive. Bartel Leendert van der Waerden would later become one of Emmy Noether's star protégés, but during this initial visit before returning to Amsterdam to take his doctorate, it was Kneser who mainly took the talented Dutchman under his wing. Kneser's Innsbruck lecture was not yet in print, but topology was one of the many topics they discussed during their walks after meeting for lunch.

Five years later, it was van der Waerden's turn to review recent research on combinatorial topology for the German mathematical community. This was the topic of his lecture at the 1929 DMV conference held that year in Prague [van, der Waerden 1930], at which Kneser was present. In fact, van der Waerden began by referring back to Kneser's Innsbruck lecture before describing where things now stood. Emmy Noether reported briefly about this lecture in a letter to Alexandrov:

> Prague showed that there is great interest in topology. In no lecture were there as many auditors as those who heard v. d. Waerden's report. ...he will send you the manuscript or page proofs, so that Moscow will be quoted correctly. As for topologists belonging to the guild, only H. Kneser was there, whom you don't even count as a guild member! [Rowe, 2021, 100–101].

Noether's last amusing comment ought not to deflect attention from the fact that Alexandrov was deeply interested in another aspect of combinatorial topology, namely its relevance for bridging the gap between point-set topology in the style of Hausdorff and the classical approach of geometric topology. During the summer of 1930, he offered a lecture course in Göttingen on his new combinatorial theory of dimension. Two years earlier, on 20 April 1928, he wrote about this to Hausdorff:

> I'm personally more and more interested in a combinatorial construction of dimension theory; it is truly surprising how many – by appearances set-theoretic properties of geometric figures – are ultimately of combinatorial origin and in a different sense than I thought just a short time ago; completely unexpected connections arise along this path, sometimes as new results, often as problems, with ties even to the theory of knots. [Rowe, 2021, 149]

In his talk, van der Waerden described various proposals in combinatorial topology, focusing especially on competing and possibly incompatible ideas for a theory of manifolds. He called this a "battlefield of different methods," in which he distinguished between those that were purely combinatorial and others that employed a *méthode mixte*, just as Kneser had done five years before. Since van der Waerden said nothing about the older combinatorial approaches of [Dehn and Heegaard, 1907] and [Tietze, 1908] or the more recent axiomatic proposal put forward by Weyl, it seems the former had by now served its purpose, whereas the latter probably never found much favor. Given his strong interests in algebraic geometry, it comes as no surprise that van der Waerden's report provided ample coverage of Lefschetz's works, but he also discussed the contributions of Vietoris and van Kampen to homology theory as well as work by Brouwer, Veblen, and J.W. Alexander.

That left Kneser's axiomatic treatment of combinatorial topology, which depended on proving the *Hauptvermutung*. Since van der Waerden considered this far out of reach, he instead put his faith in Alexander's mixed approach using singular homology. As he summed up the present situation:

> I believe one can now be quite certain that the difficulty [in proving the *Hauptvermutung*[14]] is and will remain so great that one should give preference to a *direct* proof of the invariance properties of homology groups, fundamental groups, etc., by means of the *méthode mixte* rather than taking the detour through the combinatorial equivalence and invariance concepts.[15]

At the end of his lecture, van der Waerden also emphasized that no progress had been made on "famous problems," like the generalized Poincaré conjecture or the triangulability of manifolds, except in dimension two. Indeed, the course of developments in topology had gradually taught mathematicians how to make progress by consciously avoiding these problems, recognizing that they could only be successfully attacked when entirely new methods became available. Perhaps he thought the same applied with Dehn's Lemma, as he also announced – for the first time in public, though he said that insiders had long been aware of this – that the proof Dehn had given in 1910 was invalid. Whether the result itself was correct thus now became an important open problem [van, der Waerden 1930, 133]. When Kneser returned to Greifswald after attending the Prague conference, he wrote to Dehn letting him know about van der Waerden's announcement.[16]

Dehn's Lemma

That letter marks the end of a dramatic episode in the lives of Dehn and Kneser. Without attempting to describe what happened in detail, this section gives a summary account of the key events based on the extant correspondence in the Dehn Papers at the Dolph Briscoe Center for American History at the University of Texas in Austin. The background related, once again, to a talk Kneser gave, this time at the annual meeting of the DMV held in September 1928 in Hamburg. This led to the publication of [Kneser, 1929], an important paper in the history of 3-manifolds (see [Gordon, 1999, 466–469]). In April of the following year, after he had received proofs, Kneser discovered a serious problem. He then wrote to Dehn from Greifswald on 22 April, informing him that "to my despair, I find something in your proof . . . that appears to be a gap" (Figure 8). Kneser referred here to a specific example where continuing the switch over process through a triple point leads to the double curve being diverted.

Dehn was away on vacation, but when he returned to Frankfurt six days later he answered Kneser in a postcard that evinced remarkable equanimity and optimism. He was apparently none too worried when he wrote: "I'm in less despair than you since I've known of this gap for a long time. I've pointed this out numerous times to people who were interested and to whom I sent an offprint." He added that he

---

[14] The *Hauptvermutung* turned out to be false in dimensions four and higher, though these efforts led to a number of astonishing new results (see [Scholz, 1999, 57–58]).

[15] van der Waerden, B.L.: Kombinatorische Topologie, *Jahresbericht der Deutschen Mathematiker-Vereinigung* 39(1930): 121–139, especially pages 126–127.

[16] Kneser to Dehn, 7 October 1929, Dehn Papers, Dolph Briscoe Center for American History, University of Texas at Austin.

Greifswald, d. 22.4.29.
St. Georgsfeld 44.

Lieber Herr Dehn!
Zu meiner Bestürzung finde ich in dem Beweise Ihres Lemmas (M. Ann. 69) etwas, was mir als eine Lücke erscheint. Zunächst die Widerlegung: Fängt man in dem Beispiel der beiliegenden Figur (hoffentlich ist sie deutlich) die Umschaltung bei a an, so müssen nach Ihrer Regel die mit p und q bezeichneten Flächenteile verschmolzen werden. Setzt man die Umschaltung in der Pfeilrichtung fort, so erhält die Doppelkurve nacheinander die in den beiden weiteren Figuren gezeigten Gestalten: es trennt sich ein Teil ab; man kommt nicht zu allen Strecken der (einzigen) Gruppe von Doppellinien, wie Sie es auf S. 150-151 behaupten. Der Fehler scheint mir

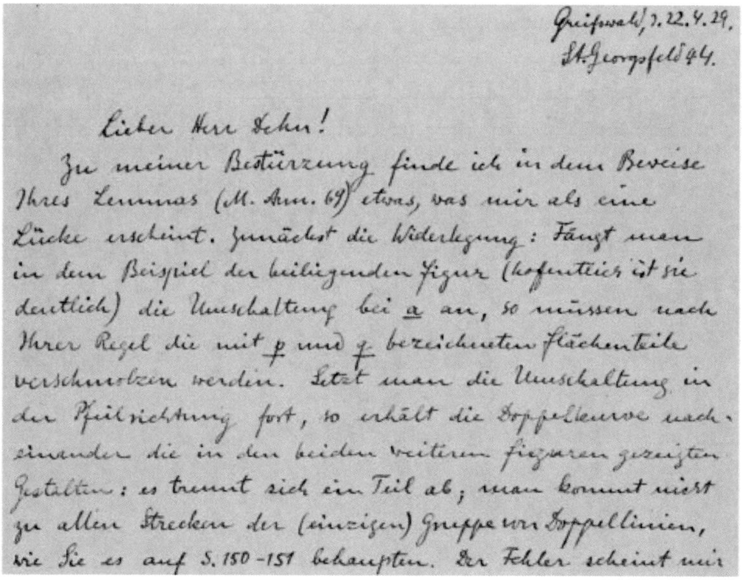

FIGURE 8. Excerpt from Kneser's first letter to Dehn, 22 April 1929.

could vaguely recall how he had encountered the problem of triple points, and that he had no difficulty dispensing with them. Since the summer semester was now about to begin, though, he had no time to ponder these matters, but remarked that he would be very happy, if Kneser succeeded in cleaning the proof up soon.

During the weeks that followed, Dehn began pondering Kneser's counter-example while looking for a way to fill the gap in his proof. On 29 May 1929, thus about a month after he mailed the postcard, he wrote a letter confirming that his original argument in the 1910 paper overlooked the problem that Kneser was now pointing out. Still, he seemed not at all worried: "the matter is, in fact, pretty harmless," he claimed. According to his present thinking, one needs to take into account two situations before carrying out the final switch over. If the standard operation failed, then performing the opposite switch over would ensure that the resulting surface remained connected.

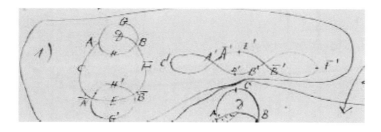

FIGURE 9. Dehn's depiction of Kneser's "counter-example"

Dehn first pointed out this problem before describing how he proposed to handle Kneser's counter-example, which he redrew in order to simplify the visualization process. As he explained: "Now to your example. Here I remark that it seems most purposeful to use the singularity-free image since the processes involved are completely independent of the embedded manifold. This way one doesn't need to be tortured with 3-D pictures (darstellend-geometrischen Zeichnungen abzuquälen). Your very beautiful example is then represented by the following figure" (Figure 9). This letter may well represent the first instance in which Dehn employed what came to be called Dehn diagrams. These show the singularity set $S(f)$ in the disc's interior, which consists of pairs of points with the same image under f. A simple example that Dehn himself later considered is given below (Figure 10):

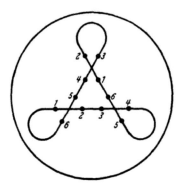

FIGURE 10. Example of a singularity set due to Dehn (Johansson 1935, 314)

Traversing the curve in Figure 10 by following points 1, 2, ... , 6, we see that it crosses itself between points 3 and 4, which meets again between points 5 and 6. These crossings arise when moving through the second set of points 1, 2, ... , 6. Schematically, this corresponds to a triple point where the images of the segments 12, 34, and 56 cross one another in the singular image. However, this particular diagram cannot be realized as a map of the disc into a 3-manifold. The 3D picture that Kneser sent to Dehn had two such triple points, as in Figure 11.

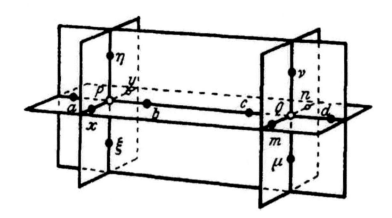

FIGURE 11. Two triple points P and Q situated on the segments ab and cd (Johansson 1935, 312)

In Dehn's drawing of Kneser's example, the singular set has two components. The preimages of triple points are those marked A and B, whereas the double lines are given either by traversing the two components – CADBFBEA – or by following the two cycles: AGBH. Dehn described how one could apply a switch over to both components at once, thereby removing the triple points in two steps. He then drew pictures to illustrate the process he had in mind. In this case, one could begin by removing triple points in a similar manner, but Dehn emphasized that the final switch over had to be done in the opposite way in order to preserve connectivity. He thus realized that one needed a general criterion to decide which of the two *Umschaltungen* would be needed in the end.

A week later, on 1 June 1929, Dehn wrote again: "Please don't ridicule me because I write you again about the Lemma. But the matter must indeed be put in order as quickly as possible, and today that will hardly happen." This time, Dehn wanted to explain the original motivation behind his earlier work, part of which aimed to make progress on the Poincaré conjecture. He was also curious to know something about Kneser's talk from the Hamburg meeting. Dehn then noted: "For ordinary space there are many simple proofs for the Lemma. In the paper from 1910, I only applied the Lemma for E^3, but I originally wanted to use the result for an arbitrary M^3 in order to prove something about spaces with trivial fundamental group (see the end of the paper). Have you used the Lemma for a general M^3?"

Kneser answered in a 4-page letter written on 16 June 1929. He explained that he has shown that every closed surface in S^3 can be built from an S^2 by adding handles (which may be knotted and linked). For an arbitrary M^3 the analogue involves replacing S^2 by surfaces whose curves are null homotopic if they have this property in M^3. Kneser went on to describe his method of splitting an M^3 along 2-spheres, a technique that led to a unique decomposition into irreducible sub-3-manifolds. (This does not require the Lemma.) Kneser's main theorem then states that M^3 can be constructed as the connected sum of two 3-manifolds with fundamental groups A and B if and only if its fundamental group is the free product of A and B [Gordon, 1999, 466–468]. Referring to these results, Kneser then added: "All that I presented in Hamburg – I believe with success – and have now given to the DMV-Berichten for publication, and all that stands or falls with your Lemma. You can measure accordingly how much I owe to it and how concerned I am to see it stabilized. If I manage to find something I'll write to you of course."

This letter, in fact, marks the climax of the Dehn-Kneser correspondence, even though Dehn continued to press ahead afterward. Kneser, who had a heavy teaching load in Greifswald, was also thinking hard about how he might be able to fill the gap, and for a brief time he thought he had succeeded. In his letter to Dehn he briefly described a strategy for removing triple points systematically, an approach he presented in the Greifswald colloquium. Afterward he noticed, however, that the problem was still unresolved, and so he explained to Dehn what he had tried to achieve and why the argument breaks down.

Both Dehn and Kneser were by now aware that the crux of the problem came down to showing that all the triple points could be removed systematically by means of *Umschaltungen*. Each such switch over also had to ensure two things: 1) control over any changes in the Euler characteristic, and 2) ensuring that the resulting surface remained connected. Kneser was not at all convinced by the arguments Dehn had sent him, writing:

> Unfortunately I don't understand your last letter, in which you say "such-and-such works". When applying Umschalten one has to ensure two things: 1) that at the end of the process the final Umschalten raises the Euler characteristic by two; 2) that the removal of the final segment on the double curve does not separate the surface. I absolutely cannot see how this can be done if the Umschaltung starts at two different places from the beginning. At any rate, after various unsuccessful attempts in this direction I have given this effort up. I will be thankful to you for any communique; as a user of the Lemma I'm just as interested in it as you are as its inventor.

Dehn was greatly disappointed when Kneser later sent him page proofs for his Hamburg lecture, to which he now proposed to add a brief remark indicating that the original proof in [Dehn, 1910] was faulty. In a letter from 6 September 1929, Dehn returned these proofs while requesting that Kneser omit his remark that Dehn had long been aware of this gap. Clearly, Dehn had not taken this to be a serious problem, but he now insisted that he had only been aware that "something wasn't quite in order," not that the proof was "completely false." Otherwise, he would have, of course, called attention to his mistake. Kneser accordingly omitted this remark when he returned the proofs on 7 November. The note he appended read simply:

> During the printing I noticed that the proof of Dehn's Lemma contains a gap. The deliberation in the second paragraph on [Dehn, 1910, 150–151] is unsound and cannot be put in order, as shown by counterexamples. Until this gap is filled, my results too are not firmly grounded, though they throw light on the significance of the lemma for this circle of questions. [Kneser, 1929, 260]

The fertility of Max Dehn's topological and group-theoretic ideas is discussed in Chapters 6-8, whereas the theory of 3-manifolds remained seriously stuck until the 1950s. For a survey of how it got unstuck, see [Gordon, 1999], which also briefly describes how the problem Dehn and Kneser fought with back in 1929 was finally resolved. Christos Papakyriakopoulos, better known as Papa, had also long struggled to prove "Dehn's Lemma" before his breakthrough in [Papakyriakopoulos, 1957]. Like Dehn, Papa hoped this would open the door to proving the Poincaré conjecture [O'Shea, 2007, 168–170]. Already in the 1940s, he sent a purported proof to Princeton's Ralph Fox, who was apparently undismayed when the argument turned out to be faulty. Fox afterward invited Papa to Princeton and managed to arrange financial support that kept this venture going. John Milnor later celebrated the final breakthrough with a limerick that exploited Papa's complicated name:

> The perfidious lemma of Dehn
> Was every topologist's bane
>   'Til Christos D. Pap-
>   akyriakop-
> oulos proved it without any strain.[17]

---

[17]This is a slight variant on the version given in [O'Shea, 2007, 169]. Milnor later remarked that he had heard many variations over the years [Milnor, 2009, 18].

Insiders knew that the final line referred to Papa's "tower construction" using covering spaces (on which, see [Gordon, 1999, 478–480]). This "relieved the strain" of trying to prove the lemma by standard cut-and-paste methods. So "Dehn's conjecture" was, indeed, true, though obviously far harder to prove than its author had originally imagined. Nor did he live to witness this vindication: Dehn died quite suddenly in 1952. He had been teaching for some years at Black Mountain College in Western North Carolina, a place he and his wife Toni came to love (see Part III).

In 1948, the year Ralph Fox invited Papa to Princeton, Max Dehn turned 70. His birthday brought a flood of mail from well-wishers both in the United States as well as from former friends in Germany. One of those who wrote to him was Hellmuth Kneser, who took the opportunity to recall the inspiration Dehn had given him when he was a young man studying in Breslau:

> This is the proper occasion to acknowledge what I owe to you. When in the springtime of my youth I chose topology as my field of study, I learned from your papers that there is a topology of 3-dimensional space and how and with what one can grasp it. Then a number of conversations provided further motivation. ...It must have been around 1920; then we met in the Tiergartenstrasse in Breslau and discussed properties of the two-dimensional homotopy group, that it's commutative and usually not finitely generated. Recently I had your letters from 1929 in hand that you wrote in the matter of the Lemma. That proved to be indeed a really hard nut (ein recht harter Knochen).

References

Alexander, James W.: On the subdivision of 3-space by a polyhedron, *Proceedings of the National Academy of Sciences* 10: 6–8.

Cannon, James W., Starbird, Michael (1987): A diagram oriented proof of Dehn's Lemma, *Topology and its Applications*, 26(2): 193–205.

Dehn, Max: Beweis des Satzes, dass jedes geradlinige geschlossene Polygon ohne Doppelpunkte die Ebene in zwei Teile teilt, undated manuscript ca. 1899, Max Dehn Papers, Briscoe Center for American History, University of Texas, Austin.

Dehn, Max: Die Legendréschen Sätze über die Winkelsumme im Dreieck, *Mathematische Annalen* 53: 404–439.

Dehn, Max: Über den Rauminhalt, *Mathematische Annalen* 55: 465–478.

Dehn, Max: Berichtigender Zusatz zu III AB 3 Analysis Situs, *Jahresbericht der Deutschen Mathematiker-Vereinigung* 16(1907): 573.

Dehn, Max: Über die Topologie des dreidimensionalen Raumes, *Mathematische Annalen* 69: 137–168; English translation in [Dehn, 1987, 92–126].

Dehn, Max: Über unendliche diskontinuierliche Gruppen, *Mathematische Annalen* 71: 116–144; English translation in [Dehn, 1987, 133–178].

Dehn, Max: Die beiden Kleeblattschlingen, *Mathematische Annalen* 75: 402–413; English translation in [Dehn, 1987, 203–223].

Dehn, Max: *Papers on Group Theory and Topology*, Translated and introduced by J. Stillwell, New York: Springer.

Dehn, Max and Heegaard, Poul: Analysis situs, *Enzyklopädie der Mathematischen Wissenschaften* Ill, AB 3, Leipzig: Teubner, 153–220.

Epple, Moritz: *Die Entstehung der Knotentheorie. Kontexte und Konstruktionen einer modernen mathematischen Theorie*, Wiesbaden: Vieweg.

Epple, Moritz: Geometric aspects in the development of knot theory, in [James, 1999, 301–358].

Gordon, Cameron McA.: 3-dimensional topology up to 1960, in [James, 1999, 449–490].

Gordon, Cameron McA.: Dehn Surgery and 3-Manifolds, in [Mrowka and Ozsváth, 2009, 21-72].

Gray, Jeremy; *Henri Poincaré. A Scientific Biography*, Princeton: Princeton University Press.

Guggenheimer, Heinrich: The Jordan curve theorem and an unpublished manuscript by Max Dehn, *Archive for History of Exact Sciences* 17: 193–200.

Guggenheimer, Heinrich: The Jordan and Schoenflies theorems in axiomatic geometry, *The American Mathematical Monthly* 85(9): 753–756.

Hales, Thomas: Jordan's Proof of the Jordan Curve Theorem, *Studies in Logic, Grammar and Rhetoric* 10: 45–60.

Hausdorff, Felix: *Felix Hausdorff: Gesammelte Werke, Grundzüge der Mengenlehre*, Bd. II, E. Brieskorn et al., Hrsg., Heidelberg: Springer.

Heegaard, Poul: Forstudier til en topologisk Teori for de algebraiske Flladers Sammenhaeng, Dissertation, Copenhagen.

Hilbert, David: Grundlagen der Geometrie, in *Festschrift zur Einweihung des Göttinger Gauss-Weber Denkmals*, Leipzig: Teubner (2. Aufl., 1903); *Grundlagen der Geometrie (Festschrift 1899)*, Klaus Volkert, ed., Heidelberg: Springer.

James, I.M., ed. (1999): *History of Topology*, Dordrecht: North Holland.

Johansson, I. (1935): Über singuläre Elementarflächen und das Dehnsche Lemma, *Mathematische Annalen*, 110: 312–330.

Jordan, Camille: *Cours d'Analyse*, 2nd ed., Gauthier-Villars, Paris, pp. 90–99.

Kneser, Hellmuth: Die Topologie der Mannigfaltigkeiten, *Jahresbericht der Deutschen Mathematiker-Vereinigung* 34: 1–13.

Kneser, Hellmuth: Geschlossene Flächen in dreidimensionalen Mannigfaltigkeiten, *Jahresbericht der Deutschen Mathematiker-Vereinigung* 38: 248–260.

Lennes, N.J.: Theorems on the Simple Finite Polygon and Polyhedron, *American Journal of Mathematics* 33 (1911): 37–62.

Magnus, Wilhelm (1978/79): Max Dehn, *Mathematical Intelligencer*, 1(3): 132–143.

Milnor, John: Fifty Years Ago: Topology of Manifolds in the 50's and 60's, in [Mrowka and Ozsváth, 2009, 7-20].

Mrowka, Tomasz S. and Peter S. Ozsváth, eds.: *Low Dimensional Topology*, IAS/Park City Mathematics Series, Vol. 15, AMS.

Munkholm, E.S. and Munkholm, H.J.: Poul Heegaard, in [James, 1999, 925–946].

O'Shea, Donal (2007): *The Poincaré Conjecture: In Search of the Shape of the Universe*, New York: Walker.

Papakyriakopoulos, C. D. (1957): On Dehn's Lemma and the Asphericity of Knots, *Annals of Mathematics*, 66(1): 1–26.

Poincaré, Henri: Analysis Situs, *Journal de l'École Polytechnique* (2) 1: 1–121.

Poincaré, Henri: Complément à l'analysis situs, *Rendiconti del Circolo Matematico di Palermo* 13: 285–343.

Poincaré, Henri: Second complément à l'analysis situs, *Proceedings of the London Mathematical Society* 32: 277-308.

Poincaré, Henri: Cinquième complément à l'analysis situs, *Rendiconti del Circolo Matematico di Palermo* 18: 45–110.

Rowe, David E.: *Emmy Noether: Mathematician Extraordinaire*, Cham: Springer Nature Switzerland.

Sarkaria, K.S.: The Topological Work of Henri Poincaré, in [James, 1999, 123–168].

Schoenflies, Arthur: Ueber einen Satz aus der Analysis situs, *Nachrichten der Königlichen Gesellschaft der Wissenschaften zu Göttingen. Math.-phys. Klasse*, 79–89.

Schoenflies, Arthur: Beiträge zur Theorie der Punktmengen III, *Mathematische Annalen* 62: 286–328.

Scholz, Erhard: *Geschichte des Mannigfaltigkeitsbegriffs von Riemann bis Poincaré*, Basel: Birkhäuser.

Scholz, Erhard: The Concept of Manifold, 1850-1950, in [James, 1999, 25–64].

Steinitz, Ernst: Beiträge zur Analysis situs, *Sitzungsberichte der Berliner Mathematischen Gesellschaft* 7: 29–49.

Stillwell, John: Letter to Editor, *Mathematical Intelligencer* 1(3): 192.

Stillwell, John: *Classical Topology and Combinatorial Group theory*, New York: Springer

Stillwell, John: Max Dehn, in [James, 1999, 965–978].

Stillwell, John: Max Dehn and geometry, *Mathematische Semesterberichte* 49(2): 145–152.

Stillwell, John: Poincaré and the early history of 3-manifolds, *Bulletin of the American Mathematical Society* 49(4): 555–576.

Tietze, Heinrich: Über die topologischen invarianten mehrdimensionaler Mannigfaltigkeiten, *Monatschrift für Mathematik und Physik* 19(1908): 1–118.

Tietze, Heinrich and Vietoris, Leopold: Beziehungen zwischen den verschiedenen Zweigen der Topologie, *Enzyklopädie der Mathematischen Wissenschaften*, III-2, pp. 141–236.

van der Waerden, B.L.: Kombinatorische Topologie, *Jahresbericht der Deutschen Mathematiker-Vereinigung* 39: 121–139.

Volkert, Klaus: *Das Homöomorphieproblem inbesondere der 3-Mannigfaltigkeiten in der Topologie, 1892–1935*, Habilitationsschrift, Universität Heidelberg.

Volkert, Klaus: The early history of Poincaré's conjecture, in J.-L. Greffe, G. Heinzmann, and K. Lorenz, eds., *Henri Poincaré: Science and Philosophy*, Berlin/Paris: Akademie Verlag, Albert Blanchard, pp. 241–250.

Wiener, Hermann: Über Grundlagen und Aufbau der Geometrie, *Jahresbericht der Deutschen Mathematiker-Vereinigung* 1(1891): 45–48.

CHAPTER 5

Golden Years in Frankfurt

David E. Rowe

The early history of Frankfurt University closely parallels the university founded in Hamburg, though the latter was established after the Great War and the former immediately before. Frankfurt was the first instance in Germany of a university financed by funding donated by its wealthy citizenry. Private universities with large endowments, like Harvard and Yale in the United States, had never existed before in Germany. Jewish donors played a large role in helping to launch this project, and many of those who gained academic appointments at the new university were of Jewish background. Originally the university had five faculties, but none for theology; instead there were distinct faculties for natural sciences as well as for economics and social sciences. Max Dehn's former teacher, Arthur Schoenflies, assumed the first chair in mathematics at the new university. Schoenflies already resigned his chair in Königsberg in 1911 in order to take a professorship at the Academy for Social and Commercial Sciences in Frankfurt, an institution that passed over to the University of Frankfurt three years later.

As the first Dean of the Science Faculty, Schoenflies appointed Ernst Hellinger as associate professor and in 1915 Ludwig Bieberbach assumed the second full professorship in mathematics. During the academic year 1920/21, Schoenflies also served as the university's rector. Two additional associate positions were then created for Otto Szász and Paul Epstein. It was around this time that Bieberbach left Frankfurt after accepting a professorship in Berlin. This led to complicated negotiations within the natural sciences faculty, but eventually opened the door for Max Dehn. One year later, in 1922, Schoenflies himself retired, a circumstance that led to the appointment of Carl Ludwig Siegel.

A Darkhorse Candidate

The circumstances that led to Max Dehn's appointment in Frankfurt were indeed complex, as can be seen from correspondence exchanged between Arthur Schoenflies and the Dutch topologist L.E.J. Brouwer, who were on friendly terms. Brouwer had benefited greatly from his relationship with Otto Blumenthal, managing editor of *Mathematische Annalen*, and felt Blumenthal had been unjustly treated in the past. As the journal's workhorse, he had found little time to publish work of his own, a circumstance that had left him stuck in Aachen. Nor could Blumenthal point to any major result connected with his name, whereas Dehn's career was launched early when he solved Hilbert's third Paris problem. Knowing from personal experience how invaluable Blumenthal had been, both to him and others, Brouwer wrote to Schoenflies on 21 January 1921 to enter a plea for considering his

---

©2024 by the author

candidacy as Ludwig Bieberbach's successor:

> I'm thinking of Blumenthal, with whom I have standing relations for a decade, during which time I have come to appreciate him more and more from a number of sides. In particular, I am convinced that he has hardly an equal among our colleagues when it comes to broad mathematical knowledge, work ethic, helpfulness, honesty and decency. [Rowe/Felsch, 2019, 129]

Schoenflies answered on 14 February by giving Brouwer a frank assessment of the difficulties this vacancy posed. Although he knew Dehn ever since the latter's student days and surely held him in high esteem, at this stage he was not one of the main candidates under consideration.

> We are thinking primarily of [Leon] Lichtenstein and [Georg] Pólya.[1] I'm afraid, though, that the government will ignore Lichtenstein, who is now on his way to Münster, whereas for Pólya, there may be a personal obstacle. If neither Lichtenstein nor Pólya can be called, we are clueless. We have still thoughts of [Johann] Radon and [Arthur] Rosenthal, and might seriously consider the name Blumenthal, admittedly only as a last resort. But I will be leaving soon, since I will reach retirement age according to the law on the 1st of October. Then, I believe, Blumenthal would be an excellent choice. But we cannot completely pass over Hellinger and Szász, so you see that the situation is involved in every way. [Rowe/Felsch, 2019, 130–131]

Aside from Rosenthal, who was best known for his work in geometry, all the others named were analysts with various areas of expertise. This may account for why Dehn's name had not yet come up during this early stage in the deliberations, even though Schoenflies was well aware of his accomplishments. The latter's concern with regard to Hellinger and Szász, both of whom were analysts, should not be overlooked either. Ernst Hellinger and Richard Courant had both studied under Hilbert in Göttingen, but Hellinger was five years older. His seniority was surely a major reason for why he was chosen over Courant in 1914, when both were candidates for the associate professorship in Frankfurt. In the meantime, Courant had cultivated his relationships in Göttingen, not least with his future father-in-law Carl Runge, paving the way for his appointment to Klein's former chair there. Still, Courant had been brought over from Münster, whereas elevating Hellinger in Frankfurt would have been a *Hausberufung* (in-house appointment), a far more delicate undertaking.

In view of his later notoriety as an advocate of "Deutsche Mathematik," it is interesting to follow how Ludwig Bieberbach behaved in this situation.[2] Schoenflies held Bieberbach in high esteem and for this reason he largely acquiesced to his views during these preliminary negotiations. Bieberbach accordingly wrote to the Austrian geometer Wilhelm Blaschke, at present a colleague of Johann Radon in Hamburg, asking for his opinion of the two leading candidates, Lichtenstein and

---

[1] Lichtenstein took the professorship in Münster briefly held by Richard Courant, who in 1920 succeeded Wilhelm Killing. Pólya joined the faculty of the ETH Zurich as a *Privatdozent* in 1914; he became a full professor there in 1928 and in 1940 came to the United States, ending his career at Stanford University.

[2] On Bieberbach's career as a leading Nazi mathematician, see [Mehrtens, 1987].

Pólya. Blaschke reacted with outrage that these two Jews might be favored over his former mentor in Vienna, Wilhelm Wirtinger. Bieberbach also made an inquiry with Hermann Weyl, a colleague of Pólya in Zurich. Weyl ranked both Pólya and Lichtenstein higher than Radon and Rosenthal.

As it turned out, however, neither of the two candidates, whom Schoenflies originally characterized as favorites, ended up on the "short list," which now included four names. Schoenflies explained what had happened in another confidential letter to Brouwer, dated 4 April, 1921.

> Our proposals (replacement of Bieberbach) went to Berlin; I want to share them with you confidentially. Blumenthal is, as it turned out, not among them. There was a desire within the entire faculty to find an energetic personality, with proven or hopefully potential leadership qualities, for Bieberbach, who is unfortunately irreplaceable overall. Not all the names I shall give you correspond to this, perhaps none do completely – but Blumenthal was from this point of view really not possible.
>
> Here is our list: 1) Wirtinger; 2) Dehn and Radon; 3) Nielsen. The list is a compromise based on long deliberations, which unfortunately were also drenched with nationalistic motives. We mathematicians had no such motivations, but almost all of the other members of the faculty made these their first consideration! So we were forced to give in. Otherwise we would have put Pólya and Lichtenstein on the list, too, especially since Bieberbach put a lot of effort into promoting Pólya. But in view of the circumstances it was impossible to get him through. And a minority vote by the mathematicians and their few friends would have been unsuccessful given the prevailing conditions in Berlin.
>
> Wirtinger is an excellent personality; if he decided to come, we would have what we most need. With Dehn we would at least know exactly what we are getting; whether he is willing or able to build up a school, I do not know. Radon and Nielsen are still developing and, at least for me, it is difficult to say how they will turn out. In the whole matter, I left the decision to Bieberbach. We, of course, also obtained all kinds of outside expert opinions, and the above list, resp. the names Radon and Nielsen, was partly based on these. [Rowe/Felsch, 2019, 131–132]

Schoenflies also added that he would have loved getting either Hermann Weyl from Zurich or Richard von Mises, who had recently been appointed to a professorship in Berlin. Weyl had even been approached, but declined for health reasons.[3] Schoenflies evidently decided that von Mises, too, would be unattainable. He again held out the possibility that Blumenthal might be considered when his own chair came open the following year; he would soon reach the mandatory retirement age of 68.

Wirtinger, as expected, declined the offer from Frankfurt, thereby opening the door for Dehn, who obviously belonged at a university rather than a *Technische*

---

[3]Weyl's received numerous offers during these years, but turned them all down. He thus remained at the ETH Zurich from 1913 to 1930, when he accepted the offer to succeed Hilbert in Göttingen.

Hochschule. When he received the official confirmation from the Prussian Minister of Education, Carl Heinrich Becker, on 19 July 1921, Dehn was informed of the various conditions of employment as successor to the chair previously held by Ludwig Bieberbach. His professorship also carried an appointment as co-director of the Mathematics Seminar, i.e. an institution with its own budget for expenses such as purchasing books, covering costs for teaching special courses, travel expenses to attend conferences, or invitations to guest speakers. His base salary, starting on 1 October, was 19,000 Marks, in addition to which he would receive funds for each of the three children plus a supplement to cover higher living costs in the city of Frankfurt. Furthermore, he was guaranteed a minimum of 6,000 Marks per year in student fees (which traditionally were paid directly to the lecturer). After moving from Breslau, he was to submit expense receipts for reimbursement. Two decades after his Habilitation in Münster, Max Dehn's long wait for a full university professorship was over.

Congenial Colleagues

Dehn's longstanding relationship with Arthur Schoenflies, one of the founding figures of Frankfurt University, had surely worked in his favor. Moreover, this appointment only shortly before his former teacher's retirement meant that he would become the newly anointed leader and director of the Frankfurt Mathematical Seminar. Yet Dehn had no ambitions to build a traditional mathematical school, much less the kind of massive operation that Richard Courant sought to continue, following in the footsteps of Klein or Hilbert. What unfolded under his leadership in Frankfurt was a community of scholars who cooperated with one another in an atmosphere largely free from the typical rivalries common at other leading universities. That atmosphere was later memorialized by Carl Ludwig Siegel, its last surviving member in [Siegel, 1965], a lecture he delivered on 13 June 1964 in the Frankfurt Mathematics Seminar.

The term "seminar" in this context refers to the typical institutional arrangements founded at German universities throughout the nineteenth century with the primary purpose of supporting curricular reforms aimed at prospective teaching candidates, particularly at the higher levels of the secondary schools. A mathematics seminar in this sense is essentially synonymous with a mathematics institute, a term commonly used today. Beginning in 1924, Max Dehn was appointed director of the Frankfurt Seminar, which made him the formal head of its teaching program. In the following account, we will often draw on Siegel's lecture as translated in [Siegel, 1979], beginning with those mathematicians who were already members of the Frankfurt faculty when Dehn arrived.

Ernst Hellinger was born in Silesia in 1883 and grew up in Breslau. There he befriended Max Born and Otto Toeplitz during their days together at the Gymnasium. Afterward, Hellinger studied in Heidelberg and Breslau before finishing in Göttingen under Hilbert. This was during the time when Hilbert began concentrating on the theory of integral equations, the topic of Hellinger's doctoral dissertation. He and Toeplitz would later become especially well known for their survey article on integral equations, published in 1927 in the *Enzyklopädie der Mathematischen Wissenschaften*. Siegel considered it to be still of great value in the 1960s, despite developments based on the later ideas of John von Neumann.

After taking his degree in 1907, Hellinger spent the next two years in Göttingen as Hilbert's assistant. He then habilitated in Marburg, where he taught as a *Privatdozent* until Frankfurt University was founded in 1914. He joined its faculty in that year as an *Extraordinarius*, a position akin to an associate professor. Siegel thought of Hellinger as the most gifted teacher in the group, especially because of his deep sense of dedication to his calling. He likened him to a "Prussian official of the old school," by this time a vanishing breed.

> Whereas Dehn's lectures could become difficult to follow for moderately gifted students when he got carried away, Hellinger was able, by means of careful preparation and clearly detailed presentation, to arouse interest even in those for whom initially mathematics meant little. He always took the welfare of his students to heart even outside his lectures and study groups. For many years he held an honorary position on the board of the student aid society, and thereby contributed greatly to the success of the university in areas other than the purely scientific. [Siegel, 1979, 224]

Paul Epstein was born in 1871 and grew up in Frankfurt. His father, Theobald Epstein, was a professor of mathematics and physics at the Philanthropin Academy, one of the most prominent schools for Jewish students in Germany. He was also an amateur astronomer and in charge of the Frankfurt observatory, which was first built in 1838 in the tower of the Paulskirche. His son Paul attended the Philanthropin, graduating in 1890. During this time he struck up a friendship with Karl Schwarzschild, whose love for astronomy first brought him to the attention of Paul's father. Schwarzschild came from a wealthy Frankfurt family and attended the same Gymnasium as Otto Blumenthal, from which time they became close friends [Rowe, 2018b, 9–11].

Like Schwarzschild, Epstein later studied at the Kaiser-Wilhelm University in Strassburg, which opened in 1872 as an outpost for Prussian culture after the annexation of Alsace-Lorraine following the Franco-Prussian War. However, unlike Karl Schwarzschild, who soon transferred to Munich, Paul Epstein stayed. In 1895 he completed his dissertation on a topic in Abelian functions under Elwin Bruno Christoffel, and afterward he taught from 1899 to 1919 at the local technical school and also as a *Privatdozent* at the university. As a result of Germany's defeat in the Great War, the Alsatian region became French once again, forcing Epstein to leave. He then returned to his native city, where in 1919 he took a temporary post at the university before gaining an appointment as a non-tenured associate professor two years later.

Siegel remembered Epstein as a multi-talented, artistic type, very much like Max Dehn:

> His mathematical work was primarily in the field of number theory, and he will be remembered in generations to come by the zeta functions named after him. In addition, he was interested in pedagogical problems as well as in the history of mathematics, making his work in the seminar of great value. It is widely believed that the talent for mathematics and the talent for music are closely related. I know full well, however, that there are many decidedly non-musical mathematicians: I myself am one.

Epstein, on the other hand, was extremely gifted in this respect and participated actively in the cultural life in Frankfurt. [Siegel, 1979, 224]

Otto Szász was born in Hungary in 1884. During his early career he was highly itinerant, beginning with studies in Göttingen and Budapest, where he took his doctorate in 1911 under Leopold Fejér. As a post-doc, he visited the universities in Munich, Paris, and Göttingen, before joining the Frankfurt faculty in 1914 as a *Privatdozent*. Szász was known primarily as an expert in real analysis, especially Fourier series and continued fractions. As a consummate problem solver, he had much in common with his fellow countryman Georg Pólya, whose candidacy for the vacancy in Frankfurt had been favored by Bieberbach. Siegel recalled the playful manner in which Szász tried to hide the fact that in conversation he was quite often a little slow to catch on [Siegel, 1979, 224].

Together with the senior mathematician, Arthur Schoenflies, these three comprised the Frankfurt teaching staff when in 1921 Max Dehn was called to Bieberbach's chair. At that time, Schoenflies succeeded in having both Szász and Epstein appointed as associate professors, thereby stabilizing the positions they would hold throughout the era of the Weimar Republic. As noted above, Schoenflies had imagined at this time that Otto Blumenthal, or perhaps some other senior mathematician, would be nominated to take his place the following year. Instead, however, a far younger man was chosen, Carl Ludwig Siegel.[4]

Landau's Prize Discovery

Initially, Siegel planned to study astronomy when he enrolled at the University of Berlin in 1915. This was during the midst of the "war to end all wars," and since Siegel had a deep aversion not just to warfare but to all forms of authority, he thought that studying the heavens would help distract him from the mundane affairs that preoccupied everyone else. This plan, however, soon went awry, as he recalled many years later:

> The person who was to lecture in astronomy gave notice that he would start fourteen days after the beginning of the semester. ...At the same time, ...Frobenius was to lecture on number theory. Because I hadn't the slightest idea what number theory might be, I attended the lectures for two weeks out of pure curiosity. This, however, determined my scholarly direction for the entire rest of my life. I did not attend the astronomy lectures once they began.... [Yandell, 2002, 176]

Siegel also studied under the other great algebraist in Berlin, Issai Schur. In his third semester, he learned from Schur about a 1909 paper on Diophantine equations by the Norwegian mathematician Axel Thue (on Thue, see Chapter 8). Siegel studied the argument, reorganized its structure, and thought he saw a way to strengthen the main result. Convinced that his reasoning was sound, he wrote up a four-page proof and submitted it to Schur. The reaction he got was deeply disappointing: "a couple of weeks later, Schur handed me a manuscript, yellowed by the sun from sitting on his desk, with the brief remark that I had merely calculated

---

[4][Tobies, 1996] gives many details about Siegel's preceding years in Göttingen; this serves as the principal source for the account below.

identities, from which no conclusion could be drawn" [Yandell, 2002, 177]. Number theory did not save Siegel from being drafted either, a legendary chapter in his early life. He spoke about it later to his former student and one-time companion, Hel Braun, who described how Siegel's rich and cunning friend, Egon Schaffeld, devised a plan that led to his dismissal from military service. After spending several weeks under close examination by doctors at a military hospital, Siegel was declared mentally unfit to be in the army and released [Braun, 1990, 19–21]. This ordeal, along with his mother's death, left him shattered. An only child, he returned home to his father, who worked as a delivery man for money orders in Berlin.

As it turned out, Siegel senior knew the father of Edmund Landau, a distinguished gynecologist to whom he made deliveries. The latter heard about the young man's mathematical abilities and invited him to his home, where he showed him his son's two-volume study devoted to the distribution of prime numbers. Edmund Landau wrote it as a *Privatdozent* in Berlin, not long before he was called to Göttingen as Hermann Minkowski's successor. Through his father, Landau learned about young Siegel's despair; he also found out that Siegel claimed to have proved a deeper result than Thue's. It took some time before Siegel could clean up his proof, but he eventually did, after which Landau wrote a glowing report on his dissertation results (reproduced in [Tobies, 1996, 29–30]):

> It was very surprising and astonishing for me to learn that a Berlin student (in his 3rd semester!), this very Herr Siegel, found on his own that by strengthening Thue's method one can reduce the *order of magnitude* of n to $\sqrt{n}$.... His proof is long, and he was able to correct all the original mistakes. This alone would have been a highly significant dissertation. The Berlin colleagues were unable to understand Siegel's argument, so I had him come here to do his doctorate.[5] The present dissertation, nevertheless, goes far beyond this. Above all, the author investigates approximations of algebraic numbers by algebraic (not just rational) numbers with equal success and draws the most important conclusions from them for the previously unexplored field of diophantine equations whose coefficients and unknowns belong to any algebraic number field. ...
>
> Our faculty can be proud to have such a study appear under its auspices as a dissertation. [Tobies, 1996, 29–30]

Richard Courant only knew about Siegel's immense talents secondhand through Landau. During the summer of 1920 when Siegel completed his studies in Göttingen (see Figure 1), Courant was teaching in Münster, where he had taken the chair long held by Wilhelm Killing. This postwar period saw a great deal of turnover in professorial chairs, which opened the door for Courant to return to Göttingen after just one semester in Münster, by which time Siegel was already in Hamburg. During his brief stay at the Georgia Augusta, Siegel had also studied number theory under Erich Hecke, who took up a newly established professorship in Hamburg in 1919. That was the year Hamburg University opened its doors with Hecke and two Austrians, Wilhelm Blaschke and Johann Radon, representing mathematics on

---

[5]Landau's reference to his "Berlin colleagues" may well have carried a hint of *Schadenfreude*, since he, as Frobenius' star pupil, surely had imagined himself as his mentor's successor when the latter died in 1917; instead, Carathéodory received the call.

FIGURE 1. Carl Ludwig Siegel (in handcart) after passing his doctoral exam on 9 June 1920. Left to right: Karl Grandjot, Erich Bessel-Hagen, Werner Wolfgang Rogosinski, Wilhelm Ness, Willi Windau, Arnold Walfisz, Wolfgang Krull, Otto Emersleben, Hans Kopfermann, Hedwig Wolff, Boskowits and Hellmuth Kneser. (Photo by G.F. Hund - Own work, CC BY 3.0, https://commons.wikimedia.org/w/index.php?curid=11968671)

the faculty. Hecke arranged a temporary assignment for Siegel teaching a course in integral calculus. This did not go well, however, and he soon became frustrated over not being able to advance his own work.

Courant was in regular contact with the geometer Blaschke, who served as coeditor for the so-called "yellow series," textbooks Courant edited for the publisher Ferdinand Springer. Possibly it was through Blaschke that Courant learned about how unhappy Siegel was in Hamburg, and luckily Courant was in a position to help. In his new position as Hecke's successor to the chair once held by Felix Klein, he was able to arrange for Siegel to return to Göttingen as his second assistant, the first position having gone to Hellmuth Kneser. One can sense from these few names how quickly Courant was spinning an impressive web of contacts; indeed, the Kneser name already had a distinguished pedigree. After studying in Breslau, where he first got to know Max Dehn, Hellmuth Kneser continued his studies in Göttingen, obtaining his doctorate under Hilbert with a dissertation on the mathematical foundations of quantum theory. Soon thereafter, he became a leading authority on combinatorial topology in the spirit of Dehn (see Chapter 4). Carl Siegel, who only a short time before had no connections and no academic degree, now found himself in exalted mathematical company, as these fortuitous circumstances now paved the way for him to habilitate in Göttingen under Landau in December 1921.

Since he had already praised Siegel to the skies in his report from the previous year, Landau's sequel merely alluded to some of the candidate's more recent achievements. He underscored how "to an astonishing extent, Siegel had found the few really good results published in the vast mathematical literature, and in a very short time he had united and deepened them." Here Landau mentioned ideas of Hardy and Littlewood on additive number theory as well as recent papers by Erich Hecke. As for his teaching ability, the candidate's performances both in seminars as well as before the mathematical society confirmed that he was a brilliant lecturer. In Landau's eyes, Siegel had far surpassed the expectations he had placed in him only a year earlier. Indeed, he playfully praised himself for discovering Siegel and bringing him to Göttingen in the first place. In the meantime, Landau and a handful of other experts familiar with Siegel's work had "come to regard this 24-year-old youth as the coming man of his generation and the pride of German science."

Landau's enthusiasm for Siegel was surely unsurpassed, but others, in particular Courant, clearly recognized his talent as well. In commenting on his unusual promise, Courant wrote:

> If his very delicate mental constitution is not too strongly affected by the stresses of professional life, we can expect many extraordinary achievements from him. His main field of research, number theory, is not the only subject in which he takes a deep interest. I believe that he will later turn to other branches of our science, as I know from him that he is strongly drawn to celestial mechanics. Our colleagues may be interested to know that Siegel is an extraordinarily well-educated person, who, for example, reads Russian literature in the original language. [Tobies, 1996, 33]

Courant had in the meantime evidently gotten to know Siegel fairly well. He probably also knew that the young man's main hobby during these years was painting, especially impressionistic landscapes. After he was called to Frankfurt, Siegel lived at first with Fritz Wucherer and his family in Kronberg, a wealthy town in the idyllic Taunus region northwest of Frankfurt (see Figure 2). Wucherer owned an impressive villa and belonged to an artists' colony in Kronberg. He was well known for his landscape paintings, and for some time Siegel took lessons from him. According to Hel Braun, who knew Siegel well during the 1930s, he gave up painting soon before he fled to the United States in 1940 [Braun, 1990, 21–22].

The testimonials cited above surely speak for themselves, but they also throw considerable light on the situation in 1922 when Siegel was called to fill Schoenflies' chair in Frankfurt. Little seems to have been written about who else might have been considered for this coveted position, but one must naturally assume that Schoenflies and Dehn turned to Göttingen for advice. By this time, Hilbert had just turned 60 and Klein was in the twilight of his long career. Both now deferred to Richard Courant, who had now taken up the reins of power as director of the Mathematics Institute.

Siegel was only 25 when he became an *Ordinarius* in Frankfurt in 1922. This turn of events delighted Edmund Landau, who always had an eye for significant dates in history. At the end of that year, he wrote to congratulate his colleague Felix Klein, who 50 years earlier had been appointed *Ordinarius* in Erlangen when he was only 23. Landau wished to let Klein know that not even Gauss had lived to

FIGURE 2. Siegel rented an apartment in this villa in Kronberg owned by the painter Fritz Wucherer. (Photo by Karsten Ratzke - Own work, CC0, https://commons.wikimedia.org/w/index.php?curid=33764343

experience the fiftieth anniversary of his *Ordinariat*, but he also offered a prognosis: Siegel would live to see that day. As it turned out, Landau was right.

Mathematics Education, 1925–1932

Cooperation and harmony were watchwords for the atmosphere cultivated by Max Dehn and his colleagues in Frankfurt. With virtually no turnover in the teaching staff for more than a decade and a relatively small student population, this group gradually worked out a set of rules and procedures for Frankfurt's academic program that cut through all unnecessary red tape. Teacher training was still the dominant task for those who taught mathematics at the German universities, a duty that some took very seriously, despite the fact that their research interests rarely had any direct relevance for the standardized mathematics taught in secondary schools. University professors, then as now, often cultivate an idealized picture of who should teach mathematics, what should be taught, and how it should be done. In Dehn's time, other groups were beginning to have their say, which added to the sense of urgency. In a lecture to the German Mathematical Society, "Problems in Post-Secondary Teaching of Mathematics" [Dehn, 1932a], Dehn described some of the idiosyncrasies of the program in Frankfurt before turning to general issues confronting mathematical educators throughout Germany.

Ernst Hellinger – certainly far more than Siegel but to some extent also more than Dehn – took a strong interest in pedagogical matters. This was reflected in his deep concern for the general welfare of all his students. Hel Braun confirmed that impression in her memoirs, beginning with her very first encounter with the mathematics faculty in Frankfurt. This took place when she attended an orientation session at the beginning of the turbulent summer semester in 1933 [Braun, 1990, 4–5]. These gatherings may well have been a Frankfurt novelty, since Dehn took the time to describe their purpose in [Dehn, 1932a, 73]. Each semester began with such a general meeting of all staff and students to discuss specific course offerings, but also other matters of general concern. These ranged from standards of performance to professional opportunities for those who wished to pursue more specialized studies, as for example insurance mathematics. Most of the students, though, wished to teach mathematics at a Gymnasium, which ultimately meant they would need to pass the state examination required before they could apply for a probationary position as a student teacher. On that particular occasion, as Braun remembered it long thereafter, Hellinger spoke especially to the beginners, offering them friendly advice. This included tips about how to go about studying the lecture notes they would soon be taking in their courses. He was also frank with them about the need to scrutinize their own progress, including the possibility that they might want to drop out and pursue a different course of study. Above all, he counseled beginners to spend a good amount of time every day working through the course material, studying theorems and solving related problems – and to do this work on their own, without the help of others. Clearly, the Frankfurt faculty wanted not only to impart knowledge to their students; above all, they sought to instill intellectual independence.

Since there were no regular exams, the instructors relied heavily on how well students performed in the problem-solving sessions attached to the introductory courses in analysis, algebra, and geometry. These courses were assigned to different faculty members each semester, so the teaching staff gradually reached a general consensus with respect to standards of accomplishment. Those who fell far short of those standards were advised not to continue with their studies, whereas those who wished to go on were required to pass an oral exam conducted by the entire teaching staff. If they passed, they were then allowed to enroll in the introductory seminar (*Proseminar*); a similar procedure governed admission to the advanced seminar (*Oberseminar*).

This emphasis on oral testing was, until quite recently, the norm throughout Germany, especially at the universities. Many students never had to pass a written exam (*Klausur*) during their entire education; they were evaluated instead by means of direct oral interrogation. Dehn flatly opposed written exams, but he wanted to raise a more general issue. Roughly one-third of the incoming students in a given year were destined to drop out at some point along the way. Since this outcome, at least in many cases, was predictable, how should faculties help to ensure that attrition takes place smoothly and as quickly as possible? From a practical standpoint, quality teaching demanded that students possess the requisite abilities and determination to learn. The overall success of the Frankfurt program thus depended on an efficient selection process in which the seminars served as the key vehicles for pursuing deeper or more creative work.

Siegel regarded the general level of the students in Frankfurt as quite high, as evidenced by the quality of several outstanding dissertations. He further noted that five graduates of the program from the period 1926 to 1930 later went on to successful academic careers. Wilhelm Magnus and Ruth Moufang were both students of Dehn, as was Ott-Heinrich Keller (see Chapter 6). Kurt Mahler and Wilhelm Maier both studied under Siegel, who was especially proud of a third former student:

> A few years later we were gratified to learn that in a highly original dissertation one of our students had solved another of the problems posed by Hilbert in 1900. This was Theodor Schneider, now a professor in Freiburg im Breisgau, and the problem he solved was described by Hilbert himself as being inaccessible and extremely difficult. Hilbert even thought this problem was more difficult than the Fermat problem or the proof of the Riemann hypothesis. [Siegel, 1979, 226]

In 1934 Schneider, along with A.O. Gelfond, succeeded in solving Hilbert's seventh Paris problem by showing that for any algebraic number $\alpha \neq 0, 1$ and for any irrational algebraic number β, α^β is a transcendental number.

Max Dehn's low-key leadership style clearly appealed to his colleagues, none of whom had a great passion for networking with star players at the more famous research centers. One of the major hubs of the era was Courant's Göttingen, where by the late 1920s one could encounter a whole galaxy of mathematical stars. Frankfurt was far more relaxed, which made its atmosphere less appealing to those who wanted to feel like they were in the middle of the action. When the Rockefeller Foundation began promoting mathematical research by funding visits to leading outposts, no one in Frankfurt took the initiative to invite a foreign visitor [Siegmund-Schultze, 2001]. On the other hand, in 1925 Siegel characteristically jumped on this opportunity to fund his own 8-month stay in Copenhagen and Cambridge, where he visited with Harald Bohr and G.H. Hardy, respectively.

During the summer of 1930, Siegel took another leave of absence from Frankfurt in order to teach in Göttingen, an arrangement that allowed Emmy Noether to fill in for him during that semester. Noether's favorite student, Max Deuring, attended Siegel's course on analytic number theory and was assigned the challenging task of writing up a readable manuscript from these lectures using the notes he took. On learning this, Emmy Noether sent him a postcard in which she remarked: "it is very nice that you are preparing the Siegel lectures; then I can read his breakneck proofs in peace in the winter, which I much prefer to hearing them" [Lemmermeyer/Roquette, 2006, 92]. Siegel's classical style was in the mold of his mentor, Edmund Landau, whereas Noether was the leading champion of abstract algebra, a trend Siegel would later vehemently oppose.

During Noether's stay in Frankfurt, Pavel Alexandrov came to speak in the mathematics seminar. One of those who heard his lecture was Paul Dubreil, a young French mathematician. He also heard lectures delivered by two other guests: Louis Mordell spoke on number theory and Wolfgang Krull on non-noetherian rings. Dubreil found the relaxed atmosphere in Frankfurt much to his liking. He got to know Magnus and Moufang and generally felt "supported by a kind of sympathetic prejudice that seemed to float in the air!" After giving his presentation, Emmy Noether commented that he had treated a problem of her father with the methods of his daughter, and Dehn joked about the terminology Dubreil had introduced (*Unterresultante*). The weather that day was splendid, and so the group made an *Ausflug* to the Taunus, where Dehn loved to hike. They surely made a stop at the Fuchstanz restaurant for cake and coffee; perhaps this was the day Dehn's daughter Maria remembered, when Emmy ordered a second piece of Torte, laughing that it couldn't hurt her figure anyway [Dubreil, 1983, 66].

Frankfurt's Historical Seminar

In 1924 Max Dehn became Director of the Frankfurt Mathematical Seminar, a post he thenceforth would hold until his professorship was terminated in 1935. Around the time he was first called, though, Dehn decided to launch a private reading circle, a *Lesekränzchen* that would long be remembered by all who attended. This group devoted its attention to the study of classical mathematical texts in their original languages, in particular Greek and Latin works written by, among others, Euclid, Bombelli, Cavalieri, Kepler, Roberval, Wallis, Huygens, Barrow, Newton, Leibniz, and Euler. The entire mathematics faculty took part in these gatherings as a truly communal undertaking, even though Max Dehn was the acknowledged *spiritus rector*.

In [Dehn, 1932a], his lecture to the DMV cited above, Dehn offered some interesting remarks about Frankfurt's most distinctive pedagogical innovation, its history of mathematics seminar. Among its several qualities, he laid stress on one of its moral virtues, namely, the sense of humility one gains through a deeper appreciation of the intellectual achievements of one's forebears. "Studying the development of mathematics, steadily, deeply, and without haste together with close colleagues," he wrote, "makes every mathematician more mature and fills him with a more humane love of his science." At the same time, Dehn had no illusions about the effectiveness of this special seminar as a teaching tool. Most students lacked the necessary linguistic skills, but even more, the intellectual patience required to delve into difficult texts. He also noted very aptly that mathematical and historical thinking tend to run in opposite directions. Over the course of ten years, he doubted whether more than a half-dozen students had gained anything of lasting value from the seminar. This telling remark clearly suggests that its true purpose was *Fortbildung*, i.e., cultural enrichment for the faculty and a few older teachers from the surrounding community.

Carl Siegel spoke of these informal meetings as particularly joyful occasions, though it took him some time to overcome his initial awkwardness. One can easily imagine how important the social aspect was for Siegel, who loved good food and drink. The history seminars took place in the late afternoon on Thursdays, and they invariably ended with convivial chatter in the Dehns' home over coffee and cake. Siegel took on the responsibility of bringing a delicious *Torte* from Müller's, a local confectionery. When Emmy Noether taught in his place in the summer of 1930, he wondered if she had been charged with the same task.

Students were naturally encouraged to participate in the Frankfurt historical seminars as well, though not many possessed the requisite language skills to do so. Siegel claimed, nevertheless, that there were never less than six students over the course of the thirteen years that it met.[6] A young astronomy student named Willy

---

[6]Dehn, on the other hand, said there were rarely more than five, and that only a few truly profited from the experience; see his comments as noted above.

Hartner, who later founded Frankfurt's Institute for History of Science, recalled in 1981 how much he regretted never having participated regularly in Dehn's seminar. Hartner possessed the necessary prerequisites – that unusual mixture of philological and mathematical talents – but he admitted that in 1922, the year he first met Dehn, he had not yet discovered his interest in history (he was only 17 at the time). Nevertheless, he shared some vivid memories of the contrasting styles of Dehn and Hellinger as teachers:

> Anyone who, like me, ever heard Ernst Hellinger's differential and integral calculus and other lectures will have remembered well into old age his almost unequaled mastery. Today educational methods are very much in fashion, but I am sure Hellinger never bothered with such theories; with him it was as if a friendly fairy had put that in his cradle.
>
> Max Dehn embodied a completely different type of brilliance. In contrast to Hellinger, he loved to improvise and abandon himself to the overflow of thoughts storming through him. With all due acknowledgment of his mastery, this proved a bit difficult for us, his inexperienced listeners. Feeling very despondent, I asked him for a brief interview. It lasted a good two hours spent in the professors' cafeteria, where one drank miserable inflationary coffee at a price of about a billion marks a cup. I was pleasantly surprised that Dehn responded to my request without any sign of annoyance. The rest of the conversation was about very different things – art, music, languages, classical and modern, about history, and finally also about the political situation. It was the beginning of a lifelong friendship that we preserved in even more difficult times. [Burde/Schwarz/Wolfart, 2002a, 23–24]

Hartner's role in helping the Dehns during the dramatic events of November 1938 is described in Chapter 9.

Occasionally, visitors attended the Frankfurt historical seminar, one of whom was André Weil, who vividly recalled the impression Dehn left on him:

> A humanistic mathematician who saw mathematics as one chapter – certainly not the least important – in the history of human thought, Dehn could not fail to make an original contribution to the historical study of mathematics, and to involve his colleagues and students in the project. This contribution, or rather this creation, was the historical seminar of the Frankfurt mathematics institute. Nothing could have seemed simpler or less pretentious. A text would be chosen and read in the original, with an effort to follow closely not only the superficial lines but also the thrust of the underlying ideas. ... It was only later that I attended it, on subsequent visits to Frankfurt, a place I made a point of visiting as often as I could. I am not sure whether it was already in the summer semester of 1926 that, during a seminar session devoted to Cavalieri, Dehn showed how this text had to be read from the viewpoint of the author, taking into account both what was commonly accepted in his lifetime and the new

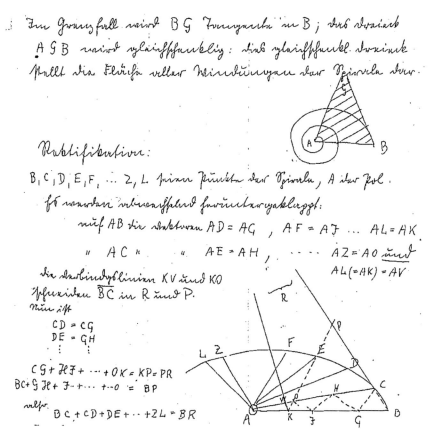

FIGURE 3. A page from Epstein's presentation of Torricelli's work on spirals (Universitätsarchiv Frankfurt am Main.).

> ideas that Cavalieri was trying to the best of his ability to implement. Everyone participated in the discussion, contributing what he could to the group effort. [Weil, 1992, 52]

Paul Epstein was one of those who took an intense interest in this undertaking, as can be seen from the overview of sixteen notebooks that still survive today [Bergmann/Epple/Ungar, 2012, 128]. In a discussion of Evangelista Torricelli's work on spirals (Figure 3), Epstein began with the observation that one needed to distinguish different cases, which Gino Loria, editor of Torricelli's *Opere*, had failed to do. Epstein thus separated the case of the logarithmic spiral, which Descartes had studied in 1638, from algebraic spirals. In both cases, Torricelli dealt with their quadrature and rectification, whereas in the latter he also investigated the problem of finding tangents to the curve (Mathmatisch-historisches Seminar, WS 1927/28, Notebook V).

Epstein was particularly interested in the aesthetics of language and sought to clarify what he considered to be a misunderstanding on the part of Goethe, who considered mathematics to be sterile and unaesthetic. In an essay entitled "Goethe and mathematics," Epstein wrote:

> Thus, the language of the mathematician is free from all of the imperfections and shortcomings of colloquial language, which Goethe himself vividly felt, deploring that language as we know it, which had developed from obvious human needs, human activities and general human sentiments and views, is not adequate to correctly express perceptions of the functioning and workings of nature as something which is indeed far removed from common human affairs. The mathematical sign language, however, – correctly understood and used in the fields to which it is suited – is the "Language of the Spirits" which Goethe sought; according to a famous dictum of Galileo, it is the language in which the Book of Nature is written. (Quoted from [Bergmann/Epple/Ungar, 2012, 192])

Commenting on this, Birgit Bergmann noted that Epstein

> ...attempted to equate Goethe's classical poetry with the clarity and precision of mathematical language. He was convinced that Goethe's rejection of mathematics was due to his failure to recognise its true beauty. Epstein sought to eliminate the most obvious and the best known contradiction between Goethe and mathematics by showing that ultimately both approaches to understanding the world had their source in the same motivation and led to the same goal. With these texts, Epstein joined the ranks of German-Jewish intellectuals who attached great importance to the classical German educational ideal known as "Bildung". [Bergmann/Epple/Ungar, 2012, 191]

Among the students who regularly attended this mathematics history seminar, one in particular stood out from all the rest – Adolf Prag, whose later career in many ways mirrored Dehn's. Not only did Prag's life crisscross with those of Max Dehn and his two daughters, Eva and Maria, but he also went on to play a singular role in historical studies devoted to the mathematics of the seventeenth century. When he died in 2004, his life and work were later recounted by the historian of mathematics Christoph Scriba in his obituary article [Scriba, 2004], the basis for much of what follows.

Prag was born in 1906 in a small village on the edge of the Black Forest, but soon thereafter his family moved to Frankfurt, where he attended the humanistic Goethe Gymnasium. There he acquired a solid grounding in classical languages that he would cultivate throughout his life. From 1925 to 1929 he studied mathematics at Frankfurt University, where he became a mainstay in Dehn's history of mathematics seminar. As Scriba well imagined the situation:

> Dehn, with his wide historical and philosophical interests, must have sparked a congenial vein in Prag. In addition, the outstanding linguistic abilities of this student, who was able to translate Latin and even Greek texts fluently into the German language, were a welcome asset for the discussions of this circle.

During this time, a lifelong friendship began between Prag and two of his fellow students, Ruth Moufang and Wilhelm Magnus.

After completing his studies, Prag still needed to pass the state examination for teaching candidates and submit a thesis (*Staatsexamenarbeit*). For a topic he went to Dehn, who suggested that he write about the Oxford mathematician John Wallis, whose work Prag had studied in the seminar. The resulting thesis was so impressive that Dehn sent it to Otto Neugebauer, who published it in his new series *Quellen und Studien zur Geschichte der Mathematik* [Prag, 1929].[7]

During the 1920s, the field of history of mathematics had begun to attract interest among some of the leading mathematicians of the day. One of the most prominent of these was André Weil, whose approach to historical studies can be found in [Weil, 2001], among other works. This has been aptly called the heritage approach to history, later canonized in Bourbaki's *Élements d'histoire de mathématique* (1960). Weil famously defended it as the only legitimate viewpoint when it came to understanding the larger historical development of mathematical ideas.

In the case of Dehn's seminar, one can easily see that the choice of texts was largely confined to classical antiquity and the period in early modern Europe leading up to the emergence of the calculus in the works of Newton and Leibniz. Siegel thus recalled spending a number of semesters studying works by Euclid and Archimedes. Another block of texts dealt with developments in algebra and geometry from Leonardo of Pisa and Cardano to Viète, Descartes, and Desargues. Finally, the seminar looked carefully at texts documenting the emergence of infinitesimal calculus over the course of the seventeenth century, especially key authors associated with the British tradition: Wallis, Gregory, Barrow, and of course Newton. This overall plan was thus entirely conventional; yet even so, knowing in advance what one expected to find in an older mathematical text was usually of little help when it came to reading and *actually understanding* such works *in detail*.

Some three decades later, when he returned to Frankfurt to speak about the times he shared with his former colleagues there, Siegel had this to say about their history of mathematics seminar:

> As I look back now, those communal hours in the seminar are some of the happiest memories of my life. Even then I enjoyed the activity which brought us together each Thursday afternoon from four to six. And later, when we had been scattered over the globe, I learned through disillusioning experiences elsewhere what rare good fortune it is to have academic colleagues working unselfishly together without thought to personal ambition, instead of just issuing directives from their lofty positions. [Siegel, 1979, 226]

History as Inspiration for Mathematical Research

Dehn's seminar proved to be deeply inspirational for Siegel, whose singular ability to attack truly formidable problems in number theory was becoming legendary [Yandell, 2002, 208]. He was surely long intrigued with the mysterious results Riemann had communicated in his 8-page paper on the zeta-function, which no one had been able to unravel. With the assistance of his friend, Erich Bessel-Hagen, he set to work studying Riemann's unpublished notes related to the distribution

---

[7]The published version, however, omitted a chapter on the Pell equation.

of primes, a question that Riemann's teachers, Gauss and Dirichlet, had studied before him. On 6 November 1927, Siegel composed a 10-page manuscript on this topic. Figure 4 shows his portrait of Riemann along with a famous poem "Friede mit der Welt" (Peace with the World) by Friedrich Rückert,[8] which he found among Riemann's notes:

> Lebe von der Welt geschieden,
> Und du lebst mit ihr in Frieden.
> Willst du dich mit ihr befassen,
> Höre, was dir widerfährt!
> Du musst lieben oder hassen;
> Keines ist der Mühe wert.

> (Live apart from the world,
> And you live with her in peace.
> Should you want to engage with her,
> Hear, what shall befall you!
> You must love or hate;
> Neither is worth the effort.)

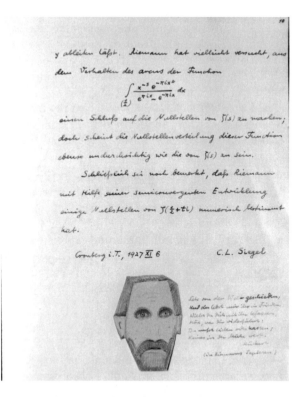

FIGURE 4. The final page from Siegel's manuscript on Riemann's unpublished work on the zeta-function

---

[8]Rückert's poetry was set to music by numerous famous composers; best known among these works are the "Kindertotenlieder" in the composition by Gustav Mahler.

Siegel's research project eventually led to his reconstruction of the Riemann-Siegel Formula, published in *Quellen und Studien* [Siegel, 1932]. H.M. Edwards summed up his accomplishment with these words:

> The difficulty of Siegel's undertaking could scarcely be exaggerated. Several first-rate mathematicians before him had tried to decipher Riemann's disconnected jottings, but all had been discouraged either by the complete lack of explanation for any of the formulas, or by the apparent chaos in their arrangement, or by the analytical skill needed to understand them. One wonders whether anyone else would ever have unearthed this treasure if Siegel had not. [Edwards, 1974, 136]

Max Dehn's contributions to the literature on history of mathematics came mainly in the form of essays and occasional articles. His single most impressive writing was a six-part appendix to the third edition of Moritz Pasch's classic monograph *Vorlesungen über die neuere Geometrie* [Pasch, 1882/1926]. The second edition of Pasch's book, published by Teubner in 1912, had long been out of print. During the postwar era, after Springer assumed Teubner's former role as the leading German publisher of mathematical texts, Courant's "yellow series" often published older standard works in an updated form. Pasch was already approaching 80, so he was in no position to produce a substantially new edition, but Courant was surely more than pleased when Dehn agreed to write an appropriate supplementary appendix.

Some five years later, Courant turned once again to Dehn to request a supplement for a new edition of Arthur Schoenflies' textbook on analytic geometry. Dehn's six appendices to [Dehn, 1931, 298–411] not only offered an overview of foundations and a modern treatment of linear algebra, it also contained a brief historical overview as well as a section on still unsolved problems in analytic geometry. In short, this material made the book far more than simply an elementary textbook. Here, as well as in the case of Pasch's book, Dehn drew on material he had developed for his courses in Frankfurt. This circumstance is reflected in his preface to [Pasch, 1882/1926], where he wrote: "The appendix corresponds approximately to a two-hour, one-semester lecture course, in which the instructor reports on what he considers to be all the more important questions, discussing the most important problems in detail, and above all seeking to stimulate independent study and the reading of classical works" [Dehn, 1926, viii].

Among the classics in the history of geometry that Dehn had in mind, two were preeminent: Euclid's *Elements* [Heath, 1908] and Hilbert's *Grundlagen der Geometrie* [Hilbert, 1899/2015], which in 1922 was published in a 5th edition containing several new supplements. As for the significance of Pasch's original text from 1882, Dehn described this as marking the end of a quest to derive projective geometry from purely elementary principles, formulated in a complete system of axioms that avoids appealing to congruence properties or notions of continuity, such as the Axiom of Archimedes [Dehn, 1926, 188]. Hilbert's axiom system, on the other hand, stood closer to the original system of Euclid, which made it possible to analyze which parts of geometry were susceptible to an elementary treatment and which were not.

Dehn's approach in this survey was largely systematic, though he added footnotes containing brief historical remarks coupled with references. The first question he raised was the role of the parallel postulate in ancient Greek geometry, a problem compounded by philological difficulties. In most of the extant manuscripts this postulate appears under the "common notions," which textbook authors usually referred to as axioms to distinguish these from the strictly geometrical postulates. The "parallel postulate" was then given as Axiom 11 in these texts (in the English tradition, following Robert Simson, it was Axiom 12). The Danish historian of mathematics, Johan Heiberg, argued that this was due to the editorial intervention of Theon of Alexandria who, according to Heiberg, had removed the fifth postulate from its original position and placed it under the "common notions." In preparing the modern Greek/Latin edition, Heiberg restored the postulate to what he believed was its original place. He found it listed as the fifth postulate in an older non-Theonine manuscript housed in the Vatican Library, which he took as his principal *Urtext* in preparing the modern edition. The English translation published afterward by T.L. Heath [Heath, 1908] follows Heiberg's edition almost without exception. In Dehn's day, these were very recent events, though today few realize that the parallel postulate has only been called Euclid's fifth for little more than a century. In several places, when discussing Greek mathematics, Dehn made similar comments about difficulties arising from a dearth of historical source material.

The logical or mathematical status of the parallel postulate long remained one of the most famous of all geometrical mysteries. Pasch's work put the last touches on projective geometry, a theory in which the properties of parallel lines play no role. Alongside those developments, however, more subversive thinkers – Lobachevsky and Bolyai – staked out arguments for a new theory of geometry in which parallel lines no longer satisfy Euclid's fifth postulate. Although it took several decades for mathematicians to embrace non-Euclidean geometry, once they did so, the contingent status of the parallel postulate became clear: Euclidean geometry was only a special case. Indeed, among the infinitely many possible spaces of constant curvature, Euclidean geometry was the one in which that constant was zero. Dehn's discussion took up the connection between non-Euclidean geometries and projective geometry, an insight Klein recognized once he learned about the possibility of obtaining a general projective metric, a technique Arthur Cayley used to derive Euclidean geometry. Dehn also briefly noted how Riemann's notion of a manifold with local curvature properties led to the natural question of the various possible global extensions, a problem that led to Clifford-Klein space forms.

Dehn sketched these various topics quite rapidly before turning to problems underlying the foundations of projective geometry. Here he focused on the difficulty of providing a logically sound and complete construction of coordinate systems in projective geometry. Dehn distinguished between an older, more intuitive approach that depended on the Archimedean axiom and the purely projective methods developed by Pasch. From Desargues' theorem – which follows immediately from the incidence axioms for points, lines, and planes in space – one can easily generate a network of rational points in the plane by iterating the construction of a fourth harmonic point for every triple. Pasch then found a way to extend this construction to irrational points by invoking a projective substitute for the Archimedean axiom.

These brief remarks then led over to Dehn's main topic, which began with a modernized account of Hilbert's segment arithmetics based on the two-line Pascal theorem (Pappus's theorem) and the theorem of Desargues. His treatment of these, however, draws on elementary group theory for geometric transformations, leading to a proof that the fundamental theorem of projective geometry entails both theorems, Pascal as well as Desargues. Dehn also gave a proof of Hessenberg's theorem, namely that the planar theorem of Desargues follows from Pascal. The individual achievements of others (Staudt, Wiener, Hilbert) are only mentioned in a footnote, and Dehn caps off this section with a schematic chart providing an overview of the relative dependence of the various axioms of fundamental theorems. All of this reflects Hilbertian interests, except for the appeal to group theory, where for details he points to [Schwan, 1919]. The author of this study was a Gymnasium teacher in Düsseldorf, who went on to write his dissertation under Max Dehn.[9]

Following this overview, Dehn presents a section containing proofs of the key theorems. He emphasizes that one must first prove the Pascal theorem without recourse to continuity, and he begins with a synopsis of the original proof given by Friedrich Schur in [Schur, 1898]. This proof made essential use of a beautiful idea first discovered by Germinal Pierre Dandelin in connection with conics that lie on a hyperboloid of one sheet, thus a quadric surface generated by two systems of lines. Dandelin showed that a spatial hexagon obtained by connecting 6 points along corresponding generators, as these alternate between the two families, leads to a so-called Brianchon point, the common intersection point of the 3 diagonals.[10] The dual incidence relation follows as well, and taking a plane section of the quadric then leads to a conic with an inscribed hexagon that satisfies Pascal's theorem. Dehn not only credited Hermann Wiener with having brought out the significance of the theorems of Desargues and Pascal for foundations of geometry, he also emphasized how Schur's proof of Pascal's theorem was inspired by Dandelin's older ideas. These enabled Schur to prove the two-line version of Pascal's theorem, the case required for a commutative segment arithmetic [Dehn, 1926, 228–232].

Turning back to his earlier discussion of Euclid's *Elements*, Dehn underscored what Schur had achieved, namely the very first purely synthetic introduction of a segment arithmetic without any appeal to continuity or the parallel postulate. He thought this work, and not Saccheri's, could more fittingly have borne the title "Euclidis ab omni naevo vindicatus" (Euclid freed of every flaw). A century earlier, the English mathematician Henry Saville had pointed out two major flaws in the classical presentation: the opaque use of the parallel postulate in Book I and the glaring break in Book V, where Euclid inserted a general theory of ratio and proportion before applying it to develop the theory of similar rectilinear figures in Book VI. The cornerstone concept in Book V was the famous Definition V.5 that provides a theoretical criterion for determining when two ratios will be equal. Euclid merely needed to invoke that definition once, in the first proposition of Book VI, after which everything fell easily into place.

Dehn seemed to be saying that this historical development – from Saville to Saccheri and Lambert, passing through the discovery of non-Euclidean and Pasch's grounding of projective geometry, and then the rigorous coordinatization of elementary synthetic geometry with Schur's work – represents a story that was already essentially closed when Hilbert stepped onto the scene. What he wrote immediately afterward, though, fully clarifies why Hilbert's *Grundlagen der Geometrie* occupies such a significant place in this chain of developments. Indeed, in surveying what

---

[9]Wilhelm Schwan, "Extensive Größe, Raum und Zahl," Diss. Frankfurt University, 1923.

[10]Pictures illustrating this argument for this case of Brianchon's theorem can be found in [Hilbert/Cohn-Vossen, 1932, 92–93].

had transpired up until 1899, Dehn described the series of highways and byways that led to important stations, but in such a complicated fashion that one could hardly view these as more than a collection of significant results that fell well short of constituting a unified theory. Hilbert, on the other hand, was the first to recognize the validity of "exotic geometries," as for example, plane geometries in which the theorem of Desargues fails to hold. This finding went hand in hand with one of his central insights: *The validity of the plane theorem of Desargues is the necessary and sufficient condition for deciding whether the plane can be embedded in space.* Schur's proof of the Pascal theorem made essential use of spatial geometry, whereas Hilbert sought to reveal the possibilities for building a theory on geometry in the plane by exploiting the power of the parallel postulate. After spelling out this motivation, Dehn proceeded to give Hilbert's planar proof of Pascal's theorem.

In the closing section on projective geometry, Dehn describes some of the simple consequences of arithmetization, illustrating the theorems of Desargues and Pascal by means of incidence configurations for points and lines in the plane. Hilbert sometimes called these closure theorems, since they lead to closed figures that lie in special position in the plane. The Desargues theorem leads to a $(10, 3)$ configuration, whereas Pascal is a $(9, 3)$ (thus 9 points and 9 lines that are incident in triples). In the first case, one has 30 linear equations, three for each of the 10 lines whose equations are satisfied by substituting the coordinates of the 3 points that lie on them. But since these linear relations are not independent, translating the theorem into algebra leads to the result that one can deduce the final relation from 29 of them. Similarly for the Pascal theorem, as both are examples of *Schnittpunktsätze*, as Hilbert described in *Grundlagen der Geometrie*. In this setting, duality follows immediately from the fact that point and line coordinates enter symmetrically in systems of linear equations.

In the remaining parts of his survey, Dehn took up several topics closely related to Hilbert's researches as well as his own. Some key points were touched on earlier in Chapter 2 in connection with Hilbert's general views on foundational issues. The focal point of the Hilbert-Bernays program to formalize mathematics was their effort to prove the consistency of the axioms for arithmetic as a first step toward solving Hilbert's second Paris problem, which required doing the same for the real numbers. By the mid-1920s, if not earlier, Dehn had begun to doubt the feasibility of this formalist program. Like Henri Poincaré before, he doubted that the principle of complete induction could ever be reduced to a consistency argument [Dehn, 1926, 260–262].

Pasch had no time to study Dehn's text in any detail when he received the page proofs; his failing eyesight likely hindered him from doing more that glancing through the text. Still, he sent his congratulations to Dehn, while expressing his delight over the sheer volume of material his survey contained as well as the careful handling of it.[11] He only added his wish that Dehn somewhere mention the term "Pasch's Axiom" in his text, since several writers had used this terminology. Hilbert himself had acknowledged in a footnote that Axiom II.4, the last among his axioms of order, was first introduced by Pasch in 1882.

---

[11] Moritz Pasch to Max Dehn, 7 July 1926, Dehn Papers, Dolph Briscoe Center for American History, University of Texas at Austin.

Frankfurt as a Mathematical Counter-Culture

In his influential study of mathematical modernity and its discontents – highlighting proponents of modernism and those who opposed some of its manifestations – Herbert Mehrtens described how Göttingen became the prototype for a modern mathematical research center [Mehrtens, 1990]. Under the leadership of Klein and Hilbert, the dominance of Göttingen within the German mathematical world was embodied in the journal *Mathematische Annalen*. The *Annalen* also served as a flagship publication for the Leipzig firm of B.G. Teubner, which dominated mathematical publishing in Germany up until the end of World War I. At that point, owing to the precarious economic situation, Teubner largely withdrew from the field, thereby enabling the Berlin firm of Julius Springer to assume that role and soon go far beyond it. This sudden turn of events had much to do with the special relationship Richard Courant cultivated with Ferdinand Springer, which began during the war years [Reid, 1976, 69–72].

Courant's Göttingen was a multi-faceted enterprise, but at its heart flourished a "publish or perish" culture that stood as the antithesis of the one cultivated in Frankfurt. Indeed, one of the striking features of the latter was how little Dehn and Siegel chose to publish once they began working together. This hardly meant that they were unproductive, however; nor did they lack ambition. In fact, their decision to withdraw from this arena stemmed from a shared understanding that "more was not better" – real progress would take place outside the "mathematical factories." which were producing and disseminating such an abundance of new results that contemporary mathematicians found themselves drowning in their own literature.

André Weil remembered Dehn invoking just this image when he visited Frankfurt around Christmas of 1926. Mathematics, Dehn told him,

> was in danger of drowning in the endless streams of publications; but this flood had its source in a small number of ideas, each of which could be exploited only up to a certain point. If the originators of such ideas stopped publishing them, the streams would run dry; then a fresh start could be made. To this purpose, Dehn and his colleagues refrained from publishing. [Weil, 1992, 53]

This view probably comes closer to Siegel's attitude than to Dehn's, if only because the latter was a born teacher and collaborator, famous for his generosity in spending fresh ideas to help others. Dehn was highly gregarious, whereas his younger colleague was a loner who vacillated between periods of incredible productivity and long patches of depression. As noted earlier, Dehn took an active interest in conferences and large international congresses, whereas Siegel only rarely participated in such events. Although he loved to travel and visit with colleagues at home and abroad, Siegel never even became a member of the German Mathematical Society. He published 13 major papers during the two years prior to his arrival in Frankfurt on August 1, 1922, having "gone from a military mental hospital to full professor in five years" [Yandell, 2002, 181]. Afterward, he kept much of his work out of the public eye, though he gave away polished manuscripts as presents to Dehn and others. His musings over Riemann's work on the zeta-function, described above, constitutes a striking example.

In Dehn's estate one also finds a 47-page manuscript entitled "On Diophantine Equations," which Siegel signed and dated 13 April 1928; he dedicated it to Schoenflies on his birthday, which took place four days later.[12] The text begins with a brief recapitulation of the main result in Weil's thesis, before going on to state and prove a related finiteness theorem for a certain class of diophantine equations. Weil learned about this text during a stay in Frankfurt, and he received permission from Dehn to read it, though under condition that Weil do so in Dehn's home. He was even allowed to take notes, since he agreed not to show them to anyone else [Weil, 1992, 53]. One can easily imagine a playful grin on Max's face as he explained the rules of the game.

André Weil, who later became the unofficial leader of the group that published under the pseudonym Nicolas Bourbaki, was very struck by the radically different atmosphere in Courant's Göttingen [Weil, 1992, 52–53]. He recalled, in particular, how he learned very little in conversations with those in Richard Courant's own group. Nearly every time he got talking with one of them, the exchange would end rather abruptly with a remark like, "sorry, I have to go write a chapter for Courant's book" [Weil, 1992, 51].[13] There was a distinct awareness in Göttingen that Max Dehn and Carl Ludwig Siegel, both of whom thought of mathematics as an art form, were cultivating an approach to research in Frankfurt that stood in conscious opposition to the Göttingen model.

Otto Neugebauer, who served as Courant's "floor manager" at the Mathematics Institute in Göttingen, was certainly sensitive to the implicit criticism coming from Frankfurt.[14] Neugebauer played a central role in designing the institute's new quarters, built with funding from the Rockefeller Foundation. When it opened in December 1929, Hermann Weyl delivered a lecture honoring Felix Klein, who had long dreamed of housing mathematics in such a building. Neugebauer, on the other hand, was eager to describe the physical arrangements as an inviting place for teaching staff and students to gather and meet. "We hope and believe," he wrote, "that the new mathematics institute will *not* provide new impetus for the "mechanization" of science, as so often prophesied, ... but rather will offer a workplace, where one can *enjoy* teaching and learning and, above all, the pursuit of pure science" [Neugebauer, 1930, 4].

In January 1928, Max Dehn addressed a large audience at Frankfurt University when he spoke about "The Mentality of the Mathematician" [Dehn, 1928]. This was a ceremonial occasion, namely the annual celebration of the founding of the modern German nation on January 18, 1871. For many Germans, that date conjured up feelings of nostalgia and even sorrow, as the original *Reich* existed no longer, and few felt any deep affection for the republic that replaced it after the fall of the monarchy. On this occasion, Dehn wanted to give the audience some idea of the mental world of the mathematician. Since he had to approach this topic from some higher plane, though, he chose to illumine what he hoped to convey by appealing to history, even going back to ancient times. Dehn's daughter Maria, who was 13 years old at the time, still remembered more than four decades later what her father had to say about Descartes and other famous figures that day. His text, which was

---

[12]It appears likely that Schoenflies was too ill to accept this gift; he died on 27 May 1928.

[13]Whereas this "publish or perish" mentality predominated in Courant's circle, Emmy Noether felt no such urgency to rush her work into print.

[14]On Neugebauer's early career, see [Rowe, 2016].

later translated in [Dehn, 1983], makes delightful reading even now (the quotations below are taken from this translation). Somewhat surprisingly, a leading historian of science also cited it in a paper that created a major stir in the late twentieth century, as will be discussed further below.

Certainly the views Dehn expressed in "The Mentality of the Mathematician" cast considerable light on the speaker's own quite unique way of thinking. His first and most immediate task was to assure his listeners that mathematicians were engaged in a creative activity. For "the layman often thinks that mathematics is by now a closed science, and gives little thought to the origin of the discipline he is familiar with from school." Dehn spoke of the sense of divine inspiration that ancient Greek mathematicians felt after making a profound discovery, and how "Eratosthenes and Perseus, in the manner of winners in an Olympic competition, made votive offerings out of joy at attaining their goals." Turning to early modern times, he talked about Cardano's wild urge to work out all the various types of solutions of cubic equations in his *Ars Magna*, but he also made clear that mathematical knowledge had to be clarified and communicated to have a decisive impact. This was particularly evident in the case of Descartes, who fashioned himself as having made a great new discovery – a method for systematically solving geometrical problems by reducing them to algebraic equations – when, in fact, he had mainly brought forth a known method with exceptional clarity. As Dehn put this:

> Descartes himself believed that, through an illumination, he had discovered a new science – *cum mirabilis scientiae fundamenta reperirem*. But this was hardly the case. His great contribution was not the discovery of a completely new idea – that of the unity of algebra and geometry. It is even incorrect to say that he realized the existing idea in a manner more daring than his predecessors. What is true is that the illumination came to a man of surpassing algebraic talent that enabled him to solve the most difficult particular problems and, more importantly, to a thinker endowed with incomparable shaping power, who presented and applied the idea – seized in a vision – with admirable sharpness, clarity, and terseness, with almost rhetorical brilliance. The historical significance of his contribution lies, above all, in formulation. It was this that produced such fruitful effects on his contemporaries – who, not surprisingly, had for the most part the impression of something entirely new.

Dehn's admiration for Descartes' accomplishments did not extend to his person, however, as this great French thinker was extremely impressed by his own sense of superiority and gloated over what he had accomplished. For Max Dehn, Gerolamo Cardano was a far more sympathetic figure, as can be seen from this passage:

> Cardano, who died in 1576 at the age of 75, was a typical man of the Renaissance. In view of our present topic – the creative power of the mathematician – Cardano is of special interest to us. His productivity was unbelievably extensive. Ninety years after his death, ten large folios of his work appeared, and the publisher assured readers that this was only half of what Cardano had written. There is no area between heaven and earth that he left untreated. He wrote about all the natural sciences, medicine,

> astrology, theology, philosophy and history. His autobiography – which Goethe compared to Benvenuto Cellini's – has great charm. In it he describes with touching ingenuousness a life afflicted with manifold misfortunes. At times we are strongly reminded of Rousseau's *Confessions.* Goethe writes at length about Cardano in his history of the science of color – about his talent, his passion, his wild and confused state that always comes to the fore . . . .

In connection with Cardano, Dehn noted that priority disputes had often played a major role in the history of mathematics – perhaps the most famous being the controversy between Newton and Leibniz as well as their respective followers regarding the discovery of the infinitesimal calculus. Yet his main point was that these seldom had real significance for the larger sweep of developments, since mathematical knowledge as attained by any one individual is always a fleeting thing. The larger body of knowledge is thus always a work in progress, so to speak. "For the historian," Dehn said, "the purest joy is to relish the contemplation of the ups and downs of the development, of the connections, of the breaks and transitions, to try to see the divine spark in each of the creators and to relive their productive moments."

He also spoke of the attractions certain famous problems still held for amateurs:

> Young and old, students, and especially adults living in relative isolation – country clergymen, small-town teachers, foresters – busy themselves with famous ancient problems, such as the trisection of an angle, the quadrature of a circle, the mysterious properties of whole numbers. The latter are probably the oldest playground of mathematics. Here a few results were found in the distant past, and here, to this very day, the most accomplished mathematicians battle to achieve progress with problems that are as easy to state as they are difficult to master.

In speaking of the role of rigorous thinking in mathematics, Dehn described the importance of the discovery of irrational ratios by the ancient Greeks. From the standpoint of practical measurements, Greek mathematicians clearly realized that the astronomer or land surveyor had no need for anything beyond the ordinary ratios formed from natural numbers. Yet such practical considerations did not satisfy them; instead, they erected a whole theory (Euclid's Book V) for precisely handling ratios of magnitudes, whether they be rational or not. Dehn emphasized the key role of the axiom of Archimedes, as these ancient Greek geometers

> realized that without it they could not derive their most beautiful results, such as the philosopher Democritus' insight that the volume of a pyramid is one third of the volume of a parallelepiped with the same base and height, or Archimedes' splendid discovery that the surface area of a sphere is four times the area of its equatorial cross section.

Few in the audience would have realized that the speaker was the first to prove rigorously what the ancient Greeks had only somehow divined, namely the impossibility of obtaining such results without invoking a continuity principle such as given by the axiom of Archimedes.

By the seventeenth century, rigor had largely fallen out of fashion, and mathematicians as varied as Kepler, Cavalieri, and Leibniz all relied on naive intuition. A century later, the special type of rigor cherished by the ancient Greeks was dead. As Dehn put it: "The new race of titans found the rigorous rules of the ancients too particular and inhibiting. One might say that productivity defeated strict propriety. For modern mathematicians, the existence of irrational numbers was not an intense personal experience." Yet during the nineteenth century a new wave of rigor arose, and through the advent of projective geometry it reached over and led to a strengthening of the ancient edifice of Euclidean geometry. Dehn here extolled the achievement of Moritz Pasch, whose classic text had only recently been republished together with Dehn's historical supplement in [Pasch, 1882/1926]. Here, again, he neglected to mention his own role in promoting a broader understanding of Pasch's work.

Mathematicians must be able to think carefully and rigorously, and though they obviously sometimes make mistakes, the larger body of mathematical knowledge is almost free of any error. This has much to do with internal consistencies that reinforce the solidity of newly acquired knowledge. As Dehn noted: "Most results are so involved in the general web of theorems, they can be reached in so many ways, that their incorrectness is simply unthinkable. This is the characteristic difference between mathematics and all systems of knowledge that arose before it ...." He went on to say that many mathematicians are prone to apply the same kind of rigorous thinking to other fields of knowledge or even human affairs in general. This inevitably leads to gross misunderstanding, which often enough ends in a merely dismissive attitude toward other forms of thinking. Dehn's personal experience with conversations between mathematicians and philosophers led him to conclude that:

> Most mathematicians find it especially difficult to come to terms with the thought constructions of the philosophers. The structure of knowledge in these two disciplines is so different that the mathematician arrives quickly at the conclusion that the philosopher works exclusively with magic incantations, while the philosopher thinks the mathematician superficial and simplistic.

In this connection, Dehn strikingly counts Descartes as a prime example of the hyperrational mindset, suggesting he had a detrimental effect on philosophy. His success in geometry made him overestimate the power of pure reason: "Blinded by this, he believed that he could comprehend the whole world by means of pure reason, that he could, so to say, explain it mathematically." Dehn also distanced himself from a viewpoint once expressed by Goethe, who claimed that mathematicians have no conscience: "He probably meant that they need no conscience, for they build on a solid foundation and cannot possibly be led into temptation. But he overestimates mathematicians or mathematics."

Recent developments, Dehn explained, have made much clearer some of the limitations of mathematical knowledge. "The taste for generalization, for ever more comprehensive concepts, ... has made us construct things that are very far removed from common intuition, such as the general concepts of set theory ...." He alluded to certain contradictions that had arisen in set theory, difficulties which had still not been entirely resolved. He also mentioned the controversies that broke out over Zermelo's axiom of choice, which turned out to be an almost indispensable assumption for proving important theorems in modern analysis. Dehn's general skepticism in this connection stands in sharp contrast with the famous views that Hilbert had propounded in Paris in 1900 (see Chapter 2). As Dehn well knew, his former mentor was still hoping to salvage the keystone result he needed to prop up his optimistic worldview, which L.E.J. Brouwer rejected in the name of his intuitionist philosophy of mathematics. Alluding to this by now well-known foundational crisis, Dehn noted:

> There is no doubt that it is necessary to develop the foundations of analysis with greater care. But even if this task is accomplished satisfactorily, no modern mathematician will want to claim that no contradictions will ever arise on this new foundation. It was primarily Hilbert who clearly formulated and tackled the problem of giving a direct proof of consistency. A complete resolution of the issue is very unlikely for a variety of reasons. I myself think it may be that the reason for this difficult state of affairs is that in the case of very general conceptual constructs there arise contradictions between geometry – the world of extensive magnitudes – and the world of counting, and that analysis – the bridge between the two – will surely remain completely free of contradictions as long as it is intuitive, that is, its concepts remain intuitive geometric concepts.

Dehn was most assuredly not referring to intuition in the sense of Brouwer, whose philosophical views were just as dogmatic as those of Hilbert.[15] His position does, however, bear some resemblance to Hermann Weyl's later views, though Weyl emphasized the broader importance of mathematical physics rather than geometry. It might be noted here that Dehn was speaking just two years before Kurt Gödel showed that Hilbert's second Paris problem could not be solved within the framework of proof theory as designed by Hilbert and Paul Bernays. This was the upshot of Gödel's Second Incompleteness Theorem, published in "On Formally Undecidable Propositions of *Principia Mathematica* and Related Systems I" [Gödel, 1931].

Dehn explicitly rejected the Platonic notion that mathematical thinking is rooted in pure ideas divorced from the senses. His understanding of geometric intuition derived, on the contrary, from human perception of worldly phenomena. He went on to say:

> Earlier, one regarded all of the mathematics – the basic theorems and the mode of reasoning – as logically necessary; mathematics was unique. Since the beginning of the 19th century it has become more and more apparent that different systems can be visualized, that, say, different types of spaces are compatible with experience. This is not to imply that the mathematician can now choose his assumptions at will. Not only is such arbitrariness likely to result in developments without beauty, but it is also likely to lead to contradictions that make all of the work an illusion.

---

[15]Ironically, Brouwer dubbed the belief that all problems in mathematics can be answered "Hilbert's dogma" [van, Dalen 2013].

Dehn's general views with regard to mathematical productivity have a truly stoic quality, and he takes pains to explain why he sees resignation as the appropriate mindset for the intellectually mature mathematician. He rests his case on some observations about developments in certain mathematical disciplines, beginning with projective geometry, which had been a major field of research during much of the nineteenth century. Over the course of time, however, it took on a canonical form and eventually its practitioners learned to package and solve its problems by using modern algebraic techniques. With mirthful exaggeration Dehn even claimed "that after two thousand years of work the solution of problems in projective geometry was virtually as trivial for a professional mathematician as the solution of problems of counting with whole numbers for a 10-year-old child."

Projective geometry thus represents a discipline with a long and intricate history, but which no longer holds any vital interest for mathematical researchers, nor is it likely to generate renewed interest ever again. Number theory, on the other hand, has an even longer history, but some of its oldest problems still remain unsolved. Dehn noted, as one example, the notion of perfect numbers first investigated by ancient Greek mathematicians. In Book IX of the *Elements*, Euclid proved that if $2^p - 1$ is prime, then $2^{p-1}(2^p - 1)$ is perfect. Thus, each Mersenne prime leads to an even perfect number, and the Greeks knew of four such cases. In 1928, the year Dehn spoke, twelve had been found, and today the number is 51. Whether an upper bound exists on the set of all perfect numbers is still unknown; nor is it known whether or not an odd perfect number might exist. Dehn then turned to topology, his own principal field of research, about which he said: "here the failure is not due to the fact that – as in number theory – the problems cannot be tackled, but to the fact that they are so intricate that the power of the human intellect, the ability to imagine different things at the same time, is not sufficient for mastery."

Dehn commented that other fields revealed other distinctive patterns of development, but he considered these good illustrations of the most important types. Summarizing, he referred to these three types as "the discipline that reduces to triviality; the one that is forever hindered in its development and beset on all sides with seemingly unassailable problems; and the one that – after a longer or shorter period of development – is brought to a halt by the complexity and difficulty of its problems." Dehn was particularly interested in this third type of discipline, which presented new and very daunting challenges to contemporary researchers. He likened its development with that of a tree:

> There is a limit to the growth of a tree, for the ability of the tree to transport nourishing substances from the earth to the crown does not extend to arbitrary heights. Similarly the continued development of a branch of mathematics requires connection with the ground of intuition, and man's limited intellectual power sets a boundary on the distance between abstraction and intuition. No development is possible beyond this bound.

Dehn's general remarks in this last instance clearly reflected his early struggles with difficult topological problems, in particular the Poincaré conjecture (see Chapter 4). They also suggest why his views on foundational issues were far more skeptical than those of his former mentor, David Hilbert.

Dehn contra Spengler

In his lecture on "The Mentality of the Mathematician," Dehn also alluded to contemporary developments in physics that ran parallel with those in pure mathematics. During the era of Lagrange and Laplace, when studies of lunar motion and the celestial mechanics of the solar system dominated the attention of many leading mathematicians, a certain sense of resignation prevailed as well, much like the type Dehn had in mind when he spoke about projective geometry. Some felt that the main problems had all been solved and one could do little more than simply refine what had been achieved. Dehn's views, on the other hand, were guided again by the example of topology, a new field that raised all kinds of new and very difficult problems. The situation was, indeed, comparable to the difficulties then facing physicists, who had begun to recognize the limitations of their intellectual resources.

In his oft-cited study [Forman, 1971], Paul Forman quoted Dehn's reflections on the mutually reinforcing trends toward skepticism in mathematics and physics, which he interpreted as signs of intellectual retreat by scientists during the Weimar era. One need not consider the merits of Forman's overall argument concerning the widespread influence of Oswald Spengler's cultural pessimism in scientific circles, however, in order to see that Dehn was in no sense following this line of argument. Thus Forman badly blurred Dehn's intent when he wrote that his views were "[p]ainted in pure Spenglerian style, the characteristic mental tone of the contemporary [German] mathematician is skepsis, mistrust of reason, self-inculpation, pessimism, and resignation." The passages Forman cited are these:

> This somewhat skeptical attitude of some modern mathematicians is reinforced by what is happening in the neighboring area of physics. It seems that it is no longer possible to consistently *interpret* physical phenomena in a mathematical 4-dimensional spacetime continuum. Until now we have been able to supply physics with somewhat loosely built scaffoldings for its ever bolder constructions. It may now be about to emancipate itself from mathematics in its important investigations on the finest structure of matter.
>
> All this inclines some of us to greater skepticism in more general questions as well. The fundamental belief of every philosopher that the world can be consistently comprehended by human reason is now open to doubt. ...[16] To be sure, this is not an entirely original attitude. It reminds one of the view taken by the late Eleatic philosophers at the time of the foundational crisis in ancient Greece.

---

[16] The omission reads: "Beyond the *ignorabimus*, he no longer strongly believes in man's ability to bring together different insights in a satisfying harmony." This, again, stands in stark contrast to Hilbert's views.

> This skepticism gives rise to certain resignation, a kind of distrust in man's intellectual power.[17]
>
> ...man's limited intellectual power sets a boundary on the distance between abstraction and intuition. No development is possible beyond this bound. But modern mathematics is certainly not dead, and someone may, and (we hope) will, so simplify processes in, say, topology... that a new development leads to the discovery of new connections. ...

Here Forman omitted the followng sentences, which are crucially important for the theme addressed in this section: "We must not be hampered by resignation. On the contrary, we must realize that the progress of science depends not on comfortable plodding but on perceiving and forming new ideas. The progress of our discipline depends not on mass efforts, not on a flood of papers filled with investigations of insignificant special cases or generalizations, but on individual creative achievements." Dehn then continued (following Forman):

> Such achievements can hardly come about in a factory-like setting. But if the mathematician complains that modern development has organized even the pursuit of his own science, then he must, above all, tell himself: *mea maxima culpa.* For it is through mathematics that man's constructive power, that brought forth the age of technology, first blossomed. And when he is gripped by despair at the sight of the evil he has brought about he is saved, for the third time, by resignation. ...

Immediately following this citation, Forman emphasized how Dehn's views paralleled similar ideas promoted by Richard von Mises and Gustav Doetsch. These reflected an accommodation to "the values and mood of their intellectual milieu [and] to effectively repudiate their own discipline. They show, moreover, that this process of ideological adaptation to the intellectual environment was, either explicitly or implicitly, in large measure a capitulation to Spenglerism" [Forman, 1971, 55].[18]

What Dehn actually thought about Spengler's bestseller – which he did not so much as mention in his Frankfurt address – can be seen from what he wrote about a biological model that somehow "explains" the general "decline of Greek culture" [Dehn, 1931, 383]. To claim that this accounts for why Greek geometry reached a highpoint with Apollonius, he noted, was quite absurd, in particular because other fields of Greek culture experienced no such decline at the time. Dehn then proceeded to examine subsequent developments in mathematics in order to assess the causes for this particular standstill. To these remarks, one should add an anecdote told by Dehn's former student, Wilhelm Magnus:

---

[17] Forman omits the passage that follows, in which Dehn describes such skepticism as a general feature of scientific inquiry, citing Newton as an example: "But quite apart from the modern developments just described, this attitude has always characterized the greatest mathematicians. For even the greatest researchers learn the sad truth that while they can discern an infinity of new and beautiful problems, they can only tackle the least of them. ..."

[18] Instead of lumping Dehn in with mathematicians like von Mises and Doetsch, it would be far more appropriate to cite the views of Otto Toeplitz, Hellinger's collaborator and a close associate of Dehn's historical seminar. Toeplitz explicitly repudiated the views of Spengler with regard to the history of mathematics, for which see [Bergmann/Epple/Ungar, 2012, 190–191].

> When someone made enthusiastic comments on the work *The Decline of the West* by Oswald Spengler (which postulates an inexorable "law of history" according to which civilizations, like individuals, are born, become vigorous, age and die), Dehn said: If you go to your physician and tell him that you feel tired and worn out, and if your doctor says: "Well, this is so because your vitality is ebbing," would you not reply: "At least, you should take my blood pressure." And Dehn added: There are specific reasons for the development of civilizations, and even if we do not understand their origin, we can at least say something about the causes of their end. [Magnus, 1978/79, 134]

By pulling the above passages out of context, Forman completely overlooked that Dehn's whole argument for resignation depended on his reflections on the inherent limitations of human knowledge as revealed by some of the most recent developments in mathematical research. What Dehn meant by this can also be seen from what he said immediately after the passage cited above:

> ... [The mathematician] is aware of the power of man's thought that can penetrate to the limits of the macrocosm and microcosm and back into the depths of time; but he also knows that this very same human thought is powerless to shape fate, that it is just a small force in a wondrously contorted development, a force that works blindly and has no access to the riddles of the future. Such humility induces profound religiosity in many mathematicians. In spite of the boldness and acuteness of their speculations, it was especially marked in Pascal, Newton, Euler, Cauchy, Gauss, and Riemann.

Dehn wanted his Frankfurt audience to know that, contrary to popular belief, mathematicians are not normally otherworldly eccentrics. He noted that mathematics stands between the humanities and the natural sciences, though its practitioners pursue a distinct version of the general scientific method. Mathematical research is more spiritual than the natural sciences and more sentient than the humanities, whereas mathematical knowledge is linked with the whole history of science, which has intimate connections with the history of philosophy. In summarizing his image of the mental world of the mathematician, he offered these words:

> At times the mathematician has the passion of a poet or a conqueror, the rigor of his arguments is that of a responsible statesman or, more simply, of a concerned father, and his tolerance and resignation are those of an old sage; he is revolutionary and conservative, skeptical and yet faithfully optimistic.

These qualities sometimes appear together in a single person, but Dehn cited the words of the Swiss writer Conrad Ferdinand Meyer in accounting for this apparent paradox:

> *Ich bin kein ausgeklügelt Buch,*
> *Ich bin ein Mensch mit seinem Widerpruch.*
> (I am not a clever book,
> but a human being with contradictions).

References

Bergmann, Birgit, Epple, Moritz, Ungar, Ruti, eds.: *Transcending Tradition: Jewish Mathematicians in German-Speaking Academic Culture*, Heidelberg: Springer.

Braun, Hel: *Eine Frau und die Mathematik, 1933–1940*, Heidelberg: Springer.

Burde, Gerhard, Schwarz, Wolfgang and Wolfart, Jürgen: Max Dehn und das mathematische Seminar, preprint.

Dehn, Max: Die Grundlegung der Geometrie in historischer Entwicklung, in [Pasch, 1882/1926, 185–271].

Dehn, Max: Über die geistige Eigenart des Mathematikers. Rede anlässlich der Gründungsfeier des Deutschen Reiches am 18. Januar 1928, *Frankfurter Universitätsreden* Nr. 28.

Dehn, Max: Sechs Anhänge, in [Schoenflies/Dehn, 1930, 208–411].

Dehn, Max: Probleme des Hochschulunterrichts in der Mathematik, *Jahresbericht der Deutschen Mathematiker-Vereinigung* 43(1932): 71–79.

Dehn, Max: The Mentality of the Mathematician: a Characterization, trans. of [Dehn, 1928] by A. Shenitzer, *Mathematical Intelligencer*, 5(2): 18–26.

Dubreil, Paul: Souvenirs d'un boursier Rockefeller 1929–1931, *Cahiers du séminaire d'histoire des mathématiques* 4: 61–73.

Edwards, Harold M.: *Riemann's Zeta Function*, New York: Academic Press.

Forman, Paul: Weimar Culture, Causality, and Quantum Theory, 1918–1927: Adaptation by German Physicists and Mathematicians to a Hostile Intellectual Environment, *Historical Studies in the Physical Sciences*, 3: 1–115.

Über formal unentscheidbare Sätze der Principia Mathematica und verwandter Systeme, I, *Monatshefte für Mathematik und Physik* 38(1931): 173–198.

Heath, Thomas Little, ed., *The Thirteen Books of Euclid's Elements*, 3 vols., Cambridge: Cambridge University Press.

Hilbert, David: Grundlagen der Geometrie, in *Festschrift zur Einweihung des Göttinger Gauss-Weber Denkmals*, Leipzig: Teubner (2. Aufl., 1903); *Grundlagen der Geometrie (Festschrift 1899)*, Klaus Volkert, ed., Heidelberg: Springer.

Hilbert, David: Mathematische Probleme. Vortrag, gehalten auf dem Internationalen Mathematikerkongreß zu Paris, 1900. *Nachrichten der Königlichen Gesellschaft der Wissenschaften zu Göttingen. Math.-phys. Klasse*, 253–297.

Hilbert, David und Cohn-Vossen, Stephan: *Anschauliche Geometrie*, Berlin: Springer.

Lemmermeyer, Franz u. Roquette, Peter: *Helmut Hasse und Emmy Noether - Die Korrespondenz 1925–1935*, Göttingen: Universitätsverlag.

Magnus, Wilhelm (1978/79): Max Dehn, *Mathematical Intelligencer*, 1(3): 132–143.

Mehrtens, Herbert: Ludwig Bieberbach and "Deutsche Mathematik," in *Studies in the History of Mathematics*, Esther Phillips, ed., Washington: The Mathematical Association of America, pp. 195–241.

Mehrtens, Herbert: *Moderne, Sprache, Mathematik. Eine Geschichte des Streits um die Grandlagen der Disziplin und des Subjekts formaler Systeme*, Frankfurt: Suhrkamp.

Neugebauer, Otto: Das mathematische Institut der Universität Göttingen, *Die Naturwissenschaften* 18(1): 1–4.

Pasch, Moritz: *Vorlesungen über die neuere Geometrie. Mit einem Anhang von Max Dehn: Die Grundlegung der Geometrie in historischer Entwicklung*, Berlin: Springer.

Prag, Adolf: John Wallis (1616–1703). Zur Ideengeschichte der Mathematik im 17. Jahrhundert, *Quellen und Studien zur Geschichte der Mathematik*, Abt. B, 1(1929): 381–412.

Reid, Constance: *Courant in Göttingen and New York: The Story of an Improbable Mathematician*, New York: Springer.

Rowe, David E.: From Graz to Göttingen: Neugebauer's Early Intellectual Journey, *A Mathematician's Journeys: Otto Neugebauer and Modern Transformations of Ancient Science*, Alexander Jones, Christine Proust and John Steele, eds., Archimedes, New York: Springer, pp. 1–59.

Rowe, David E., Hrsg.: *Otto Blumenthal, Ausgewählte Briefe und Schriften I, 1897–1918*, Mathematik im Kontext, Heidelberg: Springer.

Rowe, David E. und Felsch, Volkmar, Hrsg.: *Otto Blumenthal, Ausgewählte Briefe und Schriften II, 1919–1944*, Mathematik im Kontext, Heidelberg: Springer.

Schoenflies, Arthur und Dehn, Max: *Einführung in die analytische Geometie*, 2te Auflage, Berlin: Springer.

Schur, Friedrich: Über den Fundamentalsatz der projektiven Geometrie, *Mathematische Annalen* 51(1898): 401–409.

Schwan, Wilhelm: Streckenrechnung und Gruppentheorie, *Mathematische Zeitschrift* 3: 11–28.

Scriba, Christoph J.: In Memoriam Adolf Prag (1906–2004), *Historia Mathematica* 31: 409–413.

Siegel, Carl Ludwig: Über Riemanns Nachlass zur analytischen Zahlentheorie, *Quellen und Studien zur Geschichte der Mathematik, Astronomie und Physik*, Abt. B: Studien 2: 45–80.

Siegel, Carl Ludwig: Zur Geschichte des Frankfurter Mathematischen Seminars (Vortrag von Carl Ludwig Siegel am 13. Juni 1964 im Mathematischen Seminar anläßlich der Fünfzig-Jahrfeier der Johann-Wolfgang-Goethe-Universität Frankfurt), *Frankfurter Universitätsreden*, N.F. 36, Frankfurt: Klostermann.

Siegel, Carl Ludwig: On the History of the Frankfurt Mathematics Seminar, *Mathematical Intelligencer* 1(4): 223–230.

Siegmund-Schultze, Reinhard: *Rockefeller and the Internationalization of Mathematics Between the Two World Wars*, Basel: Birkhäuser.

Tobies, Renate: Carl Siegel zum 100. Geburtstag, *Mitteilungen der DMV*, 4: 29–34.

van Dalen, Dirk: *L.E.J. Brouwer–Topologist, Intuitionist, Philosopher. How Mathematics is Rooted in Life*, London: Springer.

Weil, André: *The Apprenticeship of a Mathematician*, Basel: Birkhäuser.

Weil, André: *Number theory, an Approach through History from Hammurapi to Legendre*, Basel: Birkhäuser.

Yandell, Ben H.: *The Honors Class. Hilbert's Problems and their Solvers*, Natick, Mass.: AK Peters.

CHAPTER 6

Three Students of Max Dehn

John Stillwell

Introduction

Max Dehn's work began in geometry, as a student of Hilbert during Hilbert's geometric period in the 1890s. His greatest success in this field was his negative solution of Hilbert's third problem [Dehn, 1900], obtained by showing that the regular tetrahedron is not equidecomposable with the cube (see Chapters 2 and 3). Around 1904 he carried Hilbert's axiomatic approach to topology, co-authoring a 1907 article on the subject with Poul Heegaard for Klein's *Encyklopädie der mathematischen Wissenschaften*. The article contained a rigorous proof of the classification theorem for surfaces on the basis of combinatorial axioms for topology, but was otherwise not very influential, probably because combinatorics was not by itself a fruitful idea in topology.

Dehn's research in topology took off when he encountered *group theory* around 1910, and applied it to knot theory and 3-manifolds in his paper [Dehn, 1910]. This paper also contains the germ of *geometric* group theory, in a section where Dehn applies hyperbolic geometry to construct what he called the *group diagram* of one of his 3-manifold groups. The diagram involves a tessellation of the plane by 9-gons, five of which meet at each vertex, which can be accomplished for regular 9-gons only in the hyperbolic plane.

In [Dehn, 1912] the union of topology, group theory, and hyperbolic geometry came into full bloom. The connection between hyperbolic geometry and surfaces of genus greater than one had been noticed in the 1880s by Klein and Poincaré, and [Poincare, 1904] made a sophisticated application of this connection to the homotopy of curves on such surfaces. But Dehn [Dehn, 1912] applied hyperbolic geometry for the first time to justify a purely combinatorial result about groups: the famous *Dehn's algorithm* for the solution of the word problem for fundamental groups of surfaces. A picture which encapsulates the relationship between geometry, topology, and group theory is shown in Figure 1. (See also Chapters 7 and 8.)

The picture shows a tessellation of the hyperbolic plane by congruent octagons, eight of which meet at each vertex. On the one hand, this tessellation is the *universal covering* of a genus 2 surface. Each octagon is a the result of cutting the surface along four closed curves a, b, c, d so as make it simply-connected, and passing across an edge from one octagon to another corresponds to travelling across a curve on the surface. On the other hand, the network of edges, if labelled by names for the corresponding curves on the surface, is the *group diagram* of the fundamental group of the surface, and travelling along an edge from one end to the other corresponds

©2024 American Mathematical Society

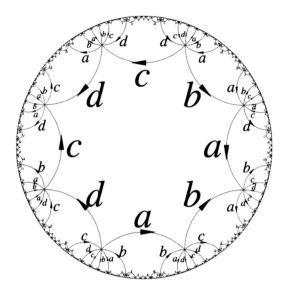

Figure 1. Tessellation for the genus 2 surface

to completing a loop on the surface. More generally, travelling around a closed path of edges correponds to completing a path on the surface that is not only closed, but *contractible to a point*. In particular, the sequence of labelled edges in the circumference of any octagon, when equated to 1, is the *defining relation* $aba^{-1}b^{-1}cdc^{-1}d^{-1} = 1$ of the fundamental group, since any closed path in the group diagram is composed from paths around octagons.

Dehn's ideas began to spread when Jakob Nielsen met Dehn as a student in Kiel in 1913. Dehn's influence can be seen in the last section of Nielsen's thesis, [Nielsen, 1913], and in almost everything that Nielsen wrote thereafter, as will be explained in more detail in the next section.

After periods of military service during the first world war – Dehn in the German army and Nielsen in the German navy – the two met again in Breslau (now Wrocław in Poland). Between them they discovered an early form of the *Dehn-Nielsen theorem*, which Nielsen published in full generality in [Nielsen, 1927]. The theorem establishes an elegant relationship between automorphisms of groups and homeomorphisms of manifolds, in the case where the manifold is a compact orientable surface S and the group is the fundamental group of S, $\pi_1(S)$. In one direction the relationship is clear: a homeomorphism f of S induces an automorphism of $\pi_1(S)$ by mapping the generating curves of $\pi_1(S)$ to another set of generating curves. Indeed, any *deformation* of the map f (that is, any homeomorphism in the same *mapping class*) induces the same automorphism.

The opposite direction of the relationship is less obvious: finding a homeomorphism of S that realizes a given automorphism of $\pi_1(S)$. To do this Dehn and Nielsen studied the large-scale nature of the automorphism, viewing it first as a map of the group diagram of $\pi_1(S)$ on the hyperbolic plane, then reinterpreting this as a map of the universal covering of S, and finally as a map of S itself.

After Breslau, the paths of Dehn and Nielsen diverged, with Dehn moving to Frankfurt and Nielsen to Copenhagen. But both of them continued to work on the topology of surfaces and the related theory of groups. For Dehn, this was one of several interests, but for Nielsen it was essentially his life's work, as we will see below.

Dehn was at the University of Frankfurt from 1922 from 1935, a particularly happy period of his career that has been described by [Siegel, 1965] and [Magnus, 1974b] (see also Chapter 5.). During this time Dehn was a fruitful source of ideas in geometry, topology, and group theory – ideas which he rarely published himself but generously allowed his students to develop. As mentioned above, Jakob Nielsen was one of these students, who had already embarked on his own career in the 1920s. The other two students who brought ideas of Dehn to fruition were Wilhelm Magnus and Ruth Moufang.

In 1928 Dehn set Magnus the problem of proving the *Freiheitssatz*, the claim that in a group with one defining relation, involving one particular generator in an essential way, the subgroup generated by the remaining generators is free. For example, in the group above with defining relation $aba^{-1}b^{-1}cdc^{-1}d^{-1} = 1$, the subgroup generated by b, c, d is free. As Magnus reported in [Magnus, 1978], Dehn outlined a proof using the diagram of the group, but was disappointed when Magnus used purely algebraic methods in his eventual proof, published in [Magnus, 1930]. Dehn evidently had a rare intuition for the geometry of groups, because later proofs of the *Freiheitssatz*, beginning with [Reidemeister, 1932], are also heavily algebraic.

Ruth Moufang was a contemporary of Magnus in Frankfurt, studying geometry with Dehn between 1925 and 1930. Her work picked up a loose thread from Hilbert's work in the 1890s, concerning the relation between algebra and the foundations of projective geometry. This thread began with work of von Staudt, who in 1847 observed that the operations of addition and multiplication can be defined geometrically, given the presence of certain projective properties, notably the *Pappus theorem*. If this theorem, stating the collinearity of certain points in a configuration of points and lines, is taken as an axiom (together with certain "obvious" geometric axioms, such as the existence of a line through any two points), then the addition and multiplication operations have the *algebraic* properties that define a field. Conversely, the presence of these algebraic properties allow *coordinates* to be introduced, so that the Pappus theorem can be proved by linear algebra.

As a byproduct of his research, Hilbert discovered that the somewhat weaker *Desargues theorem* is equivalent to the existence a weaker kind of coordinatization, by elements of a so-called *skew field*: that is, a system satisfying all the field properties except commutative multiplication. Moufang pursued this thread to its logical end, finding a projective axiom equivalent to the so-called *alternative field* properties, which include all the field properties except commutative and associative multiplication. For more on this, see the section on Moufang below.

Although much of Dehn's influence after 1920 was through his students, Dehn published one more major paper: [Dehn, 1938]. This paper sums up Dehn's later ideas on surface mappings, which had diverged from those of Nielsen since their work together on the Dehn-Nielsen theorem in the early 1920s. While Nielsen had pushed methods from hyperbolic geometry to new heights, Dehn dropped hyperbolic geometry in favor of a purely topological approach, generating homeomorphisms by what we now call *Dehn twists*. In the gathering storm of World War II, neither approach attracted much attention, and many results of Dehn and Nielsen remained unknown to English-speaking topologists until they were rediscovered in the 1960s and 1970s.

The twist generators of surface homeomophisms were rediscovered in [Lickorish, 1963] and several of Nielsen's results were rediscovered by Thurston in the 1970s, by geometric methods that went even further than Nielsen's – even bringing Dehn's twists within their scope. Thurston's work was published in [Fathi, Laudenbach and Poenaru, 1979] and brought up-to-date in [Farb and Margalit, 2012].

Jakob Nielsen

Jakob Nielsen was born in a region which is now in southern Denmark, but which was from 1866 until 1920 a part of Germany. This is why his university studies took place in Germany (in Kiel) and why he served in the German navy during World War I. His first two published papers have the address "Constantinople," where he was stationed at the time. Of his time in Turkey Nielsen later wrote:

> I walked about here as a stranger the first two years, but in the wonderful summer of 1917 and its echo in 1918 I came to love this country under the all-powerful sway of the sun, in the company of people I liked. I know that I shall often languish for the sun. I want to go there again some day.
>
> Quoted in [Fenchel, 1960, p. x].

Even after his home region was returned to Denmark in 1920, Nielsen spent some more time in Germany (at Breslau), before settling in Copenhagen for the rest of his career. By the early 1920s, Nielsen had decided that his research focus would be mappings of surfaces and he pursued it with singular depth and intensity for the rest of his life.

In the beginning Nielsen, like Dehn, was concerned as much with group theory as topology. The latter part of his 1913 dissertation, mentioned above, deals with homeomorphisms of the torus and the associated automorphisms of its fundamental group, the free abelian group on two generators. Since the group is just $\mathbb{Z}^2$, its automorphisms are easily understood – in fact they belong to the classical theory of the modular group – but [Nielsen, 1917] uncovered a surprising connection with the automorphism group of the *free* group on two generators, a much more complicated beast. This led Nielsen towards two profound results on free groups, published in [Nielsen, 1921] and [Nielsen, 1924].

The first, stating that any subgroup of a free group is free, is now known as the *Nielsen-Schreier theorem*. Nielsen proved it for finitely-generated free groups and Schreier [Schreier, 1927] subsequently proved it for arbitrary free groups. The theorem is related to the fundamental groups of graphs – a connection that was foreseen by Dehn and exploited by Schreier in his proof. Nielsen, however, proved the theorem by direct study of words in the generators of a free group, introducing in the process some generator transformations that came to be known as *Nielsen transformations*. These transformations, while not actually needed to prove the Nielsen-Schreier theorem, were shown by Nielsen to generate the automorphism group of any finitely-generated free group. This gave him a starting point for his formidably difficult determination of the defining relations for the automorphism group in [Nielsen, 1924]. The latter result was half a century ahead of its time, as is shown by the following quotation from [Magnus and Chandler, 1982]:

> An unsystematic poll taken by Magnus in 1970 seems to indicate that for about a decade after the death of Nielsen in 1959 there existed no living mathematician who had read Nielsen's paper in detail or would have been able to derive his result.[1]

In [Nielsen, 1924a] he derived another deep result from his study of free group automorphisms, generators and relations for the 3-dimensional analogue of the modular group. After that it was back to his main agenda: the study of surface mappings.

This began with the massive paper [Nielsen, 1927], which painstakingly laid the geometric foundations for Nielsen's investigation of surface mappings, starting with a careful construction of the fundamental octagon (Figure 2) for the universal covering of the genus 2 surface shown in Figure 1. The 1927 paper included his general proof of the Dehn-Nielsen theorem that any automorphism of $\pi_1(S)$, for an orientable surface S, can be induced by a homeomorphism of S. As mentioned above, Dehn's idea for proving this theorem was to view the automorphism of $\pi_1(S)$ as a map of the group diagram of $\pi_1(S)$, which lies in the hyperbolic plane, and to reinterpret the hyperbolic plane as the universal cover of S, thereby obtaining a homeomorphism of S.

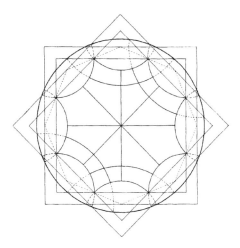

FIGURE 2. Nielsen's construction of the octagon

A key idea of the proof is to view the map of the group diagram *at infinity*; that is, on the circle that bounds the disk model of the hyperbolic plane. The view at infinity conveniently overlooks local topological variations and captures precisely the homotopy class of the map of S. This idea first emerged in [Poincare, 1904], where Poincaré observed that lifting a noncontractible curve k on S to the universal cover gives an infinite curve $\tilde{k}$ whose ends at infinity are unchanged by deformations of k. It follows, by considering all possible curves on the surface S, that a homeomorphism f of S induces a homeomorphism f_∞ of the circle at infinity depending only on the homotopy class of f. In all his subsequent work, Nielsen exploited this view at infinity for all it was worth.

---

[1] Actually, Magnus seems to have missed one eminent group theorist, Bernhard Neumann, who certainly studied Nielsen's paper enough to make some minor simplifications. This is reported by [Neumann, 1974], though Neumann describes his own contribution as "piffling."

The culmination of his research on surface mappings and automorphisms is in [Nielsen, 1943], where he proves that an automorphism of finite order can be induced by a map of finite order. To appreciate the depth of this result, one needs to appreciate the difference between realizing *one* automorphism of $\pi_1(S)$ by a map of S, and realizing a *group* of automorphisms by a group of maps. The Dehn-Nielsen theorem says that for each automorphism φ of $\pi_1(S)$ there is a homeomorphism f of S that induces φ. More precisely, the images $f(k)$ of the generating curves k of $\pi_1(S)$ are *homotopic to* the curves $\varphi(k)$ when the curves k are viewed as generators of $\pi_1(S)$ (such as the curves a, b, c, d in Figure 1). Thus if φ^n is the identity map of $\pi_1(S)$ for some n one only knows that each curve $f^n(k)$ is homotopic to k – the map f^n is not necessarily the identity map on S. The work of [Nielsen, 1943] is to construct an f such that f^n is indeed the identity. In so doing Nielsen showed that *a finite cyclic group of automorphisms of $\pi_1(S)$ can be realized by a finite cyclic group of mappings*.

For decades, this theorem of Nielsen was the high-water mark in realizing groups of automorphisms by groups of mappings. The problem of realizing an arbitrary finite group of automorphisms by a finite group of mappings became known as the *Nielsen realization problem*, and it was eventually solved in [Kerckhoff, 1983], using results of Thurston that were expounded in [Fathi, Laudenbach and Poenaru, 1979]. This raised the question of whether the ultimate result in this direction is true: can the group of *all* automorphisms of $\pi_1(S)$ be realized by a group of mappings of S? Alas, no. While realization is easy for genus = 1 (the torus), a counterexample for S of genus ≥ 5 was given in [Markovic, 2007], and Markovic later extended this result to any genus ≥ 2. (Despite the apparent depth of this result, the work of Markovic seems to have suffered the fate of many difficult counterexamples: it is not well known.)

The lengthy interval between the work of Nielsen (and also Dehn) and its full flowering in the work of Thurston and his successors may be due to Dehn and Nielsen being ahead of their time in the use of geometric methods in topology. But surely it was also due to ignorance of the German literature on the part of English-speaking mathematicians. It was not until Thurston had rediscovered some of Nielsen's and Dehn's results in the 1970s that their work became well known and properly appreciated. The classification of surface mappings discovered by Thurston is now known as the Nielsen-Thurston classification, since re-examination of the old literature has shown that it was partly known to Nielsen in the 1930s. Happily, Dehn and Nielsen now get their due in modern works on surface topology, such as [Farb and Margalit, 2012].

Wilhelm Magnus

As mentioned above, Magnus became Dehn's student at the University of Frankfurt in the late 1920s. This was a place and time that Magnus [Magnus, 1974a] later described as follows (paraphrasing Goethe's description of Baghdad in the time of Harun al-Rashid):

> Proverbially it was a time when, in a particular locality, all human endeavors interacted in such a fortunate way that the recurrence of a similar period could be expected only after many years and very different places under exceptionally favorable conditions.

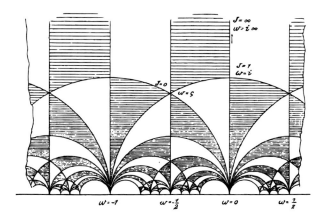

FIGURE 3. Tessellation for the modular function

Magnus, either through his contact with Dehn or by his intrinsic nature, became Dehn's natural heir. Like Dehn, he was a cultured man with a deep interest in philosophy and history, kind and generous to his students and colleagues. For the impression he made on one student read [Shenitzer, 1995]. Magnus also worked to preserve Dehn's legacy, keeping Dehn's *Nachlass* at his home in New Rochelle for many years before donating it to the Briscoe Center in Texas in the 1980s. It was at the Magnus home in 1980 that I saw Dehn's unpublished papers, which led me to make the translations of them that were eventually published in [Dehn, 1987].

The mathematical work of Magnus was remarkably varied. It included well-respected work on the special functions of applied mathematics, carried out between 1939 and 1949 after Magnus lost his academic position in Germany because of his refusal to join the Nazi party. However, Magnus is best known for his work on combinatorial group theory, which grew from questions on groups with one defining relation, posed to him by Dehn in 1928.

Combinatorial group theory studies groups from the viewpoint of *generators* (elements whose products include all members of the group) and *defining relations* (equations between particular products of generators that imply *all* equations between products of generators). Most of the motivating examples for combinatorial group theory were discovered by Klein and Poincaré in the 1880s, in their investigations of the non-Euclidean periodicity of automorphic functions, such as the modular function. For these examples, generators and defining relations can be read off a *non-Euclidean tessellation* that displays the periodicity of the function in question. Figure 3 (taken from [Klein and Fricke, 1965, p. 113]) is the tessellation that displays the periodicity of the modular function on the upper half plane of complex numbers z. The half-plane is divided into a pattern of non-Euclidean triangles that repeats when z is replaced by $z + 1$ and also when z is replaced by $-1/z$. These are the transformations that generate the modular group.

Klein and Poincaré were aware that certain non-Euclidean tessellations, such as the one in Figure 1, are related to surfaces of genus ≥ 2, and they could write down a defining relation of the corresponding group. As mentioned above, a defining relation for the genus 2 surface group is $aba^{-1}b^{-1}cdc^{-1}d^{-1} = 1$. However, it was Dehn [Dehn, 1912] who first posed the fundamental problem for such groups, and solved it for surface groups: the *word problem*. This is the problem of deciding, for

Figure 4. A braid with three threads

a given product of generators (or "word") w, whether $w = 1$. As Dehn pointed out, the word problem for the group of a surface S is equivalent to the problem of deciding, for a given curve w on S (given as a product of generating curves), whether w is contractible to point by deformation within the surface.

After Dehn's solution of the word problem for surface groups, little progress was made, even for groups with just one defining relation ("one-relator groups"). Dehn believed he could prove something general about one-relator groups – the *Freiheitssatz* – but the word problem defeated him. Thus Magnus made a major breakthrough when he not only proved the *Freiheitssatz* in [Magnus, 1930], but also solved the word problem for one-relator groups in [Magnus, 1932]. To this day, it is not known whether the word problem is solvable for all groups with as few as two defining relations. For certain groups with a large, but finite, number of defining relations the word problem is known to be *unsolvable*. This was first proved by Novikov [Novikov, 1955], shortly after Dehn's death (though he had announced the result some years earlier, so Dehn may have known about it).

In these early publications, Magnus solved problems of Dehn by algebraic methods that Dehn himself would probably not have used. His most topological paper was [Magnus, 1934], in which he applied group theory in an unexpected way to combine two of Dehn's interests: surface mappings and *knots*. Dehn made important contributions to knot theory in [Dehn, 1910] and [Dehn, 1914], but by the 1920s he had left the field to others – notably Alexander in the USA and Artin, Schreier, and Reidemeister in Germany. [Alexander, 1923] and [Artin, 1925] related knots to the more manageable objects called *braids*. Like knots and links, braids involve threads that pass under or over each other, but the threads in a braid are curve segments that are essentially parallel line segments, except at crossings. See Figure 4, which shows a braid with three threads. Braids are particularly attractive algebraically because there is a natural "product" of braids with the same number n of threads, under which they form a group B_n, and the word problem for B_n is solvable, as [Artin, 1925] showed.

The basic idea of [Magnus, 1934] is to study mappings between surfaces by connecting the points "before" and "after" the mapping by threads between two copies of the surface, as in Figure 5 (where the two copies of the surface are seen in cross section, as circles), which is from Magnus's paper. Like the theory of surface mappings, the theory of braids after the 1930s remained dormant until the 1960s. Magnus played a part in its awakening, through his student Joan Birman. Her monograph [Birman, 1974] was the first book-length treatment of the subject, and it arrived at the time when low-dimensional topology in general was awakening through the work of Thurston. However, braid theory received its biggest vindication in the early 1980s, when Vaughan Jones noticed what seemed to be braid group relations in the theory of operator algebras – a subject seemingly remote

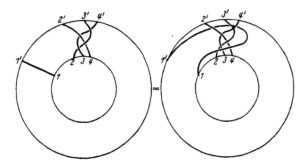

FIGURE 5. Encoding a mapping by braids

from topology. He asked Joan Birman about this and their discussions gave birth to a powerful new tool in knot theory, the *Jones polynomial*.

Among the many benefits conferred by the Jones polynomial was a new and simple proof of a theorem first proved by [Dehn, 1914]: that the left trefoil knot is not deformable into the right trefoil knot.

In the latter part of his career, Magnus helped to put combinatorial group theory on the mathematical map with the first textbook on the subject [Magnus, Karrass, and Solitar, 1966]. He also paid homage to the geometric origins of the subject (and to the magnificent illustrations in sources such as [Klein and Fricke, 1965]) with his book on non-Euclidean tessellations [Magnus, 1974a], and he coauthored the first book on its history [Magnus and Chandler, 1982].

Ruth Moufang

The work of Hilbert [Hilbert, 1899], developed at the time when Dehn was his student, included the discovery of a surprising relationship between the algebraic properties of fields and the presence of certain configurations in the projective plane. The two configurations studied by Hilbert were those named after Pappus and Desargues.

Figure 6 shows the Pappus configuration. It reflects the *Pappus theorem* stating: if a hexagon is formed from six points lying alternately on two lines (the lines through O in the picture) then the intersections of pairs of opposite sides lie on a line. In the picture the three pairs of opposite sides are colored black, gray, and dashed respectively, and they are shown meeting on a line $\mathcal{L}$ that looks like a "horizon," so that the three pairs look like "parallels." When viewed in this way, the Pappus theorem becomes the easily proved Euclidean theorem that if two pairs of opposite sides are parallel, so is the third pair.

It is useful to view lines as "parallel" if they meet on a designated "horizon line" $\mathcal{L}$ because there are natural ways (using parallels) to define "sum" and "product" of points a, b on a given line. The Pappus theorem then implies the *commutative laws* $a + b = b + a$ and $ab = ba$ because the constructions of $a + b$ and $b + a$ lead to the same point, and the constructions of ab and ba lead to the same point.

Figure 7 shows the Desargues configuration. It reflects the *Desargues theorem* stating: if two triangles are in perspective (from the point P in the picture), then the intersections of corresponding sides meet on a line. Again, corresponding sides are colored alike, and the line $\mathcal{L}$ where they meet is drawn to look like the horizon.

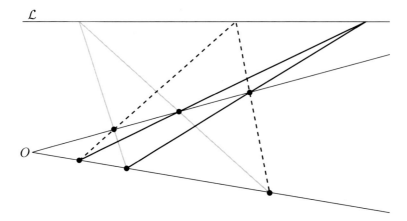

FIGURE 6. The Pappus configuration

Given the definitions of sum and product of points, the Desargues theorem makes it easy to prove the *associative laws* $a + (b + c) = (a + b) + c$ and $a(bc) = (ab)c$, and it was used in [Hilbert, 1899] for that purpose. Indeed, Hilbert showed that the Pappus and Desargues theorems together imply all the field properties, but Desargues alone does not, because it does not imply commutative multiplication. Rather surprisingly (considering that the Desargues and Pappus theorems have been known for hundreds of years), Hessenberg [Hessenberg, 1905] proved that Pappus implies Desargues, so Pappus alone implies all the field properties, and Desargues alone implies all field properties except commutative multiplication. A structure with all the field properties except commutative multiplication is called a *skew* field (rather misleadingly, since a skew field is not a specialization of the field concept, but a generalization).

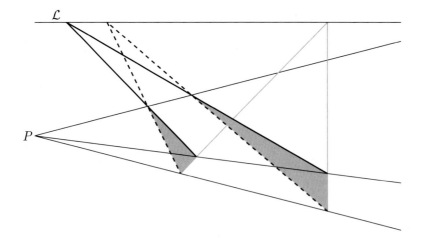

FIGURE 7. The Desargues configuration

This was the state of affairs in the 1920s when Dehn proposed the problem of investigating whether a weaker configuration theorem implies a system with weaker (but still interesting) algebraic properties. Moufang showed [Moufang, 1931] that the answer is yes, if one takes a special case of the Desargues configuration. Her proof contained slight errors which were later repaired by taking a slightly different configuration, now known as the little Desargues configuration (Figure 8). This configuration reflects the special case of the Desargues theorem in which the point P lies on the line $\mathcal{L}$.

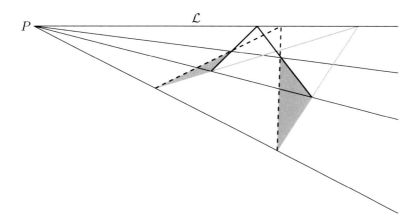

FIGURE 8. The little Desargues configuration

Moufang essentially proved that the little Desargues theorem implies all the field properties except commutative *and associative* multiplication. However, little Desargues implies a special case of associativity called *cancellation* or *alternativity*: $a^{-1}(ab) = b = (ba)a^{-1}$. A structure with these properties is called an *alternative field* (committing the same kind of terminological error as "skew field"). Dehn [Dehn, 1931] praised Moufang's discovery highly, because he viewed it as the natural end of Hilbert's investigations on the connections between projective geometry and algebra. This view was echoed in [Freudenthal, 1957], who said (my translation):

> To Hilbert, the various geometries were only a means to an end (showing the independence of the axioms), and they were all more or less pathological. He mentions neither the complex or the quaternionic geometry, which is the most beautiful example of a non-Pappus geometry ... Ruth Moufang found a new road ... She recognized the significance of an incidence theorem weaker than Desargues' theorem ... The weakening of Desarguesian ... geometry corresponds algebraically to the weakening of a skew field to an alternative field.

As Moufang realized, there are natural examples of fields, and skew and alternative fields, which serve as coordinate systems for projective planes satisfying the Pappus, Desargues, and little Desargues theorems, respectively. Namely, $\mathbb{R}$ and $\mathbb{C}$ are fields, which define the classical real and complex projective planes. The quaternions $\mathbb{H}$ are a skew field, which gives rise to the *quaternion plane*, satisfying the Desargues theorem but not Pappus. Finally, the octonions $\mathbb{O}$ are an alternative field, which gives rise to the *octonion plane*, satisfying the little Desargues theorem but not Desargues. In unpublished notes on geometry, written for a course in

1948,[2] Moufang developed some of the algebra of quaternions and octonions in parallel with the corresponding theory of projective planes. The octonions are indeed the natural end of this train of thought, since there are several theorems saying that $\mathbb{R}, \mathbb{C}, \mathbb{H}$, and $\mathbb{O}$ are the only this-or-that over the real numbers (for example, the only normed division rings).

While Moufang's results were the end of a certain kind of geometric research, they were just the beginning of others. The geometric and algebraic properties she brought to light have since been found useful for many investigations in the border zone between algebra and geometry. Today, her name occurs in Moufang planes, Moufang loops, Moufang buildings, and Moufang polygons. The latter (which include Mougang planes as the special case of Moufang 3-gons) is the subject of the deep monograph of Tits and Weiss [Tits and Weiss, 2002].

Acknowledgments

My interest in Max Dehn began when I read his inspiring paper [Dehn, 1912] around 1975. It introduced me to a new world where topology, group theory, and hyperbolic geometry all live together in harmony. As I explored further, I realized that this was a world quite different from the Cambridge-influenced analysis I had learned as an undergraduate, or the abstract algebra and algebraic topology that had discouraged me as a graduate student. Fortunately, I had learned German in high school, so I was able to explore Dehn's world by translating many of the foundational papers in geometric topology and combinatorial group theory published between 1900 and 1940.

In the process, I received guidance and encouragement from several mathematicians who were on the scene in Germany during that period, or their students. These included Wilhelm Magnus and his students Joan Birman and Abe Shenitzer, Bernhard Neumann, and Kurt Reidemeister's students Heiner Zieschang and Gerhard Burde.

I also thank Werner Fenchel, Børge Jessen, and Vagn Lundsgaard Hansen for discussions when I visited Copenhagen in 1983, and the Danish Mathematical Society for their generous support for my translations of Nielsen's works that were published in [Nielsen, 1986].

References

Alexander, J. W. (1923). A lemma on systems of knotted curves. Nat. Acad. Proc. 9, 93-95 (1923).

Artin, E. (1925). Theorie der Zöpfe. *Abh. Math. Semin. Univ. Hamb. 4*, 47-72.

Birman, J. S. (1974). *Braids, Links, and Mapping Class Groups*. Princeton University Press, Princeton, N.J.; University of Tokyo Press, Tokyo. Annals of Mathematics Studies, No. 82.

Chandler, B. and W. Magnus (1982). *The History of Combinatorial Group Theory*, Volume 9 of *Studies in the History of Mathematics and Physical Sciences*. Springer-Verlag, New York. A case study in the history of ideas.

Dehn, M. (1900). Über raumgleiche Polyeder. *Gött. Nachr. 1900*, 345-354.

Dehn, M. (1910). Über die Topologie des dreidimensionalen Raumes. *Math. Ann. 69*, 137-168.

---

[2]I found these notes in the mathematics department library at the University of Frankfurt in 2002. Thanks go to Gerhard Burde for kindly supplying me with a photocopy.

Dehn, M. (1912). Über unendliche diskontinuierliche Gruppen. *Math. Ann. 71*, 116–144. English translation in [Dehn, 1987].

Dehn, M. (1914). Die beiden Kleeblattschlingen. *Math. Ann. 75*, 402–413. English translation in [Dehn, 1987].

Dehn, M. (1931). Über einige Forschungen in den Grundlagen der Geometrie. *Matematisk Tidsskrift B*, 63–83.

Dehn, M. (1938). Die Gruppe der Abbildungsklassen. (Das arithmetische Feld auf Flächen.). *Acta Math. 69*, 135–206.

Dehn, M. (1987). *Papers on Group Theory and Topology*. New York: Springer-Verlag. Translated from the German and with introductions and an appendix by John Stillwell, and with an appendix by Otto Schreier.

Farb, B. and D. Margalit (2012). *A Primer on Mapping Class Groups*, Volume 49 of *Princeton Mathematical Series*. Princeton University Press, Princeton, NJ.

Fathi, A., F. Laudenbach, and V. Poénaru (1979). *Travaux de Thurston sur les surfaces*, Volume 66 of *Astérisque*. Société Mathématique de France, Paris. Séminaire Orsay, With an English summary. English translation in Fathi, Laudenbach, and Poenaru (2012).

Fenchel, W. (1960). Jakob Nielsen in Memoriam. *Acta Math. 103*, vii–xix.

Freudenthal, H. (1957). Zur Geschichte der Grundlagen der Geometrie. Zugleich eine Besprechung der 8. Aufl. von Hilberts "Grundlagen der Geometrie". *Nieuw Arch. Wisk. (3) 5*, 105–142.

Hessenberg, G. (1905). Beweis des *Desargues*schen Satzes aus dem *Pascal*schen. *Mathematische Annalen 61*, 161–172.

Hilbert, D. (1899). *Grundlagen der Geometrie*. Leipzig: Teubner. English translation: *Foundations of Geometry*, Open Court, Chicago, 1971.

Kerckhoff, S. P. (1983). The Nielsen realization problem. *Ann. of Math. (2) 117*(2), 235–265.

Klein, F. and R. Fricke (1890). *Vorlesungen über die Theorie der elliptischen Modulfunctionen, Vol. 1*. Teubner, Leipzig. Reprinted by Johnson Reprint Co., New York, 1965.

Lickorish, W. B. R. (1964). A finite set of generators for the homeotopy group of a 2-manifold. *Proc. Cambridge Philos. Soc. 60*, 769–778.

Magnus, W. (1930). Über diskontinuierliche Gruppen mit einer definierenden Relation. (Der Freiheitssatz). *J. Reine Angew. Math. 163*, 141–165.

Magnus, W. (1932). Das Identitätsproblem für Gruppen mit einer definierenden Relation. *Math. Ann. 106*, 295–307.

Magnus, W. (1934). Über Automorphismen von Fundamentalgruppen berandeter Flächen. *Math. Ann. 109*(1), 617–646.

Magnus, W. (1974a). *Noneuclidean Tesselations and their Groups*. Academic Press [A subsidiary of Harcourt Brace Jovanovich, Publishers], New York-London. Pure and Applied Mathematics, Vol. 61.

Magnus, W. (1974b). Vignette of a cultural episode. In *Studies in Numerical Analysis (papers in honour of Cornelius Lanczos on the occasion of his 80th birthday)*, pp. 7–13. Academic Press, London.

Magnus, W. (1978). Max Dehn. *Math. Intelligencer 1*(3), 132–143.

Magnus, W., A. Karrass, and D. Solitar (1966). *Combinatorial Group Theory: Presentations of Groups in Terms of Generators and Relations*. Interscience Publishers [John Wiley & Sons, Inc.], New York-London-Sydney.

Markovic, V. (2007). Realization of the mapping class group by homeomorphisms. *Invent. Math. 168*(3), 523–566.

Moufang, R. (1931). Die Einführung der idealen Elemente in die ebene Geometrie mit Hilfe des Satzes vom vollständigen Vierseit. *Math. Ann. 105*(1), 759–778.

Neumann, B. H. (1974). Some groups I have known. In *Proceedings of the Second International Conference on the Theory of Groups (Australian Nat. Univ., Canberra, 1973)*, pp. 516–519. Lecture Notes in Math., Vol. 372. Springer, Berlin.

Nielsen, J. (1913). *Kurvennetzen auf Flächen*. Kiel Inaugural Dissertation, in [Nielsen, 1986], pp. 14–75.

Nielsen, J. (1917). Die Isomorphismen der allgemeinen, unendlichen Gruppe mit zwei Erzeugenden. *Math. Ann. 78*, 385–397.

Nielsen, J. (1921). Om Regning med ikke-kommutative Faktorer og dens Anvendelse i Gruppeteorien. *Matematisk Tidsskrift B*, 77–94. English translation by Anne W. Neumann in [Nielsen, 1986], pp. 117–129.

Nielsen, J. (1924a). Die Gruppe der dreidimensionalen Gittertransformationen. Danske Vidensk. Selskabs Math.-fys. Meddelelser 5, Nr. 12, S. 3- 29 (1924).

Nielsen, J. (1924b). Die Isomorphismengruppe der freien Gruppen. *Math. Ann. 91*, 169–209.

Nielsen, J. (1927). Untersuchungen zur Topologie der geschlossenen zweiseitigen Flächen. *Acta Math. 50*, 189–358. English translation by John Stillwell in [Nielsen, 1986], pp. 223–341.

Nielsen, J. (1943). Abbildungsklassen endlicher Ordnung. *Acta Math. 75*, 23–115.

Nielsen, J. (1986). *Jakob Nielsen: Collected Mathematical Papers. Vol. 1.* Contemporary Mathematicians. Birkhäuser Boston, Inc., Boston, MA. Edited and with a preface by Vagn Lundsgaard Hansen.

Novikov, P. S. (1955). On the algorithmic unsolvability of the word problem in group theory (Russian). *Proceedings of the Steklov Institute of Mathematics 44*. English translation in *American Mathematical Society Translations* ser. 2, *9*, 1–122.

Poincaré, H. (1904). Cinquième complément à l'analysis situs. *Rendiconti del Circolo matematico di Palermo 18*, 45–110. English translation in [Poincaré 2010].

Reidemeister, K. (1932). Einführung in die kombinatorische Topologie. Braunschweig: Friedr. Vieweg & Sohn A.-G. XII, 209 S. (1932).

Schreier, O. (1927). Die Untergruppen der freien Gruppen. *Abh. Math. Semin. Univ. Hamb. 5*, 161–183.

Shenitzer, A. (1995). In memory of my friend Wilhelm Magnus. *Math. Intell. 17*(2), 63–64.

Siegel, C. L. (1965). *Zur Geschichte des Frankfurter mathematischen Seminars*. In Siegel's *Gesammelte Abhandlungen*, Vol. 3, Springer, Berlin. English translation in *Mathematical Intelligencer*, 1, pp. 223–230.

Tits, J. and R. M. Weiss (2002). *Moufang Polygons*. Springer Monographs in Mathematics. Springer-Verlag, Berlin.

CHAPTER 7

Max Dehn and the Word Problem

James W. Cannon

1. Introduction

Max Dehn introduced the basic structures considered in combinatorial and geometric group theory, set out an agenda for studying them, and solved the associated problems (the Word Problem, the Conjugacy Problem, and the Isomorphism Problem) in special cases important in low-dimensional topology. The original papers are [Dehn, 1910, 1912(1), 1912(2)], written in German; and they appear in English translation in [Dehn, 1987]. We will follow one of Dehn's structures, namely the *Dehn Gruppenbild*, and one problem, *Dehn's Identity or Word Problem*, through the decades to see how they have spawned a very geometric view of infinite group theory, with expanding subfields involving the *automatic groups* of Thurston [Ep, 1992]; the *word hyperbolic groups* of Gromov [Gromov, 1987, 1981]; and *CAT-0 groups* [Bridson, 1999]. The applications are central in all of 2 and 3-dimensional topology and complex variables, and they extend to a lesser degree into many other fields of mathematics.

Although Novikov [Novikov, 1955] and Boone [Boone, 1959, 1966; Britton, 1963] have shown that there are groups for which Dehn's fundamental problems are unsolvable, more recent work shows that, in a strict statistical sense, Dehn's word and conjugacy problems are solvable in almost every finitely presented group and, in fact, the basic ideas of Dehn's original algorithms are valid there.

2. The Dehn Gruppenbild

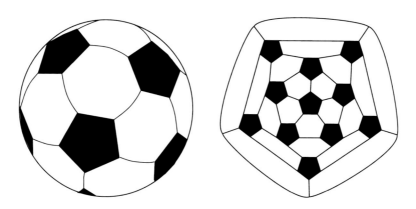

FIGURE 1. A soccerball and the Gruppenbild: $<a,b : a^2, b^5, (ab)^3>$

©2024 American Mathematical Society

We call the collection of symmetries or rigid motions of an object X its *group* $G(X)$. Max Dehn (see Paper 1 in [Dehn, 1987]) showed us how to view this collection $G(X)$ as a single geometric object that he called a *Gruppen-Bild* or *group-picture*. The Gruppenbild (plural = Gruppenbilder) is often intricate and beautiful.

It implicitly gives a complete multiplication table for the group and may reveal a good deal of geometry associated with the group. For example in Figure 1, the Gruppenbild of the group with presentation $< a, b : a^2, b^5, (ab)^3 >$ consists of the vertices and edges of the historical soccer ball.

Arthur Cayley, earlier and independently [Cayley, 1878], defined the same object, hence often called the *Cayley graph* or *group diagram*; but he emphasized the picture of finite groups, while Dehn considered as well the graphs of the infinite groups arising in topology and geometry.

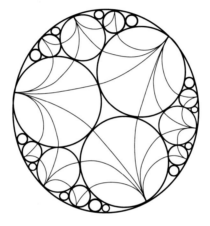

FIGURE 2. Triangle Tessellation

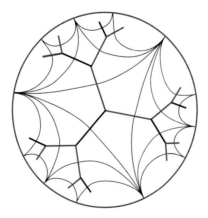

FIGURE 3. Gruppenbild: Dual Trihedral Graph $< a, b, c : a^2, b^2, c^2 >$

Given a tiling or tessellation of some geometry, such as the tiling of the non-Euclidean hyperbolic plane by ideal triangles (Figure 2) or the tiling of the Euclidean plane by squares (Figure 4), we obtain a Gruppenbild (Figure 3) or (Figure 4) as the graph dual to the tiling.

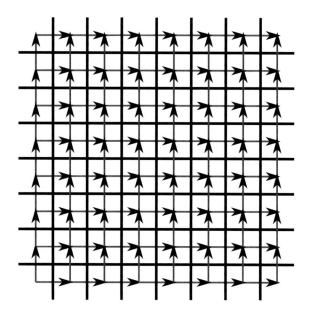

FIGURE 4. Square Tiling and Dual Gruppenbild: $< a, b : aba^{-1}b^{-1} >$

3. Defining the Gruppenbild

In combinatorial group theory, we consider a group $G =< C : R >$ defined by sets of *generators* C and *relators* R. Each element of R is a formal finite product of elements of C and their inverses. The group G is the largest group containing the elements of C and having two properties:

(1) Each element of G is a product of elements of C and their inverses.

(2) Each element of R, as a product of elements of C and their inverses, represents the identity element of G.

The associated *Gruppenbild* $\Gamma = \Gamma(C : R)$ is a graph whose vertices are the elements of G, and whose directed edges are triples $E = (v, c, w)$, with v and w denoting vertices of Γ and with label (or color) $c \in C \cup C^{-1}$ such that $v \cdot c = w$. The inverse of that directed edge is the triple $E^{-1} = (w, c^{-1}, v)$. Together, E and E^{-1} represent a single undirected edge joining v and w.

4. Constructing the Gruppenbild

Given a *finite presentation* $< C : R >$ (C and R being *finite* sets), a simple, though infinite, process can be used to construct the Gruppenbild $\Gamma(C : R)$.

The Beginning. Begin with a single vertex e representing the identity of the group $G =< C : R >$.

Then cycle through the following three steps. Each phase of the cycle will yield a graph, which then, in turn, will serve as input for the next phase or cycle. For a finite group, some unknown finite number of cycles will stabilize and yield the Gruppenbild. For an infinite group, larger and larger neighborhoods of the identity will stabilize so that the Gruppenbild will emerge as the infinite limit.

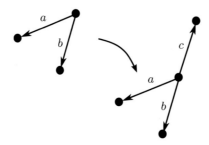

FIGURE 5. Expansion Phase: Adding a Missing Edge

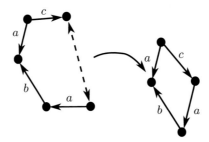

FIGURE 6. Relation Phase: Closing Up a Relator

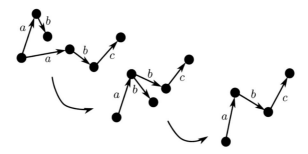

FIGURE 7. Collapse Phase: Successive Edge Collapses

Expansion Phase. For each vertex v and color $c \in C \cup C^{-1}$, if there is no edge of the form (v, c, w), create a new vertex w and new (inverse) edges (v, c, w) and (w, c^{-1}, v). (See Figure 5.)

Relation Phase. For each vertex v and relator $r \in R$, if there is an edge path $v : r$ beginning at v and labelled r, then identify the initial vertex v with the terminal vertex of $v : r$. (See Figure 6.)

Collapse Phase. For each vertex v and color c, if there are two edges (v, c, w) and (v, c, x) beginning at v and labelled by c, then identify the vertices w and x, and identify the edges (v, c, w) and (v, c, x). (Note that one identification may entail other identifications, even a large cascade of identifications.) (See Figure 7.)

5. Dehn's Fundamental Problems

The fundamental difficulty in constructing the Gruppenbild is that, at any given stage of the construction, we may not know how much of the Gruppenbild has stabilized and how much, in the future, will still collapse to a smaller subgraph. A solution to the *Word Problem* would allow us to know when a given vertex has stabilized. That is, the Word Problem is the essential difficulty in constructing and understanding at least finite portions of the Gruppenbild:

The Word or Identity Problem. Given a group $G = <C : R>$, is there a finite algorithm such that, for each word W in the alphabet $C \cup C^{-1}$, the algorithm will determine in finite time whether or not W represents the identity element $e \in G$.

The Conjugacy Problem. Given group $G = <C : R>$, is there a finite algorithm that, for each pair V and W of words in the alphabet $C \cup C^{-1}$, will determine in finite time whether or not there is a word X such that $W = X^{-1} \cdot V \cdot X$ in G. (In a rigid geometric object, V and W represent positions within the object, and the question is whether the two positions "look" the same in the object. In topology, where V and W may be represented by closed curves, the question is whether those two curves are homotopic.)

The Isomorphism Problem. Are groups $G_1 = <C_1 : R_1>$ and $G_2 = <C_2 : R_2>$ isomorphic?

6. Dehn's Solution to the Word Problem

Dehn solved his fundamental problems when they were restricted to the fundamental groups of compact, 2-dimensional surfaces, with their standard finite presentations. We will consider only the sphere with $n \geq 2$ handles and the word problem for its fundamental group

$$G = <a_1, b_1, \ldots, a_n, b_n : [a_1, b_1] \cdots [a_n, b_n]>,$$

where $[x, y] = x^{-1} \cdot y^{-1} \cdot x \cdot y$.

Dehn realized the Gruppenbild as the graph dual to a tessellation of the non-Euclidean hyperbolic plane by regular non-Euclidean polygons having $4n$ sides. An edge path in this graph is labelled by a word representing the identity element if and only if it is a closed path. Therefore, given a word W to be tested, we think of W as a circular word (closed path) which we can traverse beginning at any vertex, or even the inverse word obtained by traversing the path in the opposite direction. Likewise, we think of the relators (elements of R) as circular words since they represent the identity. We may add to R the trivial relators of the form $c \cdot c^{-1}$, for $c \in C \cup C^{-1}$, since such trivial cancellations always represent the identity.

We say that the (circular) word W *can be shortened* by the (expanded) relator set R provided W contains more than half of one of the (circular) relators of R. Thus, for example (see Figure 8), if $W = A \cdot Y$ and $r = A \cdot X$, and if A is longer than X, then we may shorten the word W to the shorter word $W' = X^{-1} \cdot Y$.

We can now describe Dehn's *greedy algorithm*.

Dehn's Greedy Algorithm. The word W represents the identity in G if and only if W can be reduced to the empty word by a finite number of shortenings.

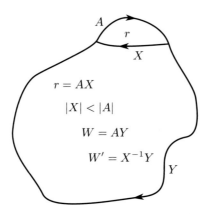

FIGURE 8. Shortening a Relator

7. Generalizing Dehn's Solution

Dehn at first used the properties of the hyperbolic metric in verifying that his greedy algorithm was effective. According to Lyndon and Schupp [Lyndon, 1977], "Reidemeister (1932) pointed out that Dehn's conclusion followed from the combinatorial properties of the tessellation, without metric considerations." The essential property was that, in the product $r_1 \cdot r_2$ of two relators in the system R expanded, there could be only small cancellation. Tartakovskii [Tartakovskii, 1949] undertook the study of relator systems in which products of defining relators admitted only small cancellation. The small cancellation conditions, as finally refined, were local and combinatorial, hence easily checked. Groups satisfying these conditions had solutions to the word problem that used greedy shortening arguments like those of Dehn. Among those contributing to the study were Britton, Schiek, and Greendlinger (Lyndon and Schupp, [Lyndon, 1977], p. 252). Lyndon [Lyndon, 1966] recognized the combinatorial conditions as essentially curvature conditions approximating the geometry of the Euclidean plane or of the non-Euclidean hyperbolic plane.

8. Logic and Unsolvability

Novikov [Novikov, 1955] and Boone [Boone, 1959, 1966], during the 1950s, shocked the world of combinatorial group theory by showing that there are groups for which the word problem is not solvable. A very nice exposition appears in [Rotman, 1973]. See also [Davis, 1965] and Chapter 8. This development required two advances in mathematical logic. The first was the formalization of the notion of algorithm. Algorithms were formalized in a number of ways during the 1930's by Church [Church, 1932, 1933, 1936], Turing [Turing, 1937, 1950, 1956], and Kleene [Kleene, 1936, 1980], with further formalizations later by Markov and Kolmogorov. Eventually, all of these approaches were proved equivalent.

The second advance was the proof that there are subsets of the positive integers that can be listed by finite algorithm but whose complement cannot be so listed. We say that a finite algorithm lists the set E of positive integers if, for each element $x \in E$, the algorithm stops in finite time when given initial input x, but for all other integers as initial input, the algorithm runs forever without stopping. Except for the phrase 'by finite algorithm', this result is not surprising. Consider an infinite sequence of drawings from the set of positive integers. It is an easy matter to record, one by one, the numbers drawn at the time they are drawn; but we will not know that a given number will never be drawn until the complete sequence of drawings is finished. That is, for a given number, it will be known in finite time that the number is drawn, if in fact it is to be drawn; but it may take the entire infinite sequence of drawings to know that a number is not to be drawn.

How might we find a set E of positive integers that can be listed by finite algorithm but whose complement E' cannot be listed by finite algorithm? Turing lists, by algorithm, all finite algorithms as a sequence of algorithms, $T_1, T_2, \ldots, T_n, \ldots$. Then, by finite algorithm, he runs all successive algorithms a bit at a time on successive positive integer inputs. If algorithm T_t stops in finite time on input t, then t is listed in set E. But consider the set E' of integers t that will not be listed by this algorithm, namely, the complementary set E'. An integer t lies in E' if algorithm T_t runs forever on input t without stopping. We may clearly assume the algorithms are listed in such an order that both E and E' are nonempty. We claim that no finite algorithm will list E'. Suppose to the contrary that algorithm T_t lists E'. The important question to ask is this: "Does T_t list t (stop on input t), or does T_t not list t (run forever on input t)?"

If T_t stops on input t, then $t \in E$ by definition of E. But $t \in E'$ since T_t lists precisely E', a contradiction.

If T_t runs forever on input t, then $t \in E'$ by definition of E'. But T_t must stop on every element of E' since T_t lists E', a contradiction.

We conclude that no finite algorithm lists E'.

How is such a set associated with unsolvable word problems? The construction of the Gruppenbild in a previous section is a finite algorithm that allows us to list algorithmically all of the finite words W that represent the identity of the group: If W has length n, then after n cycles of the algorithm, there will be an edge path $e : W$ beginning at the identity and labelled by the word W. Let v denote the terminal vertex of that path. We may then monitor the movement of v through the algorithm. If W represents the identity element of the group, v will after some unknown finite number of steps be identified with the vertex e. The solution of the word problem requires that we also be able to list those vertices v that are not ever identified with e. That is, we must be able to algorithmically list the complementary set.

Novikov based his work on the proof by Turing that there are semigroups with cancellation for which the identity problem is not solvable. Turing's idea is to encode into the structure of the semigroup a set of the type just described.

Novikov takes a semigroup S as described by Turing and shows how to embed it into a group G in such a manner that a solution to the word problem in G would solve the word problem in S.

9. The Geometric Structure of Groups

Progress in low-dimensional topology raised interest in groups as geometric objects. The rest of this paper will be influenced by, and limited by, my own experiences. In trying to fill a glaring gap in my knowledge of low-dimensional topology, I set out to learn about combinatorial group theory by reading Coxeter and Moser's "Generators and Relations for Discrete Groups" [Coxeter, 1972]. I was captivated by their description of the Cayley graph or Gruppenbild. I set out to construct explicitly the Gruppenbild of as many of the groups in the book as I could. I constructed hundreds of Gruppenbilder.

9.1. Thurston's Influence. Bill Thurston, in conversations at the International Congress of Mathematicians in Helsinki, Finland, 1978, advertised the following question:

> Suppose that G is a group with finite presentation $< C : R >$. Let $n(k)$ denote the number of elements of the group that can be expressed as a product of length $\leq k$ in elements of $C \cup C^{-1}$. If G has polynomial growth, in the sense that $n(k)$ is bounded above by a polynomial $p(k)$ (that is, $n(k) \leq |p(k)|$ for all positive integers k), is the group nilpotent?

Misha Gromov [Gromov, 1981] showed that the answer was 'yes'. We mention only a bit of his amazing argument. Gromov interpreted the group as a geometric object, the Gruppenbild. He scaled that object infinitely often to create a sequence of metric spaces that had a limiting object. The idea is well illustrated by our view of a screen door. Viewed from nearby, the screen is perforated by multitudes of small holes. But as we view a screen from further and further away (the screen door *scaled*), the screen appears more and more like a simple plane, with the holes disappearing entirely in the limit. Gromov used the famous solution to Hilbert's 5th Problem [Montgomery, 1955] to show that, in a critical basic case, the limiting space was a manifold.

At the same congress, Thurston, in his plenary address in Finlandia Hall, emphasized geometric structures on 3-manifolds, which, for example, allowed him in most cases to tessellate the universal cover, with the Gruppenbild as dual graph. He said, as best I remember, "I think of the 3-manifold not just as *having* a fundamental group but as *being* that group."

Thurston [Scott, 1983; Thurston, 1982, 1997, 1982] showed that exactly eight rigid geometries lie at the base of 3-manifold theory, with the 3-dimensional non-Euclidean hyperbolic geometry, which generalizes the 2-dimensional geometry of Lobachevskii, Bolyai, and Gauss, being, by far, the richest source of examples.

All of us who had studied 3-manifolds realized that Thurston had revolutionized the theory. Since Thurston's tools required techniques from analysis that I was at that time unprepared to understand, I was determined to study group theory by methods parallel to the techniques of Thurston, and I would do so by examining something I knew, namely, hundreds of group pictures (Gruppenbilder). As I talked to Thurston one afternoon over reindeer steaks, I was encouraged by the fact that he too liked to think of groups in terms of their Gruppenbilder.

At that same dinner, Thurston conjectured that the growth series

$$\sum_{k=0}^{\infty} n(k)x^k$$

of a cocompact, discrete hyperbolic group (with Gruppenbild dual to a tessellation or tiling of hyperbolic space) should be a rational function. I had no clue how to proceed, but learned the fundamentals of non-Euclidean geometry from Thurston's notes and learned how to recognize rational functions from Pólya. In proving Thurston's conjecture and looking at many examples, I discovered two facts that were amazing to me [Cannon, 1984]:

(1) If G is a cocompact, discrete hyperbolic group, then the Word Problem in G can be solved by a version of Dehn's greedy reduction algorithm. That is, given a finite generating set C, there is a finite relator set R such that every nonempty word representing the identity can be shortened by means of one of the relators in R. This fact greatly extended the class of groups for which it was known that Dehn's solution to the word problem was valid. (There is a solution to the Conjugacy Problem that also extends Dehn's results.)

As opposed to the combinatorial small cancellation arguments mentioned earlier, which are explicitly local in nature, the new arguments use the fact that, not in the small but in the large, the geometry of the Gruppenbild closely matches the geometry of hyperbolic space. Paths that are straight in the Gruppenbild are not at all straight in hyperbolic space but are *globally almost straight*. And straight paths in hyperbolic space are very restricted. This observation requires combinatorial arguments that have to be very forgiving in the small but can be approximately precise in the large. The arguments are explicitly geometric, just as Dehn's original proof had been.

(2) We may view the Gruppenbild as a partially ordered structure by defining vertex x to precede vertex y if there is a shortest possible path from the identity vertex e to y that passes through x. We define the *cone or shadow* of x to be the partially ordered subset consisting of those elements y with $x \leq y$. For cocompact, discrete hyperbolic groups, there are only finitely many isomorphism types of vertex shadows.

We think of each isomorphism type as being a *cone type*. Each vertex shadow type (or cone type) can therefore be recognized by a sufficiently large finite initial subset of the group graph. Consequently, once we have recognized the cone types, we can build the Gruppenbild recursively bit by bit, with no cancellations at all. Another consequence of the fact that there are only finitely many cone types is that the growth function is rational (Thurston's conjecture).

Finally, Thurston [Ep, 1992] formalized the cone structure I discovered to define what has come to be called *automatic groups*. We will discuss automatic groups below.

9.2. Gromov's Translation from Differential Geometry to Group Theory. At roughly the same time, Gromov [Gromov, 1987, 1981] was in the process of carefully translating the global facts he knew about Riemannian manifolds into facts about discrete groups. In particular, he used a property of hyperbolic space apparently suggested by Eli Rips as the characteristic property of hyperbolic space, namely the *thin triangle property*, which we will discuss again later. A group is called *word hyperbolic* if its geodesic triangles are uniformly thin (defined below). In a Gruppenbild satisfying this property, all of the combinatorial arguments previously discovered as applying to cocompact discrete hyperbolic groups could be carried out almost without change. In particular, Dehn's solutions to the word and conjugacy problems applied.

Gromov's papers were long and full of riches. His translations from differential geometry were very careful so that his theorem statements were almost invariably correct. His stated proofs using discrete means were much less carefully prepared, so that a very large cottage industry was created in finding complete combinatorial and geometric proofs of Gromov's theorems.

10. Spin-offs from the Work of Thurston and Gromov

10.1. Automatic Groups. Thurston noted that, if we assign a vertex to each of the finitely many cone types in a hyperbolic group, then there is a natural transition graph describing the passage from one cone type to another. For example, if (x, c, y) is a directed edge from x to y, with $x < y$, then the cone from y is contained in the cone from x. Hence, the cone type of x determines the cone type of y, and the corresponding vertices can be joined by a directed edge labelled c. We can interpret the resulting finite graph of cone types as a finite state automaton. Thurston showed that, in a hyperbolic group, there are finite state automata describing all geodesic paths, canonical representatives for the elements of the group, and multiplication by generators within the group. He called a group for which such automata exist an *automatic group*. The basic properties of automatic groups were presented in [Ep, 1992], a book primarily prepared by D. B. A. Epstein.

David Epstein and colleagues at the University of Warwick created beautiful computer programs to find and realize each of the appropriate finite state automata, when such automata exist. They called their program 'Aut'.

10.2. Word Hyperbolic Groups. Consider paths ab, bc, and ca in a metric space X. We say that the triangle abc with these three edge paths is δ-thin, if each of the three paths lies in the δ-neighborhood of the union of the other two paths. We say that a space X is δ-hyperbolic if each triangle in X having geodesic edges is δ-thin. A group is *word hyperbolic* if its Gruppenbild is δ-thin for some fixed δ. In hyperbolic space, every geodesic triangle is δ-thin, for $\delta = \ln(1 + \sqrt{2}) \approx 0.88$. Thin triangles in non-Euclidean hyperbolic space look like a small central core from which three long thin arms extend. (See Figure 9.)

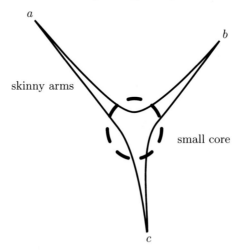

FIGURE 9. A Thin Triangle

Gromov stated many wonderful theorems about word hyperbolic groups, how to recognize them, and how to construct them. In particular, all of them admit solutions to word and conjugacy problems that are geometric extensions of Dehn's.

10.3. Ol'shanskii. Are most finitely presented groups word hyperbolic? Gromov suggested ways to measure the density of word hyperbolic groups among all finitely presented groups. Ol'shanskii [Ol'shanskii, 1992] managed to carry out Gromov's suggested analysis so as to prove that most groups are word hyperbolic, hence admit Dehn solutions to word and conjugacy problems. Thus, the groups of Novikov and Boone for which these problems are unsolvable and for which Dehn's solutions are not valid are the rare exception. Dehn's analysis applies to almost all groups.

11. Groups Involving Other Geometries

As mentioned earlier, Thurston showed that, in low dimensional topology, there are other geometries than hyperbolic geometry to consider. People spent a good deal of time trying to understand the appropriate results on groups and their Gruppenbilder from each of Thurston's eight geometries.

The groups associated with spherical geometry are all finite, hence potentially very complicated, but in a sense uninteresting in the large-scale analysis that worked so well in hyperbolic space.

The next geometry to consider was obviously Euclidean geometry. Because of Bieberbach [Bieberbach, 1911, 1912], the general structure of Euclidean groups is known. In the fundamental group of the torus, whose Gruppenbild is the square lattice in the plane, the Dehn greedy algorithm for the word problem is not valid. There are, however, exactly nine cone types, one the cone from the identity, four associated with the four axes, and four associated with the four open quadrants. One obvious conjecture would be that there are only finitely many cone types for each Euclidean group. Thurston [Ep, 1992] showed that each Euclidean group is automatic, but the automatic structure he suggested did not rely on cone types. The Euclidean cone-type problem led to an amusing experience.

I very much wanted to get to know Gromov. At a mathematical conference, I saw him sitting at the lunch table with no one else around. Here was my chance. I took my tray to the table and said, "Hi! I wanted to get to know you. My name is Jim Cannon." "I know," he said, "and I've solved your problem." "Oh, and what problem is that?" I asked. "In my paper, I prove that each Euclidean group only has finitely many cone types," he explained. "That's interesting," I said. "I have a counterexample." "Then I'll have to change my paper," he replied. It is always wonderful in mathematical research to have both a proof and a counterexample to a theorem.

Despite this difficulty, it seems that Euclidean groups are not too difficult to understand.

In dimension 2, hyperbolic space, spherical space, and Euclidean space are the three spaces with constant curvature. They also seem the most easily understood. The geometric difficulty in understanding geometric groups arises when there are both positive and negative sectional curvatures.

11.1. CAT-0 Groups. In summary, the groups most easily understood are those associated with spaces of constant sectional curvatures. Algorithmically, among those, the spaces of constant negative curvature are simplest. The most successful geometric generalization beyond the groups of constant curvature spaces are the CAT-0 groups; these are the groups whose geometry models sectional curvatures ≤ 0. At the boundary of this collection are those of curvature 0, modelled by the flat or Euclidean spaces. The algorithms tend to be like those of Euclidean groups and are, in general, more complicated than those associated with groups of negative sectional curvatures bounded away from 0.

Each geodesic triangle in any metric space can be compared with the triangle in Euclidean space having the same side lengths. If the triangles in the space in question are at least as thin as the comparison triangles in Euclidean space, then the space is said to be a CAT-0 space.

The major developments in the study of CAT-0 groups is reported in the wonderful book by Martin Bridson and André Haefliger [Bridson, 1999].

Thanks. The original papers of Max Dehn have brought me great pleasure and satisfaction, as have my friendship and encounters with Bill Thurston, Misha Gromov, Martin Bridson, and others. I give special thanks to Bill Floyd and Walter Parry with whom I have spent many years discussing these topics. Finally, thanks to Bill Floyd for help with the figures.

References

Bieberbach, Ludwig: Über die Bewegungsgruppen der Euklidischen Räume. (German) Math. Ann. 70 (1911), no. 3, 297-336.

Bieberbach, Ludwig: Über die Bewegungsgruppen der Euklidischen Räume (Zweite Abhandlung.) Die Gruppen mit einem endlichen Fundamentalbereich. (German) Math. Ann. 72 (1912), no. 3, 400-412.

Boone, W. W.: The word problem. Ann. of Math. 70 (1959), 207-265.

Boone, William W.: Word problems and recursively enumerable degrees of unsolvability. A sequel on finitely presented groups, Annals of Mathematics, second series, vol. 84, no 1 (Jul., 1966), pp 49-84.

Bridson, Martin R.: Geodesics and curvature in metric simplicial complexes. Group theory from a geometrical viewpoint (Trieste, 1990), 373-463, World Sci. Publ., River Edge, NJ, 1991.

Bridson, Martin R.: Non-positive curvature in group theory. (English summary) Groups St. Andrews 1997 in Bath, I, 124-175, London Math. Soc. Lecture Note Ser., 260, Cambridge Univ. Press, Cambridge, 1999.

Bridson, Martin R.; Haefliger, Andre: Metric spaces of non-positive curvature. Grundlehren der Mathematischen Wissenschaften [Fundamental Principles of Mathematical Sciences], 319. Springer-Verlag, Berlin, 1999. xxii+643 pp.

Britton, John L.: The word problem, Annals of Math., second series, Vol. 77, no. 1 (Jan., 1963), pp. 16-32.

Cannon, James W.: The combinatorial structure of cocompact discrete hyperbolic groups. Geom. Dedicata 16 (1984), no. 2, 123-148.

Cayley, A.: The theory of groups: graphical representations. Amer. J. Math. 1, 174-176 (1878).

Church, Alonzo: "A set of postulates for the foundation of logic," Annals of Mathematics, 2s., vol. 33 (1932), pp 346-366.

Church, Alonzo: "A set of postulates for the foundation of logic (second paper)," Annals of Mathematics, 2s., vol 34 (1933), pp. 839-864.

Church, Alonzo: A note on the Entscheidungsproblem, J. Symb. Logic 1 (1936).

Coxeter, Harold Scott MacDonald; Moser, William O. J.: Generators and relations for discrete groups. Third edition. Ergebnisse der Mathematik und ihrer Grenzgebiete Band 14. Springer-Verlag, Heidelberg-New York, 1972. ix + 161 pp.

Davis, Martin: The Undecidable: Basic Papers on Undecidable Propositions, Unsolvable Problems and Computable Functions. Raven Press, Hewlett, New York, 1965.

Dehn, M.: Über die Topologie des dreidimensionalen Raumes. Math. Ann. 69, 137-168 (1910).

Dehn, M.: Über unendliche diskontinuerliche Gruppen. Math. Ann. 71, 116-144 (1912).

Dehn, M.: Transformation der Kurve auf zweiseitigen Flächen . Math. Ann. 72, 413-420 (1912).

Dehn, Max: Papers on Group Theory and Topology, Translated and Introduced by John Stillwell, Springer-Verlag, New York, Berlin, Heidelberg, London, Paris, Tokyo, (1987), iv+ 396 pp.

Epstein, David B. A.; Cannon, James W.; Holt, Derek F.; Levy, Silvio W. F.; Paterson, Michael S.; Thurston, William P.: Word processing in groups. Jones and Bartlett Publishers, Boston, MA, 1992. xii + 330 pp.

Gromov, M.: Groups of polynomial growth and expanding maps. Inst. Hautes Etudes Sci. Publ. Math. No. 53 (1981), 53-73.

Gromov, M.: Hyperbolic groups. Essays in group theory, 75-263, Math. Sci. Res. Inst. Publ., 8, Springer, New York, 1987.

Gromov, Mikhael: in Riemann surfaces and related topics: Proceedings of the 1978 Stony Brook Conference (Stony Brook, NY, 1978), 183-213, Ann. of Math. Stud., 97, Princeton Univ. Press, Princeton, NJ, 1981.

Haken, H.: Zum Identitätsproblem bei Gruppen. Math. Z.56, 335-362 (1952).

Haken, W.: Connections between topological and group theoretical decision problems. In: Word Problems, pp. 427-441. Amsterdam: North-Holland 1973.

Kleene, S. C.: General recursive functions of natural numbers, Math. Ann. 112 (1936), 727-742.

Kleene, Stephen: Introduction to Metamathematics, North-Holland, Amsterdam; P. Noordhoff, Groningen; D. Van Nostrand, New York and Toronto, 1952. Eighth reprint (1980), North-Holland, Amsterdam and New York; Wolters-Noordhoff, Groningen.

Lyndon, Roger C.; Schupp, Paul E.: Combinatorial Group Theory. Springer-Verlag, Berlin, Heidelberg, New York, 1977, xiv+339 pp.

Lyndon, R. C.: On Dehn's algorithm. Math. Ann. 166, 208-228 (1966).

Montgomery, D.; Zippin, L.: Topological Transformation Groups, Interscience, N. Y. (1955).

Novikov, P. S.: On the algorithmic insolvability of the word problem in group theory, American Mathematical Society Translations, Ser. 2, 90 (1958), 1-152. (English translation of the Russian: Trudy Mat. Inst. Steklov 44 (1955).)

Ol'shanskii, A. Yu.: Almost every group is hyperbolic. Internat. J. Algebra Comput. 2 (1992), no. 1, 1-17.

Post, E. L.: Recursive unsolvability of a problem of Thue, J. Symb. Logic, 12 (1947), 1-11.

Rotman, Joseph J.: The Theory of Groups, An Introduction, Second Edition, Allyn and Bacon, Inc., Boston, x+342 pp. (See Chapter 12: The Word Problem.)

Scott, Peter: The geometries of 3-manifolds. Bull. London Math. Soc. 15 (1983), no. 5, 401-487.

Tartakovskii, V. A.: Solution of the word problem for groups with a k-reduced basis for $k > 6$. Izv. Akad. Nauk SSSR Ser Math. 13, 483-494 (1949).

Thurston, William P.: Three-dimensional manifolds, Kleinian groups and hyperbolic geometry, Bull. Amer. Math. Soc. (N.S.) 6 (1982), no. 3, 357-381.

Thurston, William P.: Three-dimensional geometry and topology. Vol. 1. Edited by Silvio Levy. Princeton Mathematical Series, 35. Princeton University Press, Princeton, NJ, 1997. x + 311 pp.

Thurston, W.: Hyperbolic geometry and 3-manifolds, low dimensional topology (Bangor, 1979), pp. 9-25, London Math. Soc. Lecture Note Ser., 48, Cambridge Univ. Press, Cambridge-New York, 1982.

Turing, Alan: On computable numbers, with an application to the Entscheidungs-Problem, Proceedings of the London Mathematical Society, s. 2, vol 42 (1937), pp. 230-265. A correction, to the preceding, vol. 43, pp. 544-546. Both reprinted in [6], pp. 115-154.

Turing, A. M.: The word problem in semi-groups with cancellation, Ann. of Math. (2) 52 (1950), 491-505.

Turing, A. M.: On computable numbers, with an application to the Entscheidungsproblem, Proc. London Math. Soc. (2) 42 (1956), 230-265.

CHAPTER 8

Max Dehn, Axel Thue, and the Undecidable

Stefan Müller-Stach

Introduction

Dehn was not the only mathematician to pose and develop what has come to be known as the word problem (see Chapters 6 and 7). In addition to Dehn's approach through geometric group theory, the word problem was formulated independently by Axel Thue for general tree structures in 1910 and for semigroups in 1914.

In his book with Bruce Chandler, Dehn's student Wilhelm Magnus remarked that Dehn and Thue knew each other and mentioned the amazing parallel between their discoveries:

> What appears to be incidental or, if one prefers, miraculous, is the fact that independent of Dehn and independent of topology, a contemporary mathematician had begun to ask questions of the type of the word problem in combinatorial group theory, but in an even more general and highly abstract setting. We are referring to the work of Thue, who may be considered as the founder of a general theory of semigroups. With one widely quoted exception, this work of his is largely forgotten nowadays. We do not know whether Dehn was influenced by Thue, and we have reasons to doubt it. We know that Dehn knew Thue personally, but only very superficially. Dehn mentioned Thue's work on occasion, observing that Thue's papers dealt with combinatorial problems. But he never used them, and indeed there is no known direct application of Thue's work to Dehn's group-theoretic problems.[1]

A similar statement occurs in [Magnus, 1978, Footnote 5]. There Magnus also remarked that Dehn's wife Toni did not recall that he had ever mentioned Axel Thue in her presence although Dehn had visited Norway quite a few times and was skilled in the Norwegian language. Moreover, there are no known personal relations among the students of Dehn and Thue's only student, Thoralf Skolem.

While Dehn's work had spread quickly, we do not know when mathematicians became aware of Thue's work on semigroups and what we now call Thue systems. In a paper from 1947 Emil Post mentions that he learned of Thue's 1914 paper from Alonzo Church. Around 1935–1955, the word problem became an attractive challenge for people working in the theory of computation (alias recursion theory). Aside from the *Entscheidungsproblem* (see below), it is one of the first genuine

---

©2024 American Mathematical Society
[1] See [Chandler and Magnus, 1982, p. 54].

mathematical problems which appeared to be potentially undecidable[2]. Our investigations indicate that Alonzo Church, Emil Post and others at Princeton were instrumental in bringing the word problem and the theory of computation together, thus placing the heritage of Dehn and Thue in the right historical context.

Many problems in mathematics are accessible through computation and algorithms. The Euclidean algorithm is an example of how effective mathematical thinking can lead to powerful algorithms. Gottfried Wilhelm Leibniz was the first scientist to express in a precise way how a device (which he called calculus ratiocinator) can decide the truth of all reasonable statements, not necessarily restricted to mathematics, by a sort of logical computation. Although Leibniz's thoughts remained in an abstract realm, he worked on the realization of an arithmetic calculating machine throughout his life. A later attempt, The Analytic Engine, by Charles Babbage and Ada Lovelace – albeit unsuccessful – was a harbinger of a programmable computer. Finally, from about 1940 on, Konrad Zuse at Berlin and John von Neumann at Princeton started to construct the first fully universal (Turing complete) computers, the Z3 and ENIAC. This was, of course, the beginning of a success story of incredible impact.

In the early 20th century, the notion of algorithmic computability still needed an underlying mathematical theory. Nevertheless, people had a pragmatic idea what computability was supposed to mean, that is, to reach a result in a finite number of computational steps. An example is the formulation of Hilbert's 10th problem:

> Given a diophantine equation with any number of unknown quantities and with rational integral numerical coefficients: To devise a process according to which it can be determined by a finite number of operations whether the equation is solvable in rational integers.[3]

After 1917 Hilbert started his program in proof theory (Hilbert's program) that sought a solid foundation of mathematics with a method he called finitistic.[4] Hilbert showed a lot of optimism[5] that all metamathematical questions could be settled within a mathematical proof theory. In 1928, together with Ackermann, he posed his famous *Entscheidungsproblem* (decision problem). It asks for an algorithm in the spirit of Leibniz that decides the provability of statements in first order axiomatic theories. By Gödel's completeness theorem for first order logic, proved in his dissertation from 1929, this is equivalent to asking for satisfiability in all possible set-theoretic models.

---

[2] A yes/no decision problem for an infinite set of mathematical objects is called decidable (alias algorithmically solvable, or recursive) if the set S of Gödel numbers of the involved mathematical objects is a decidable subset of $\mathbb{N}$. This means that there is an algorithm which has output 1 if $n \in S$ and output 0 otherwise.

[3] Eine diophantische Gleichung mit irgendwelchen Unbekannten und rationalen ganzen Zahlenkoeffizienten sei vorgelegt: Man soll ein Verfahren angeben, nach welchem sich mittels einer endlichen Zahl von Operationen entscheiden läßt, ob die Gleichung in ganzen rationalen Zahlen lösbar ist. [Hilbert, 1900].

[4] The finitistic approach somehow rejects the use of infinitely many steps. This is related but not the same as the intuitionistic and constructivistic approach of Brouwer, Kronecker, and Weyl which Hilbert disliked. Gödel's system T, or equivalently Gentzen's proof for the consistency of arithmetic, may also be considered as finitistic in some sense.

[5] See Hilbert's famous words: "Wir müssen wissen, wir werden wissen."

In 1931, Gödel's two famous incompleteness theorems were discovered.[6] The first theorem shows that first-order Dedekind-Peano arithmetic is incomplete in the sense that there are statements which are neither provable nor disprovable but true in the standard model. The second theorem states that the consistency of a theory at least as rich as Dedekind-Peano arithmetic cannot be proved as a syntactic formula within the theory. Hilbert's proof theoretic program subsequently had to be modified. Gödel's proof used primitive recursive functions and the technique of Gödel numbering of arithmetic statements and proofs. Tarski had independently shown that arithmetic truth predicates exist only outside the realm of the theory, which implies Gödel's incompleteness theorem.[7] From his correspondence with John von Neumann we know that Gödel was aware of this result. However, he was not able to solve the *Entscheidungsproblem* at that time; the theory of computable functions was developed in full depth after 1936.

That a given mathematical problem like Hilbert's 10th problem or, as we will see, the word problem might be undecidable was probably considered unlikely by most people before 1931. But after Gödel's achievements this became a more realistic possibility. We will see, however, that Dehn and Thue had already realized the difficulty of the word problem around 1910.

A full-fledged theory of computation emerged around 1936 based on the work of Church, Gödel, Herbrand, Kleene, Markov, Post and Turing. Using this, the undecidability of the *Entscheidungsproblem* and of the related *Halteproblem* was shown. See [Davis, 1965] for the precise history of these developments. In addition, Church showed the undecidability of the word problem for finitely generated semigroups[8] in 1937. It took until 1947 before Post and Markov gave the first proof of the undecidability of the word problem for finitely presented semigroups. Another five years passed before the word problem for finitely presented groups was shown to be undecidable by William Boone and Pyotr Novikov in 1952. Two decades later, the undecidability of Hilbert's 10th problem was shown by Yuri Matiyasevich, building upon the work of Martin Davis, Hilary Putnam and Julia Robinson (see [Matiyasevich, 1970]).

In this chapter, we describe the impact that both Dehn and Thue had on researchers in recursion theory, that is, the theory of computation, and we shed some light on the period between 1936 and 1955 during which many people worked on proving the (un)solvability of the word problem. I would like to thank Steve Batterson, Martin Davis, Catherine Goldstein, Jemma Lorenat, John McCleary, Carl-Fredrik Nyberg-Brodda, Edmund Robertson, David Rowe, Marjorie Senechal, Reinhard Siegmund-Schultze, Jörn Steuding and Marcia Tucker for helpful remarks and assistance.

---

[6]Gödel only published the first theorem, see [Gödel, 2003, Vol. I] and [Plato, 2020] for the full story. Here, completeness has a different meaning than in Gödel's dissertation.

[7]This means that the set of Gödel numbers of true statements in the standard model of the natural numbers is not an arithmetic set, hence not even recursively enumerable.

[8]Church's example was not finitely presented, as it had infinitely many relations.

1. Max Dehn and the Word Problem for Groups

In his paper [Dehn, 1911] on the word problem, Dehn used the presentation of a (finitely presented) group G by generators and relations.[9] As he remarks, this concept had been studied before in detail by Walther von Dyck [von Dyck, 1882]. It came up even earlier in the work of William Rowan Hamilton.

Dehn posed the word problem, the conjugacy problem, and the isomorphism problem for groups in [Dehn, 1911]. The word problem asks to decide whether a given word, a product of generators to integer powers, w in G is equal to 1. The conjugacy problem extended this question to determine whether two words w, w' in G are conjugate and if they are, find $u \in G$ such that $w' = uwu^{-1}$. The conjugacy problem implies the word problem, since a word w is equal to 1 if and only if it is conjugate to 1. Finally, the isomorphism problem aims to determine whether two given groups G and G' are isomorphic.

Dehn formulated the word problem, *das Identitätsproblem*, as follows:

> Let an arbitrary element of a group be given by its composition out of generators. One shall provide a method which decides in a finite number of steps whether this element is equal to the identity or not.[10]

Dehn was aware that the word problem might turn out to be difficult for a general group. He wrote:

> Here we have three fundamental problems whose solution is very important and probably not possible without a thorough study of the subject.[11]

In fact, Dehn may have been aware of the potential unsolvability of the problem.[12] Dehn's main tool was what he called the *Gruppenbild* (now known as the Cayley graph): Given a finitely presented group $G = \langle S \mid R \rangle$ with set of generators $S = \{s_1, \ldots, s_n\}$, the vertices are all elements of G, and the (directed) edges connect g and gs for every $g \in G$ and $s \in S \cup S^{-1}$. The edges are usually colored. In Figure 1 this is illustrated for the dihedral group[13] D_5. Dehn solved the word problem for infinite surface groups [Dehn, 1912], that is, groups that are the fundamental groups of orientable closed 2-manifolds.[14]

The idea of Dehn's proof is described in modern language, that one needs to prove that any non-trivial closed loop in the Cayley graph of a surface group G

---

[9]This is usually denoted by $G = \langle \underbrace{s_1, \ldots, s_n}_{\text{generators}} \mid \underbrace{r_1, \ldots, r_m}_{\text{relations}} \rangle$. A relation r is given by a word r, and the notation amounts to identifying every occurrence of r with the trivial word, i.e., setting $r = 1$. Here, a word w (of length ℓ) is a finite combination of generators, possibly with repetition: $w = g_1^{\pm 1} \cdots g_\ell^{\pm 1}$. The length of a word w is denoted by $|w|$. The inverse w^{-1} of a word w is obtained by inverting all g_i involved and reversing the order, e.g., $(g_1 g_2)^{-1} = g_2^{-1} g_1^{-1}$.

[10]1. Das Identitätsproblem: Irgend ein Element der Gruppe ist durch seine Zusammensetzung aus den Erzeugenden gegeben. Man soll eine Methode angeben, um mit einer endlichen Anzahl von Schritten zu entscheiden, ob dies Element der Identität gleich ist oder nicht. [Dehn, 1911].

[11]Hier sind es vor allem drei fundamentale Probleme, deren Lösung sehr wichtig und wohl nicht ohne eindringliches Studium der Materie möglich ist. [Dehn, 1911].

[12]Magnus in [Chandler and Magnus, 1982, p. 55] cites Dehn as follows: Solving the word problem for groups may be as impossible as solving all mathematical problems.

[13]$D_n = \langle \sigma, \tau \mid \sigma^n = \tau^2 = 1, \tau\sigma = \sigma^{-1}\tau \rangle$.

[14]These are of the form $G = \langle a_1, b_1, \ldots, a_g, b_g \mid \prod_{i=1}^g a_i b_i a_i^{-1} b_i^{-1} \rangle$ with one relation.

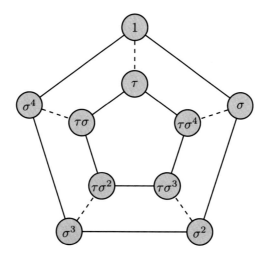

FIGURE 1. (Undirected) Cayley graph of D_5 with two generators $\sigma = (1\,2\,3\,4\,5)$ (rotation) and $\tau = (2\,5)(3\,4)$ (reflection, dashed). All figures in this text were prepared by the author using tikz.

of genus $g \geq 2$ contains more than half of the defining relations, or can be freely reduced.[15] In this way, Dehn provided an algorithm (now called Dehn's algorithm) to solve the word problem for G and some other groups. The algorithm can be presented in a very general form:

(1) Let any freely reduced word $w = w_0$ be given. We construct a finite sequence $w_0, w_1, \ldots, w_n$ of freely reduced words by recursion such that $w = w_0$ and the lengths decrease $|w_0| > |w_1| > \cdots > |w_i| > \cdots$.
(2) If w_i is already constructed and empty, i.e., $w_i = 1$, then terminate.
(3) If w_i contains a subword a such that for some relation $r = ab$ and $|a| > |r|/2$, then replace a by b^{-1} in w_i and obtain w_{i+1}.
(4) If not, terminate at step i.

A group is said to have a Dehn presentation when it is sufficient for Dehn's algorithm to work, see [Miller, 2014, p. 345]. An example where Dehn's algorithm does not succeed is the genus one case, that is, the fundamental group of the torus, and – more generally – the free abelian group $\mathbb{Z}^n$ for $n \geq 2$ [Miller, 2014, p. 345]. Note that the word problem is, on the other hand, easy to solve for free abelian groups of finite rank.

There are large classes of groups beyond surface groups for genus $g \geq 2$ to which Dehn's algorithm can be extended. One direction where this was successful is the focus of small cancellation theory, which deals with (finitely presented) groups where the relations have small overlap. We refrain from presenting any definitions and refer to the books [Lyndon and Schupp, 1977] and [Sims, 1994] for an account of this theory. Historically, small cancellation theory was mainly developed in [Tartakovskii, 1949], [Greendlinger, 1960], [Lyndon, 1966] and [Schupp, 1968]. For example, in [Greendlinger, 1960] it is proved that a group satisfying a small cancellation property denoted by $C'(1/6)$ has a solvable word problem.

---

[15] Freely reduced words have no substrings of the form $x^{-1}x$ or xx^{-1}.

Small cancellation is not a geometric concept. A geometric class of finitely presented groups where Dehn's algorithm works are word-hyperbolic groups[16] which satisfy certain metric conditions on the Cayley graph (see [Gromov, 1987]). Small cancellation groups satisfying the $C'(1/6)$-condition are examples of word-hyperbolic groups. It is a theorem due to Gromov and Olshanskii that for a general group G – in the sense that G is in some way chosen randomly – the Dehn algorithm solves the word problem for G, see [Gromov, 1987; Olshanskii, 1992].

For other algorithms related to the word problem see [Knuth and Bendix, 1970] and [ToddCoxeter, 1936]. The Knuth-Bendix algorithm[17] for completing term rewriting systems can be used to solve the word problem for the large class of automatic groups [Epstein et al., 1992] which contains word-hyperbolic groups and braid groups. The Todd-Coxeter algorithm, which is primarily a coset enumeration method for finite index subgroups, can also be applied to the word problem.

The historical survey of John Stillwell [Stillwell, 1982] on the word problem contains many examples of finitely presented groups with a solvable word problem. In the following table we list some of them:

Type of group	Reference
Surface groups	[Dehn, 1912]
Trefoil knot group	[Dehn, 1914]
Subgroups of free groups (abelian or not)	[Nielsen, 1921]
Braid groups	[Artin, 1925]
One-relator groups	[Magnus, 1932]
Residually finite groups[18]	[McKinsey, 1943]
Hypo-abelian groups	[Engel, 1949]
Linear groups	[Rabin, 1960]
Knot groups	[Waldhausen, 1968]
Hyperbolic groups	[Alonso et al., 1991]
Automatic groups	[Epstein et al., 1992]

Among these people, Engel and Magnus were students of Dehn (see chapter 6).[19] Magnus proved his *Freiheitssatz*[20] in 1930 to treat the one-relator case. Amazingly, the word problem for one-relator semigroups is still open, see the survey [Nyberg-Brodda, 2021].

Other finitely presented groups for which the word problem has been solved are finite groups, polycyclic groups, Coxeter groups and finitely presented simple groups. We refer to the textbooks [Lyndon and Schupp, 1977] and [Sims, 1994] for these and other cases.

---

[16]Hyperbolic groups are defined as follows. Consider the Cayley graph of G and endow it with its graph metric. Then G is word-hyperbolic, if the resulting topological space is hyperbolic in the sense of [Gromov, 1987], i.e., there is a constant $\delta > 0$ such that any triangle is δ-thin.

[17]Donald Knuth was a great-great-grandstudent of Thue via Thue-Skolem-Øre-Hall-Knuth.

[19]A list of students of Dehn is contained in [Magnus and Moufang, 1954].

[20]The Freiheitssatz asserts that leaving away at least one generator appearing in the relation induces a free subgroup in any one-relator group G.

2. Axel Thue and the Word Problem for Semigroups

Axel Thue was a number theorist with broad interests and he was well-known far beyond Norway for his work in arithmetic. He held a chair in applied mathematics at Oslo from 1903 on. Some of Thue's most important work in number theory is concerned with diophantine equations. For example, in [Thue, 1977] he considered the integer solutions of equations $f(x,y) = c$ for a homogenous polynomial f with integer coefficients and showed that the number of such solutions is finite, provided certain conditions on f are valid. In particular the degree of f needs to be at least three.[21] Such results were later extended by Carl Ludwig Siegel and are the basis of finiteness conjectures in modern arithmetic geometry (on Siegel, see chapter 5). In the same paper of 1909, Thue looked at generalizations of Liouville's result which bounds the approximation of irrational algebraic numbers by rational numbers from below. Thue's results were strengthened by Siegel in his 1929 dissertation under Edmund Landau, and in 1955 by Klaus Friedrich Roth who obtained an optimal estimate. Today the final result is known as the Thue-Siegel-Roth theorem.[22] For this work, Roth received the Fields Medal at the 1958 ICM in Edinburgh.

Thue claimed that he often discovered results which were previously obtained by others. For example, he wrote in a letter to Elling Holst from 1902 [Thue, 1977, p. xxi] that he had discovered the transcendence of e and π independently of Hermite and Lindemann during his time as a teacher at the technical college in Trondheim, that is, between 1894 and 1902.

Among Thue's many papers are four quite abstract works about trees, words, semigroups and term rewriting (see below) which were written in German and belong to mathematical logic:

- Über unendliche Zeichenreihen [Thue, 1977, p. 139–158] from 1906.
- Die Lösung eines Spezialfalles eines generellen logischen Problems [Thue, 1977, p. 273–310] from 1910.
- Über die gegenseitige Lage gleicher Teile gewisser Zeichenreihen [Thue, 1977, p. 413–477] from 1912.
- Probleme über Veränderungen von Zeichenreihen nach gegebenen Regeln [Thue, 1977, p. 493–524] from 1914.

The papers from 1906 and 1912 present the general theory of trees and words. In the 1906 paper Thue proved that there are infinitely long sequences consisting of three or four letters which are square-free, that is, no finite length word B occurs twice as BB in the sequence. Such sequences are called irreducible. The 1906 paper continues by showing that there is an infinite sequence

$$01101001100101101001011001101001\cdots$$

in two letters which is cube-free, that is, no finite word B occurs as BBB in the sequence. Thue's 1912 paper elaborates on the case of two and three symbols even more and classifies irreducible sequences on two letters. See [Hedlund, 1967] for a discussion of this work.

---

[21]The equation defines a plane curve in the projective plane of the same degree with equation $f(x,y) = c \cdot z^{\deg(f)}$. The other conditions on f which we did not mention take care that this curve is not rational, i.e., the image of a projective line.

[22]The theorem asserts that for every algebraic number α and every $\varepsilon > 0$ the inequality $\left|\alpha - \frac{p}{q}\right| < q^{-2-\varepsilon}$ has only finitely many solutions in coprime integers p,q.

Such sequences had been discovered before Thue's work by Eugène Prouhet in 1851 (solving the Tarry-Escot problem) and later, independently, by others.[23] The sequences of these types show that an infinite chess game is possible without violating certain chess regulations [Morse, 1938; Morse and Hedlund, 1944].

Thue's paper from 1910 introduced a very general philosophical (or logical) problem which he phrased in a metamathematical language. In modern terminology, he introduced term rewriting systems for tree-like structures. Term rewriting means that certain trees are transformed into other trees in single steps by replacing (rewriting) parts according to certain rules. The 1914 paper is concerned with words (*Zeichenreihen*) instead of binary trees. The underlying algorithmic problems in the case of words are known as (semi-)Thue systems or as term rewriting systems [Büchi, 1989, p. 181].

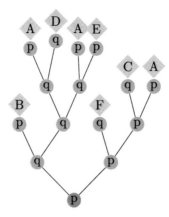

FIGURE 2. A copy of Axel Thue's figure.

Thue considered finite, binary, rooted trees as in Figure 2 (a copy of Figure 3 from [Thue, 1977, p. 275]). The outer leaves correspond to variables *A-F* of a certain type (either of type p or q in Figure 2). Thue explains that for him there is a theory of a certain logical kind behind such diagrams (called *Begriffe* and *Begriffskategorien* by Thue). In the inner nodes going to the root, each time two values (of type p resp. q in figure 2) are combined by a binary operation into a new value of the indicated new type. Hence, going all the way to the root corresponds to the computation of a tree automaton that computes a value of type p from the given values of the entry variables *A-F*.

These trees may be viewed as objects representing certain algebraic or logical terms, as in figure 3 illustrating the associativity of addition:

$$(A + B) + C = A + (B + C).$$

Conversely, a binary tree corresponds to a term. Thue had already imagined the celebrated correspondence between trees and terms, generalizations of which occur in Post's work on canonical systems [Post, 1921].

---

[23]Notably, Marston Morse (1921), see [Morse, 1921], Kurt Mahler (1929), and the chess player Max Euwe (1929).

FIGURE 3. Two trees symbolizing $(A+B)+C$ resp. $A+(B+C)$.

In his 1910 paper Thue viewed this term rewriting problem as an algorithmic problem about the relation between two given trees A and B:

> ...so we ask in other words, whether one can find trees $C_1 C_2 \cdots C_h$, such that $A \sim C_1 \sim C_2 \sim \cdots \sim C_h \sim B$.[24]

Thue suggested its possible undecidability by continuing:

> A solution of this problem in the most general case may perhaps be connected with unsurmountable difficulties.[25]

Thue's 1910 paper essentially contains the word problem without any relations [Büchi, 1989, p. 235]. In the 1914 paper, Thue reduced this problem from binary to unary trees, that is, to words or strings of letters and he introduced (semi-)Thue systems consisting of a finite set of words $a_1, \ldots, a_n$ over a given countable alphabet together with a finite set of operations (called productions) given by pairs of words (g, h). Any word of the form xgy with possibly empty words x, y may be replaced by the word xhy for any given production (g, h). A (semi-)Thue system is called Thue system, if for any production (g, h), the inverse production (h, g) is also contained in the set of productions. By composing words via concatenation, Thue systems can be viewed as semigroups with a finite presentation. This paper represents a remarkable point in history where the idea of a semigroup was born.

With this setup, in [Thue, 1977, p. 494], the term rewriting problem becomes the word problem for (finitely generated) semigroups. Notice that one replaces the question $w = 1$ in the word problem for groups by $w_1 = w_2$ for two words w_1, w_2 in the case of semigroups. Thue describes the problem as follows: assuming an arbitrary choice of given words A and B, to find a method through which one can always decide after a computable number of operations whether any two given words are equivalent with respect to A and B.[26]

The 1914 paper is cited frequently in the literature, for example, by Post in [Post, 1947], while the three other papers are mostly unknown. Richard Büchi speculates in [Büchi, 1989, p. 235] that Post might have known Thue's papers already in 1921. His paper [Post, 1921] on canonical systems may be seen as a

---

[24]...so fragen wir mit anderen Worten, ob man solche Bäume $C_1 C_2 \ldots C_h$ finden kann, sodass $A \sim C_1 \sim C_2 \sim \ldots \sim C_h \sim B$. [Thue, 1977, p. 280].

[25]Eine Lösung dieser Aufgabe im allgemeinsten Falle dürfte vielleicht mit unüberwindlichen Schwierigkeiten verbunden sein. See [Steinby and Thomas, 2000], [Thue, 1977, p. 280].

[26]Bei beliebiger Wahl der gegebenen Zeichenreihen A und B eine Methode zu finden, durch welche man nach einer berechenbaren Anzahl von Operationen immer entscheiden kann, ob zwei beliebige gegebene Zeichenreihen in Bezug auf die Reihen A und B äquivalent sind oder nicht. [Thue, 1977, p. 494].

continuation (and extension) of Thue's ideas. However, this is not reflected in the first sentence of [Post, 1947]:

> Alonzo Church suggested to the writer that a certain problem of Thue might be proved unsolvable.[27]

3. Dehn, Thue, and the Princeton Community

Dehn's student Wilhelm Magnus was a faculty member in Frankfurt from 1933 to 1938 (see Chapters 5 and 6). He rejected the Nazi government in public and was suspended from office for this reason. During the second world war he had to work in a private company. In 1947 he was appointed to Göttingen but moved to the United States one year later and finally became a member of the Courant Institute in 1950. In the academic year 1934/35, Magnus visited Princeton. This fact alone implies that in the mid 30's the Princeton community was fully aware of the word problem. This applies in particular to the prominent topologists Solomon Lefschetz, James Alexander, Ralph Fox and Marston Morse (who arrived in 1935). The books by Lefschetz [Lefschetz, 1930] (1930) and Kurt Reidemeister [Reidemeister, 1932] (1932) refer to Dehn. Alexander and Fox were experts in knot theory at Princeton.

We do not know much about the dissemination of the work of Axel Thue. Although his work on number theory, in particular the paper from 1909, was well-known to many people, his four articles on logic were probably not. On the other hand, with the help of Princeton librarians we found out that the journal[28] in which Thue had published his papers had been on the shelves in Princeton university between 1894 to 1960.

Princeton University and the Institute for Advanced Studies (IAS) play a major role in the development of the theory of computability and in the history of the word problem. While the IAS was officially independent from Princeton, there was significant overlap among the early mathematics faculty, including Oswald Veblen, John von Neumann and James Alexander (see [Dyson, 2012]).

Veblen was of Norwegian descent, born in the United States in 1880. As a professor at Princeton, Veblen spent the fall of 1913 visiting Oslo, Göttingen and Berlin [Batterson, 2007]. Veblen and Thue were both participants at the 1913 Scandinavian Congress of Mathematics, but we do not know whether they met at this occasion or at any time. After 1932, Veblen became a leading figure in the newly founded IAS at Princeton.

At Princeton and the IAS, Veblen supported the hiring of people in seemingly remote areas like mathematical logic. For example, the Polish immigrant Emil Post spent the year 1920–1921 at Princeton as a postdoctoral fellow. During this time he wrote his famous article on canonical systems [Post, 1921]. Later he spent time at Columbia, Cornell and New York University, often interrupted by periods in which he suffered from manic attacks.

Without doubt, Veblen had an important impact via his student Alonzo Church who began studying at Princeton in 1924 and finished his dissertation under Veblen in 1927. Church was then a postdoc in Göttingen and Amsterdam between 1927 and 1928. He joined the Princeton faculty in 1929 and stayed until his retirement

---

[27]See [Post, 1947].

[28]The Kristiana Videnskabs Selskabets Skrifter, Mathematisk-Naturvidenskabelig Klasse I, superseeded after 1924 by the journal of the academy Skrifter utgitt av det Norske Videnskaps-Akademi i Oslo I, Matematiske-Naturvidenskabelig Klasse.

in 1967. After that he continued to teach at UCLA until 1992. Church had an impressive roster of students. His list of students include Boone, Collins, Davis, Henkin, Kleene, Rabin, Rogers, Rosser, Scott, Smullyan and Turing who all contributed to the theory of computation, the word problem or related areas of logic in some essential way.

4. The Rise of the Undecidable

The year 1936 was the *annus mirabilis* for the theory of computation, also called recursion theory. It saw the birth of four notions of computability: the λ-calculus of Church [Church, 1936], the concept of a Turing machine [Turing, 1936], another machine concept by Post [Post, 1936], and the notion of partial recursive function (alias μ-recursive functions) by Kleene [Kleene, 1936], the latter building upon the work of Dedekind (between 1872 and 1888), Peano (1889), Skolem (1923), Gödel and Herbrand (1930–1934). Surprisingly, these four notions are equivalent. It is conjectured that there is no other feasible notion of computability beyond them; this statement is often called Church's thesis.[29]

Equipped with a notion of a (partially defined) computable function $f\colon \mathbb{N} \to \mathbb{N}$, one can define recursively enumerable sets $S \subset \mathbb{N}$ as domains, or equivalently, as images of such maps. A set $S \subset \mathbb{N}$ is called decidable, if S and its complement are both recursively enumerable, i.e., the characteristic function of S is computable. In this way, the algorithmic (un)solvability of a logical or mathematical problem, i.e., the computation of the characteristic function of the set S of Gödel numbers associated to the instances of the problem, is related to the (un)decidability of S.

Central to this theory is the existence of undecidable sets. The first examples in this direction were given by sets of natural numbers related to undecidable problems like the *Entscheidungsproblem* and the *Halteproblem*.[30] After it had been shown that these problems were undecidable (i.e., algorithmically unsolvable), people were looking for more traditional math problems for which undecidability could be shown. It turned out that the word problems for groups and for semigroups were suitable candidates. Other undecidable sets were later discovered in the negative solution of Hilbert's tenth problem.[31]

In 1937, Church announced that he could prove the undecidability of the word problem for a particular finitely generated semigroup which is not finitely presented:

> By a semigroup is meant a set in which the product of any two elements is a unique element of the set, the multiplication being associative but not necessarily obeying a law of cancellation. Consider the system of combinators, in the sense of Rosser (Duke

---

[29] To be more historically correct, it should be called Church-Markov-Post-Turing thesis. The relevant literature in this field is reprinted in [Davis, 1965].

[30] The *Halteproblem* has the following simple interpretation. If we look at computable partial functions $f\colon \mathbb{N} \to \mathbb{N}$, then it is possible to define a sequence of Turing machines T_n labeled by $n \in \mathbb{N}$ such that each computable partial function f can be computed by at least one T_n. Then the set S of all n such that T_n halts on input n is undecidable. See [Davis, 1965].

[31] This Hilbertian problem asks for an algorithm to decide whether a polynomial system of equations over the integers has a non-trivial integer solution. This problem turned out to be undecidable by showing that every recursively enumerable set is diophantine, i.e., the projection of the zero set of a system of integer polynomial equations. By applying this to an arbitrary undecidable set S, one shows that the family X_s of zero sets over every $s \in S$ has the property that one cannot decide whether X_s is empty or not.

Mathematical Journal, vol. 1 (1935), p. 336), allowing as equivalence operations r-conversions, p-conversions, and also the operations (allowed by Curry) of replacing BI by I and inversely. This system is a semigroup, with identity element I, if we take as multiplication the operation (introduced by Curry) which is denoted by Rosser as $\times$. From the relations $ab = Tb \times Ta \times B \times T$ and $T(ab) = Tb \times Ta \times B$ it follows that every element is expressible as a product formed out of the four particular elements TI, TJ, B, T. The semigroup thus has a finite set of generators, although the set of generating relations must apparently be infinite. There is, however, an effective process of writing out the series of generating relations to as many terms as desired; also an effective means of distinguishing generating relations from others. From the results of the author (American Journal of Mathematics, vol. 58 (1936), pp. 345–363), it follows that the word problem of this semigroup is unsolvable. (Received April 14, 1937.)[32]

Higman type embedding theorems, available only much later [Higman, 1961], can be used to show that Church's construction can be embedded into a finitely presented semigroup. Post proved the undecidability of the word problem for finitely presented semigroups (without cancellation) in 1947 [Post, 1947]. He mentioned that Church pointed out the 1914 paper of Thue to him. The same result was also proved in the same year (but independently) in [Markov, 1947] by A. A. Markov jr., the son of the famous mathematician who invented Markov processes. We refer to the survey article of Miller [Miller, 2014] and Rotman's book [Rotman, 1995] for more details on these and the following results.

The method Post used was to associate a (semi-)Thue system G_T to any Turing machine T [Post, 1947]. In a different way, the undecidability of the *Halteproblem* can be used to show the undecidability of the word problem for some (semi-)Thue system G_T, see [Oberschelp, 1993, §33] and [Büchi, 1989, p. 181].

Turing proved the undecidability of the word problem for semigroups admitting cancellation in 1950 [Turing, 1950] in an attempt to obtain the full result for groups. However, this result did not imply the corresponding result for groups, since the semigroups used in the proof at least a priori cannot be embedded into groups.

The word problem for groups was successfully attacked during the following years (see Chapters 6 and 7 for further details). Max Dehn died in 1952 shortly before Novikov announced his proof of the undecidability of the word problem for groups in [Novikov, 1952]. Novikov's published proof of this result in [Novikov, 1955] uses Turing's result from the 1950 paper although it employs a different method.

Church's student William Boone independently proved the undecidability of the word problem for (finitely presented) groups in his thesis. His final results were published in [Boone, 1959] after a long series of six papers [Boone, 1954–57]. Boone used Post's semigroup approach [Post, 1946]. It is known that Fox and Gödel had many conversations with Boone during this work. John Britton independently gave a proof in 1958 and later developed Boone's and other methods further [Britton, 1963]. There is a fascinating set of technical results, called Britton's lemma and Novikov's principal lemma, in Boone's, Britton's and Novikov's proofs which turned

---

[32]See [Church, 1937].

out to be related to each other; see [Miller, 2014, p. 355] and [Rotman, 1995, Ch. 12].

Much simpler proofs were later discovered in parallel with developments in group theory. One of the shortest proofs uses Higman's embedding theorem from [Higman, 1961]; see below. The same method also implies that there exists a finitely presented group G containing isomorphic copies of all finitely generated groups having solvable word problem. This group G then does not have a solvable word problem, that is, there is no uniform algorithm for all finitely presented groups that have a solvable word problem.[33]

We remark that there are many other properties of groups which cannot be recognized algorithmically, for example, the properties of being trivial, finite, abelian, nilpotent, solvable, free, torsion-free, residually finite, simple or automatic.[34] It is not difficult to prove the related undecidability of the homeomorphism problem[35] for manifolds from this.

5. Explicit Unsolvable Examples

Many construction principles are now known to yield finitely presented groups for which the word problem is unsolvable (undecidable). One particular method is quite simple and is based on work of Higman and others [Higman, 1961]. To obtain such an example, take the finitely generated (and recursively presented) group

$$G = \langle a, b, c, d \mid a^{-e}ba^e = c^{-e}dc^e \; \forall e \in E \rangle$$

where $E \subset \mathbb{N}$ is a recursively enumerable, but non-recursive set, that is, an undecidable set. Then use Higman's embedding theorem [Higman, 1961] to embed G into a finitely presented group G' with unsolvable word problem. Explicit examples are given in [Borisov, 1969] and [Collins, 1986].

An simple example of Gregory S. Tseytin [Collins, 1986; Tseytin, 1957] for a semigroup with unsolvable word problem – even in the stronger sense that on a fixed word w the decision problem $w' = w$ for any other word w' is undecidable – is given by

$$G = \langle a, b, c, d, e \mid ac = ca, ad = da, bc = cb, bd = db, ce = eca, de = edb,$$
$$cdca = cdcae, caaa = aaa, daaa = aaa \rangle$$

The word in question is $w = aaa$. There is also an example with 2 generators and 3 relations in [Matiyasevich, 1967].

References

J. M. Alonso, T. Brady, D. Cooper, V. Ferlini, M. Lustig, M. Mihalik, M. Shapiro, H. Short, in: Notes on word hyperbolic groups (Ed. A. Verjovsky), World Scientific, Singapore (1991).

Emil Artin: Theorie der Zöpfe, Abhandlungen des Math. Seminars in Hamburg, 47–72 (1925).

Steve Batterson: The vision, insight, and influence of Oswald Veblen, Notices of the AMS, Vol. 54, 606–618 (2007).

---

[33]This fact is called the Boone-Rogers theorem.
[34]This is a theorem of Adian and Rabin, see [Miller, 2014, p. 366] for a short proof.
[35]This problem asks for deciding whether two given n-manifolds are homeomorphic (for $n \geq 4$). This was treated by A. A. Markov in 1958.

William Boone: Certain unsolvable problems in group theory, Indagationes Mathematicae, Vol. 16 (p. 231–237 and 492–497), Vol. 17 (p. 252–256 and 571–577), Vol. 19 (p. 22–27 and 227–232) (1954–1957).

William Boone: The word problem, Annals of Math., Second Series, Vol. 70(2), 207–265 (1959).

Victor V. Borisov: Simple examples of groups with unsolvable word problem, Akad. Nauk., Vol. 6, 521–532 (1969).

John L. Britton: The word problem, Annals of Math., Vol. 77, 16–32 (1963).

J. Richard Büchi: Finite automata, their algebras and grammars, Springer Verlag (1989).

Bruce Chandler, Wilhelm Magnus: The history of combinatorial group theory – a case study in the history of ideas, Springer Verlag (1982).

Alonzo Church: An unsolvable problem of elementary number theory, Amer. Journ. of Math., Vol. 58, 345–363 (1936).

Alonzo Church: Combinatory logic as a semigroup, preliminary report, Bulletin of the AMS, Vol. 43(5), 333 (1937).

Donald J. Collins: A simple representation of a group with unsolvable word problem, Illinois Journal, Vol. 30(2), 230–234 (1986).

Richard H. Crowell, Ralph H. Fox: Introduction to knot theory, GTM 57, Springer Verlag (1963).

Martin Davis (ed.): The undecidable, Raven Press (1964).

Max Dehn: Über die Topologie des dreidimensionalen Raumes, Math. Annalen, Vol. 69, 137–168 (1910).

Max Dehn: Über unendliche diskontinuierliche Gruppen, Math. Annalen, Vol. 71(1), 116–144 (1911).

Max Dehn: Transformation der Kurven auf zweiseitigen Flächen, Math. Annalen, Vol. 72(3), 413–421 (1912).

Max Dehn: Die beiden Kleeblattschlingen, Math. Annalen, Vol. 75, 402–413 (1914).

Walter von Dyck: Gruppentheoretische Studien, Math. Annalen, Vol. 20(1), 1–44 (1882).

George Dyson: Turing's cathedral: the origins of the digital universe, Pantheon book, New York (2012).

Joseph Henry Engel: Some contributions to the solution of the word problem for groups (canonical forms in hypo-abelian groups), dissertation at the University of Wisconsin (1949).

D. Epstein, J. Cannon, D. Holt, S. Levy, M. Paterson, W. Thurston: Word processing in groups, Jones and Bartlett (1992).

Kurt Gödel: Collected works, Vols. I–V, Oxford University Press (1986–2003).

Catherine Goldstein: Axel Thue in context, Journal de Théorie des Nombres de Bordeaux, Vol. 27, 309–337 (2015).

Martin Greendlinger: Dehn's algorithm for the word problem, Communication on Pure and Applied Math., Vol. XIII, 67–83 (1960).

Mikhail Gromov: Hyperbolic Groups, in: Essays in Group Theory (Ed. Steve Gersten), MSRI Publ., Vol. 8, 75–263 (1987).

Gustav A. Hedlund: Remarks on the work of Axel Thue on sequences, Nordisk Matematisk Tidskrift, 148–150 (1967).

Graham Higman: Subgroups of finitely presented groups, Proceedings of the Royal Society, Vol. 262, 455–475 (1961).

David Hilbert: Mathematische Probleme, Lecture ICM Paris (1900).

Verena Huber-Dyson: The word problem and residually finite groups, Notices of the AMS, Vol. 11, 734 (1964).

Stephen Kleene: General recursive functions of natural numbers, Math. Annalen, Vol. 112, 727–742 (1936).

Donald E. Knuth, Peter B. Bendix: Simple word problems in universal algebra, in: Computational problems in abstract algebra (Ed. John Leech), Pergamon Press, London, 263–297 (1970).

Solomon Lefschetz: Topology, AMS (1930).

Roger C. Lyndon: On Dehn's algorithm, Math. Annalen, Vol. 166, 208–228 (1966).

Roger C. Lyndon, Paul E. Schupp: Combinatorial group theory, Ergebnisse der Mathematik und ihrer Grenzgebiete, Vol. 89, Reprint of the 1977 edition, Springer Verlag (2001).

Wilhelm Magnus: Das Identitätsproblem für Gruppen mit einer definierenden Relation, Math. Annalen, 295–307 (1932).

Wilhelm Magnus: Residually finite groups, Bulletin of the AMS, Vol. 75(2), 305–316 (1969).

Wilhelm Magnus: Max Dehn, Math. Intelligencer, Vol. 1, 132–143 (1978).

Wilhelm Magnus, Ruth Moufang: Max Dehn zum Gedächtnis, Math. Annalen, Vol. 127, 215–227 (1954).

Andrei Andrejewitsch Markov (jun.): Die Unmöglichkeit gewisser Algorithmen in der Theorie assoziativer Systeme (in Russian), Doklady Acad. Nauk SSSR, Vol. 55, 587–590 (1947).

Yuri Matiyasevich: Simple examples of undecidable associative calculi, Soviet Math. Dokl., Vol. 8, 555–557 (1967).

Yuri Matiyasevich: Enumerable sets are diophantine (in Russian), Doklady Akademii Nauk SSSR, Vol. 191(2), 279–282 (1970).

John C. C. McKinsey: The decision problem for some classes of sentences, Journal Symbolic Logic, Vol. 8, 61–76 (1943).

Charles F. Miller: Turing machines to word problems, in: Turing's Legacy (Ed. Rod Downey), Lecture Notes in Logic, Vol. 41, Cambridge Univ. Press, 329–385 (2014).

Marston Morse: Recurrent geodesics on a surface of negative curvature, Transactions of the AMS, Vol. 22, 84–100 (1921).

Marston Morse: A solution of the problem of infinite play in chess, Bulletin of the AMS, Vol. 44, 632 (1938).

Marston Morse, Gustav A. Hedlund: Unending chess, symbolic dynamics, and a problem in semigroups, Duke Math. Journal, Vol. 11, 1–7 (1944).

Andrzej Mostowski: On the decidability of some problems in special classes of groups, Fundamenta Mathematicae, Vol. 59(2), 123–135 (1966).

Jakob Nielsen: Om Regning med ikke kommutative Faktorer og dens Anvendelse i Gruppeteorien, Mat. Tidsskrift B, 77–94 (1921).

Pyotr S. Novikov: On the algorithmic unsolvability of the problem of identity (in Russian), Dokl. Akad. Nauk SSSR, Vol. 85, 709–712 (1952).

Pyotr S. Novikov: On the algorithmic unsolvability of the word problem in group theory, Proc. of the Steklov Inst. of Math. (in Russian), Vol. 44, 1–143 (1955).

Carl-Fredrik Nyberg-Brodda: The word problem for one-relation monoids: a survey, Semigroup Forum, Vol. 103, 297–355 (2021).

Arnold Oberschelp: Rekursionstheorie, BI-Verlag, Mannheim (1993).

Alexander Yu. Olshanskii: Almost every group is hyperbolic, Int. Journal of Algebra and Computation, Vol. 2, 1–17 (1992).

Jan von Plato: Can mathematics be proved consistent?, Sources and Studies in the History of Mathematics and Physical Sciences, Springer Verlag (2020).

Henri Poincaré: Analysis situs, Journal de l'École Polytechnique, Vol. 1(2), 61–123 (1895).

Emil Post: Introduction to a general theory of elementary propositions, American Journal of Math., Vol. 43(3), 163–185 (1921).

Emil Post: Finite combinatory processes, formulation I, The Journal of Symbolic Logic, Vol. 1, 103–105 (1936).

Emil Post: A variant of a recursively unsolvable problem, Bulletin of the AMS, Vol. 54, 264–268 (1946).

Emil Post: Recursive unsolvability of a problem of Thue, The Journal of Symbolic Logic, Vol. 12, 1–11 (1947).

Michael O. Rabin: Computable algebra, general theory and theory of computable fields, Transactions of the AMS, Vol. 95, 341–360 (1960).

Kurt Reidemeister: Knotentheorie, Ergebnisse der Mathematik und ihrer Grenzgebiete, Vol. 1, Springer Verlag, Berlin (1932).

Joseph J. Rotman: An introduction to the theory of groups, Graduate Texts in Mathematics, Vol. 148, fourth edition (1995).

Paul E. Schupp: On Dehn's algorithm and the conjugacy problem, Math. Annalen, Vol 178, 119–130 (1968).

Charles Sims: Computation with finitely presented groups, Cambridge Univ. Press (1994).

Magnus Steinby, Wolfgang Thomas: Trees and term rewriting in 1910: on a paper by Axel Thue, Bulletin of the European Association for Theoretical Computer Science, Vol. 72, 256–269 (2000).

John Stillwell: The word problem and the isomorphism problem for groups, Bulletin of the AMS, Vol. 6(1), 33–56 (1982).

Vladimir Abramovich Tartakovskii: Solution of the word problem for groups with a k-reduced basis for $k > 6$ (in Russian), Izvestiya Akad. Nauk, Ser. Mat., Vol. 13(6), 483–494 (1949).

Wolfgang Thomas: When nobody else dreamed of these things – Axel Thue und die Termersetzung, Informatik-Spektrum, Vol. 33(5), 504–508 (2010).

Selected mathematical papers of Axel Thue (Eds. Nagell, A. Selberg, S. Selberg, Thalberg), Universitetsverlaget Oslo (1977).

Heinrich Tietze: über die topologischen Invarianten mehrdimensionaler Mannigfaltigkeiten, Monatshefte für Mathematik und Physik, Band 19, 1–118 (1908).

John A. Todd, Harold S. M. Coxeter: A practical method fo enumeratng cosets of finit abstract groups, Proc. Edinburgh Math. Soc., Vol. 5, 26–34 (1936).

Gregory S. Tseytin: An associative calculus with an insoluble problem of equivalence (in Russian), Trudy Mat. Inst. Steklov, Vol. 52, 172–189 (1957).

Alan Turing: On computable numbers with an application to the Entscheidungsproblem, Proc. London Math. Society, Vol. 42, 230–265 (1936).

Alan Turing: The word problem in semigroups with cancellation, Annals of Math., 491–505 (1950).

Friedhelm Waldhausen: The word problem in fundamental groups of sufficiently large irreducible 3-manifolds, Annals of Math., Vol. 88(2), 272–280 (1968).

CHAPTER 9

Mathematics under the Sign of the Swastika

David E. Rowe

First Wave of Dismissals

When Adolf Hitler was appointed chancellor in January 1933, the vast majority of German academics assumed that the Nazis would soon be out of power and that their lives would go on as before. Few suspected that many of their colleagues, but especially younger men who aspired to promote their careers, would quickly accommodate themselves to the new political climate. Many got their first shock on April 7, 1933, when the Nazi government issued a new Law for the Restoration of the Professional Civil Service (BBG) [Siegmund-Schultze, 2009, 61–65]. The word "restoration" (*Wiederherstellung*) was, of course, a euphemism for "purging" undesirable elements from government employment. The BBG contained an array of criteria, racial as well as political, and was directed not only against Jews – defined as anyone with at least one Jewish grandparent – but also against persons who were considered politically unreliable. Even those who were merely critical of German nationalist ideology could be dismissed. While university employees had not been the principal target of BBG, but rather judges and higher civil servants, the law turned out to have devastating consequences for the university system.

Hitler seized power very quickly, but at this early stage he still had to contend with rivals in his own party as well as the powerful German military. The former general and venerable war hero Paul von Hindenburg, now 85 years old, still remained President of Germany, and he raised objections to the original version of the bill. Hitler's government then agreed to amend it so that three classes of civil servants were excluded:

1) World War I veterans who had served at the front;
2) those who had already been civil servants before the Great War; and
3) those who had lost a father or son in combat during the war.

These exemptions meant that many Jewish academics, including Max Dehn, Ernst Hellinger, and Paul Epstein, could hope to continue teaching. The Hungarian-born Otto Szász was thus the only mathematician on the Frankfurt faculty immediately affected by the BBG. Norbert Wiener afterward helped him to gain a temporary appointment at MIT, and he eventually gained a permanent position at the University of Cincinnati, where he taught from 1936 up until his death in 1952. He died while vacationing at Lake Geneva.

Quite apart from official actions coming from above, however, was the agitation of right-wing student groups operating on the ground. In Frankfurt they posed a sufficient threat to Jewish faculty that the Dean advised both Dehn and Hellinger to cancel their courses during the summer semester of 1933. This they did, though

---

©2024 by the author

they resumed teaching in the winter semester. Siegel's recollections of Hellinger accord very well with the testimonials of others who knew him.

> Hellinger was a Prussian official of the old school, though I fear this definition is not understood as well today as it was 40 years ago. Here I should point out that Hellinger was popular with everyone in student circles for his selfless dedication to duty, even during the two years before being forced to leave his position, a period when ever higher levels of officialdom were bringing their fanatical influence increasingly to bear in academic life. [Siegel, 1979, 225]

This last remark suggests that Hellinger had nothing to fear from the student body, but everything from political officials intent on interfering with academic affairs. The situation was no doubt none too clear, and it seems university authorities typically proceeded with caution. In Frankfurt they placed Hellinger's name alongside Dehn's on a document from 7 March 1934 listing faculty members whose courses might be targeted. A report of the NSDAP from this time on Dehn noted that he was not a party member, but furthermore that he had connections with Frankfurt's *Neues Theater* through his relatives [Burde/Schwarz/Wolfart, 2002a, 10]. The *Neues Theater* received no financial support from local or state governments, though the Nazis were nevertheless seeking ways to make difficulties for its management. One year later, Dehn's teaching career in Frankfurt ended; he was officially placed on retirement on 19 June 1935. His pension was more than adequate for him and Toni to continue their lives as before, but they had deep concerns over the future of their three children, all of whom were now living abroad.

Although the Nazi's virulent antisemitic rhetoric was long familiar, few imagined that the new regime would strike out so quickly against German Jews, who had long grown accustomed to enjoying full rights of citizenship. Suddenly, Jews were now singled out in a law that stripped them of those rights by claiming they were an alien people, i.e. a foreign race. For many of these Jews, the shock was all the greater since they saw themselves entirely as Germans; it many cases they or their parents had converted to Christianity. Mixed marriages between secular Jews and Gentiles had become more and more common as the country modernized. Few realized that the BBG represented merely the first of several waves of measures directed at individuals whom the Nazis portrayed as enemies of the German people.

The Nazis far more sweeping antisemitic policies later culminated with the passage of the notorious Nuremberg Laws, enacted on 15 September 1935 during a special meeting of the *Reichstag*, which was by this time completely dominated by the Nazi Party. This session, in fact, convened during the 1935 Nuremberg Party Rally, held not long after the release of Leni Riefenstahl's *Triumph of the Will*, the propaganda film based on the previous year's rally. There were two Nuremberg Laws, one regulating interracial relationships, the other bearing on legal rights. The Law for the Protection of German Blood and German Honor was aimed directly at German Jews as a racial minority, thus independent of their religious affiliation. It forbade marriages and extramarital intercourse between Jews and Germans, whereas the Reich Citizenship Law declared that only those of German blood could be citizens of the German Reich. Hitler and the Nazis often referred to their "Thousand Year Reich" as the legitimate successor of the modern German

Empire, which died with the country's defeat in World War I, but in a new form destined to reign as long as the Holy Roman Empire (800–1806).

Lonely Years Outside Academe

After being forced to give up his directorship of the Göttingen Mathematical Institute, Richard Courant began a new career at New York University. By the summer of 1935, he and other German refugees learned that Max Dehn had been forced into retirement. Although he was approaching 60, Dehn still led a very active life. Courant had been offered a position in Calcutta, but turned this down and nominated Dehn instead. Dehn may well have been interested in the job, but he expressed skepticism in a letter to Courant from 21 June. "It is very doubtful," he wrote, "whether the board there will be interested to hire an old German geometer, given that Indian interests lean toward the opposite direction" [Siegmund-Schultze, 2009, 123]. If Dehn's candidacy was considered at all, then his age was probably a more decisive factor than the field he represented. In fact, a younger German geometer, Friedrich Wilhelm Levi, was appointed to this position.

Having spent the academic year 1934/35 in Princeton, Siegel was probably not well informed about events in Frankfurt during his absence. This circumstance alone makes his later account of the atmosphere there less than convincing. "The students of that time," he wrote,

> behaved toward Dehn, Epstein, and Hellinger as would be expected of civilized humans, right down to their last hour at the university. I do not say this merely in passing, for in 1933 shameful incidents occurred at several other Universities. [Siegel, 1979, 224]

One should weigh this opinion against the recollections of Siegel's former student and later companion, Hel Braun. She was then living at home with her family on a tight budget when Werner Weber arrived in Frankfurt in the summer semester of 1935 to take Siegel's place while he was in Princeton.[1] Dehn's position had just been liquidated and word soon got round about Weber's political orientation and his role in the organized boycott of Landau's lecture course in Göttingen. Hel Braun then got involved in an attempted boycott of Weber's course, which brought her under fire from the faction of students who supported him. Braun ran for protection to Hellinger, whose days on the faculty were now numbered, and eventually she obtained her father's permission to study in Marburg. Having very little money to do so, she set off on her bicycle, and spent the next two semesters studying there under Kurt Reidemeister. When she returned to Frankfurt, Siegel greeted her with open arms, though much to her surprise. She later speculated that Hellinger had informed him about her problems with the Nazi students in Frankfurt [Braun, 1990, 23–29].

Toni Dehn remembered the brief period following Siegel's return to Frankfurt as a time when they would meet with him and Ernst Hellinger on weekly trips to the nearby Taunus Mountains. These reminiscences, however, probably relate to their earlier times together. Braun also recalled the atmosphere at the institute during this time and the troubles it caused her mentor:

---

[1]On Weber's prior political activities in Göttingen, see [Segal, 2003, 128–153].

> Siegel tried to let Dehn and Hellinger at least attend the colloquia. It took more and more courage to maintain contacts with Jewish colleagues. ...
>
> The political events and their consequences weighed heavily on Siegel. We still thought that this "thousand-year empire" must soon come to an end, but all the miseries of that time before the war dominated Siegel's world, especially in connection with Dehn and Hellinger. [Braun, 1990, 42–43]

During the summer of 1936, Dehn and Siegel both enjoyed a brief respite in Oslo, where they attended the International Congress of Mathematicians. Siegel had been invited as a plenary lecturer, and he chose to speak about a topic of recent interest that came to be known as Siegel modular forms. Here he broke new ground by connecting number theory with classes of functions in several complex variables [Hollings and Siegmund-Schultze, 2020, 169–171]. Siegel was one of 30 mathematicians whose expenses were paid by the government as a member of the official German delegation.[2] Max Dehn was invited to Norway by his friend Poul Heegaard, who played a key role in bringing the ICM to Norway. Dehn's former student, Jakob Nielsen, also gave a plenary lecture on combinatorial methods used in studying mappings between surfaces [Hollings and Siegmund-Schultze, 2020, 176–177]. Both he and Heegaard would soon re-enter the lives of Max and Toni Dehn during their flight from Nazi Germany.

Only one letter to Dehn survives from Siegel's last year in Frankfurt; it was dated 10 February, 1937. Siegel began with apologies for having not written in a long time, an indication that their pleasant wanderings together in the Taunus region were a thing of the past. He was busy preparing lectures he would deliver two weeks hence in Brussels, following up on his plenary lecture at the ICM in Oslo. He planned to spend three weeks in Brussels altogether. He ended this short letter with a personal remark: "The devil only knows why, but I miss you very much. A good guide on the rope is at times advisable, especially when there's fresh snow on the ice. Now everyone is left to go their own way."[3] This single letter strongly suggests that by now Dehn and co. were completely cut off from life at the university.

In a somewhat longer letter to Toni Dehn, written on March 6, Siegel elaborated on the time he spent in Belgium. In Brussels, he presented four lectures in French, a language he undoubtedly knew well, but spoke only occasionally. To polish these, he decided to visit Otto Blumenthal in Aachen, knowing that his French was close to flawless. For his four lectures, Siegel memorized the entire text and delivered it verbatim, an exhausting challenge. Around 25 listeners came to the first talk, but for the second the number fell to six. Georg Pólya also came and spoke to the same group, and afterward they did some sightseeing together. In Paris, Pólya spoke in Jacques Hadamard's seminar, whereas Siegel held forth in the competing seminar chaired by Gaston Julia. "The devil only knows if this is really a vacation," he concluded, "but in any case it's better than sitting in Princeton under the

---

[2] This delegation was under the leadership of Walther Lietzmann; none were of Jewish background. For the full list, see [Hollings and Siegmund-Schultze, 2020, 75].

[3] C.L. Siegel to Max Dehn, 10 February 1937, Siegel Nachlass, SUB Göttingen.

protection of Mrs. Eisenhardt [sic]."[4] The wife of the Princeton mathematician Luther P. Eisenhart had taken a dim view of Siegel's lifestyle during his stay there.

Unlike Dehn, who always seemed to have a clear direction in life, despite all that transpired, Siegel was prone to brooding and dark moods. His various wanderings also often had an air of mystery about them. According to Hel Braun, he had a tendency to run away from unpleasant surroundings. He thus imagined that in a new locality everything would be better, but "he never foresaw what awaited him, only his present situation, which he believed he could no longer bear" [Braun, 1990, 44]. Soon after he resettled in Göttingen, she, too, left Frankfurt to take a position as his assistant there.

The Dehns received a letter from Siegel, dated 11 February 1938, not long after he began teaching in Göttingen. He reported that his course on celestial mechanics was making his life difficult, though it was now going better since the astronomers were no longer attending. His new *Assistentin*, Hel Braun, was preparing an *Ausarbeitung* of these lectures, the original basis for his famous *Vorlesungen über Himmelsmechanik* (1956), later revised and translated by his star pupil, Jürgen Moser. When he first arrived in Göttingen two months earlier, Siegel was coming out of a period of deep depression. He ended his letter by expressing his hopes to see Dehn and others after the semester ended on the 28th of February, which is when he planned to arrive in Frankfurt. Except for a 10-day walking tour, he would remain throughout March. At the very end he wrote: "God punish the *katapygones*!" (spelled in Greek letters – probably meaning the weak-kneed colleagues who quickly accommodated to the Nazi policies).[5] Whether this visit actually materialized in the end seems not to be known, as no further mention of it appears in the rather scant extant correspondence from this time. What is well documented, however, concerns Siegel's visit to Frankfurt in November 1938 when he came to celebrate his friend's sixtieth birthday. The dramatic events that occurred in Frankfurt just before his arrival mark a turning point not only in their lives but also for those of millions of others as Nazi Germany descended further into barbarism.

Dehn and Hellinger during the Frankfurt Pogrom

After the Dehns' children had left Germany, Max and Toni moved out of their home on Klettenbergstrasse 16 and into a small apartment on the second floor of a private home. This belonged to a Jewish family named Rothschild, possibly descended from the famous Frankfurt banking family. Dr. Rothschild was a retired judge who had formerly presided over the juvenile court. Toni Dehn vividly recalled what happened around six o'clock on the evening of November 9, 1938 when the bell rang and two men appeared saying that her husband and Dr. Rothschild needed to come along with them. After they left, she instinctively went to Dehn's desk and began searching for any identification papers she could find.

About an hour later, much to everyone's relief, the two men had returned home. They had been taken to the police station, where they saw several large vehicles filled with people about to be sent off to detention camps. Max Dehn, however, happened to be in the company of a man well known to many of the authorities at the station, and one of them who recognized Rothschild looked at the clock and said something like: "It's 6:35, and we have orders not to accept anyone after

---

[4]C.L. Siegel to Toni Dehn, 6 March 1938, Siegel Nachlass, SUB Göttingen.
[5]C.L. Siegel to Max and Toni Dehn, 11 February 1938, Siegel Nachlass, SUB Göttingen.

6:30. So you can go home." When Dehn returned, he wasted no time calling the home of Willy Hartner (see Chapter 5). He and his wife, Else, who was Norwegian, lived in Bad Homburg, about 20 kilometers northwest of Frankfurt. She answered the phone, and Max asked her if she knew what had happened that day. She did because she knew that some teachers from the high school in Bad Homburg had taken older students to see the city's burning synagogue. The Hartners had earlier discussed the precarious situation that the Dehns might find themselves in were the Nazis to instigate a nation-wide pogrom. That day, the 9th of November, had now come. Else Hartner counseled them to leave the city at once and seek refuge with them in Bad Homburg.

Willy Hartner was still working at the university in Frankfurt; he would later bear witness to what he saw that day, both there and in Bad Homburg. He waded through streets that were strewn ankle-deep with glass and debris, while groups of rounded-up Jews stood by watching with terrified looks on their faces. Once he reached home, he called the Dehns' friend, Jakob Nielsen, in Copenhagen to tell him that they were in grave danger and needed to flee from their country. In an obituary article, Hartner later described Dehn's demeanor during this time:

> He and his wife escaped the first pogrom in November 1938 by a timely flight. They spent those days of shame at friends' homes. At that time he was not concerned about his own safety, but about the fate of those who had fallen into the hands of the henchmen. He celebrated his 60th birthday on November 13, 1938. He was unforgettable for those who saw him then, his serenity, his philosophical attitude. The discussions were not about the events of the day, but about the relationships between mathematics and art, about problems of archeology, and finally about the concept of humanity with Confucius. Thanks to the help of Scandinavian colleagues, Dehn was able to leave Germany in the spring of 1939. A helpful antiquarian from Frankfurt willingly bought his extremely valuable library of four thousand volumes for four hundred Reichsmarks with storage fees withheld. [Burde/Schwarz/Wolfart, 2002a, 30]

This condensed account of what had transpired reflects Hartner's reverential view of and fondness for Max Dehn, though some of the information he conveyed is not entirely accurate. Toni Dehn gave the 18th of January as the day they left for Copenhagen, not in the spring. During those first days, Max may well have seemed as tranquil as always, but if so, he did not remain that way. It did not take long before he came to recognize that he and his wife were in a truly desperate situation. Toni recalled how during their days with the Hartners, they discussed a plan for reaching Denmark. She and Max would go to Hamburg, soon to be joined by Jakob Nielsen and Carl Siegel. The latter had learned of their whereabouts in Bad Homburg from Ernst Hellinger, as Siegel himself related years later:

> I had already been transferred to the University of Göttingen at the beginning of 1938, where I led a somewhat retiring life. For this reason, and because there were so few Jews in Göttingen, I saw nothing of the events of the 10th of November.[6] Two days

---

[6] Siegel's remark about the number of Jews in Göttingen is misleading; over 50 were arrested the next day and taken into "protective custody," as happened all over Germany. Shortly after

> later I traveled to Frankfurt to congratulate Dehn on his 60th birthday, the 13th of November. It was then, while on my way from the station to Dehn's house, that I saw what the organized mob had done. After ringing Dehn's doorbell unsuccessfully, I went to Hellinger and learned from him what had happened. He himself had not yet been arrested, for everything that could hold prisoners was already crowded. He refused to flee, he explained, because he wanted to stay and see just how far beyond the traditional standards of justice and ethics the authorities would go in his case. He learned that the next day, while I was with Dehn in Hartner's house talking with them about the past sixty years, prompted by Dehn's birthday. [Siegel, 1979, 227]

Ernst Hellinger apparently never spoke about what he had to endure after his arrest, but his experience must have been similar to those of some 30,000 Jews, who were arrested in that period and incarcerated in concentration camps.

Many Germans, including a great number of German Jews, long held out hope that Adolf Hitler would uphold law and order in their country. The Nazis played on these hopes quite skillfully by maintaining the pretense of legality, but with the outbreak of the pogroms on 9-10 November 1938 – the infamous *Kristallnacht* (Night of Broken Glass) – they revealed to the world the true face of Hitler's regime. Within a span of less than 24 hours, mobs of rioters destroyed over 200 synagogues in cities across Germany, Austria and the Sudetenland. Thousands of Jewish businesses and homes were ransacked, store-front windows were smashed, and the streets were afterward filled with shards of broken glass. Those who never witnessed the devastation of *Kristallnacht* firsthand surely knew about it, and what they heard from the government – the Jews themselves are to blame! – should not have surprised anyone. This, after all, was the message that Hitler and his closest followers had been preaching all along. Joseph Goebbels, the Nazi Propaganda Minister, wanted the world to believe that what had happened was a spontaneous expression of the outrage of the German people over what he alleged was a Jewish plot to assassinate a high Nazi official.

For Goebbels, who had been looking for such a pretext, the timing could not have been better. A few days earlier, a stateless young Polish Jew, who was living in Paris, entered the office of the German diplomat, Ernst vom Rath, and shot him several times. The assassin, one Herschel Grynszpan, was a 17-year-old Yiddish-speaking youth, who until recently had lived with his parents in Hanover. November 9 was a high holiday in Nazi Germany when all the bigwigs gathered in Munich, this year to celebrate the fifteenth anniversary of Hitler's abortive Beer Hall Putsch. That afternoon, Goebbels received word that Rath had died from his gunshot wounds. He went to Hitler, who approved the plan to call off the police and let the Jews feel the people's fury. After sending out orders through security channels, Goebbel's informed the paramilitary forces in the Nazi Party's *Sturmabteilung* (SA) of what was afoot: they would be the ringleaders in carrying out the wanton destruction, whereas Himmler's secret police (*Gestapo*) would take

---

midnight, the synagogue was burnt to the ground, the crowning ceremony after a day of wanton violence and looting, a nationwide pogrom that spelled the end of all chances for a peaceful life for the Jewish population. In his "retiring life," Siegel simply took no notice of the events taking place around him.

care of the arrests. That evening, in his speech at the infamous beer hall in Munich where Hitler's career began, Goebbels knew how to stir up the crowd while describing this recent terrible crime, in which a German diplomat had been slain by a Jew. This, he predicted, would cause great outrage in the German people, who might even take the law into their own hands. Some ordinary civilians did just that once the SA set the pogrom in motion; others merely looked on in disbelief, wondering why the regular police were nowhere to be seen.

Willy Hartner spoke very graphically and powerfully about these events some twenty years later at the opening of a special exhibition held in Frankfurt's city hall, the Römer. This photo display, entitled "The Past Admonishes" (*Die Vergangenheit mahnt*), illustrated the crimes of the Nazi era, and Hartner's address represented an early contribution to what later became known as *Vergangenheitsbewältigung* or "confronting the past." Speaking on 28 November 1960, Hartner attacked the pervasive atmosphere of denial in West Germany that went hand in hand with a failure to atone for the horrors perpetrated against the Jews in the name of the German people. In doing so, he conceded that some few Germans might really not have been aware of the various forms of discrimination, injustice, and humiliation that the German Jews had to endure for over five years up until November 9, but after that date, how could anyone claim not to know?

> Because throughout all of Germany the synagogues had been burned, in the streets one literally waded in shards and willfully destroyed goods, and all the way to the villages one heard the piercing screams of the persecuted and abused. All Jews up to the age of 60 were driven down the streets in huge masses that were scorned, mocked, and finally carried off. Everyone who lived in Germany then had to have seen it or at least heard of it, because for a short time, the people were greatly agitated and spoke every day about nothing but this event. And for the first time in years, you heard again from the mouth of those who had previously been quiet, one is ashamed to be a German. [Hartner, 1961, 12]

Carl Siegel did not accompany the Dehns all the way to Hamburg, but rather got off the train in Göttingen to visit the Hilberts. That day Wilhelm Magnus, whose father was a chemistry professor in Frankfurt, had helped escort Max Dehn to the Frankfurt train station. In Hamburg, Max and Toni stayed with his sister Hedwig and her husband Heinrich Wohlwill at their home. Dehn's sister and brother-in-law could still move about freely, so this seemed like a safe refuge at the time. Siegel recalled how, when he checked into an older, highly respectable hotel, he had to sign a statement declaring that he was of Aryan descent. He also reported on the brutal treatment Hellinger experienced at the hands of the Nazis:

> By the time I returned to Frankfurt, Hellinger had already been taken away: first to the Festhalle along with many other innocent people, and then to the concentration camp at Dachau.
>
> He was interned there for about six weeks, until, with the help of a sister in America, he received permission to emigrate. I saw Hellinger in Frankfurt a few days after his release. He looked emaciated from the utterly insufficient diet at the camp, but maintained a strong will to live as a result of his impending

emigration. He refused to discuss his horrifying experiences and was never able to forget the humiliation that was done him.
[Siegel, 1979, 227]

Ernst Hellinger was released from the Dachau camp on condition that he emigrate immediately. He left Germany for the United States in late February 1939 to take a position as a lecturer in mathematics at Northwestern University.

Planning the Escape

A glimpse of Max Dehn's state of mind during the weeks immediately following *Kristallnacht* emerges from passages in letters he wrote to Jakob Nielsen. These were later copied by Nielsen in a letter he sent Poul Heegaard on 2 December 1938. Nielsen had heard from someone else (very likely Siegel) that "Max was in a despairing condition; he has the gravest concerns about how to survive in the future." To Nielsen, Dehn wrote on the 29th of November:

> ...that was a fast answer. You shouldn't be surprised that I was rather impatient in my last letter. This is such an incredible time. Sometimes it looks to be really difficult, and then one loses one's ease. But then a milder mood returns, and so on back and forth. What's frustrating is that I have no time or peace of mind to work. All activity, all conversation, circulates around this. Now I want to report on the little that is new. [Siegel] was here and your letter from yesterday came just as we were leaving for the station. From that as well as from recent information out of England we can now see that I can hardly hope to obtain official temporary work or any kind of stipend. It is all left to private help from friends and it will not be easy for me to burden them ....[7]

Two days earlier, Dehn had written that "a direct permit for Scandinavia or England appears to be impossible, so we are quite without hope ...."

Nielsen turned next to Heegaard, the Danish geometer who had collaborated with Dehn on their oft-cited article [Dehn and Heegaard, 1907] (see Chapter 4). Heegaard had been teaching for the last twenty years in Oslo and was therefore well informed about mathematical conditions in Norway. Furthermore, Dehn had stayed with him in Oslo both during and beyond the time of the International Congress of Mathematicians held in 1936, during which time many leading figures visited the city for the first time. Nielsen's first goal was to bring the Dehns to Denmark with the longer term goal of reaching the United States, where their son Helmut and one of Max's brothers, Karl Arnold, lived. They could submit affidavits during the time that Max and Toni stayed near their girls in England, a shorter-term stay that seemed assured if they could only obtain entry visas. After the war broke out, however, this possibility quickly vanished; had they managed to reach England, they would have been sent to a detention camp.

Since Dehn had excellent connections in Norway – he was a member of the Norwegian Academy of Science and Letters (*Det Norske Videnskaps-Akademi*) – the

---

[7]From a copy of Nielsen's letter made by Heegaard on 6 December, Dehn Papers, Dolph Briscoe Center for American History, University of Texas at Austin. This portion of the letter was written in German, whereas the rest is in Danish. The passages below were translated by Reinhard Siegmund-Schultze.

possibility of a longer stay there was also eventually considered. Thoralf Skolem had recently received news of his appointment in Oslo, which meant that the faculty in Bergen presumably had funds to hire a successor. To pursue this avenue and others, Heegaard began collecting several very impressive testimonials supporting efforts to find a suitable position for Dehn in Norway. In his lengthy letter to Heegaard, Nielsen wrote that he saw no possibility of creating more job opportunities for emigrants in Denmark than had been done already. Appealing to Scandinavian solidarity, he affirmed that Denmark had no regrets at all about having created such positions, but also mentioned how Sweden had also been helpful lately, as with the case of the probabilist William Feller. After spending some time working under Harald Bohr in Copenhagen, Feller taught in Stockholm and Lund. Nielsen thus asked, should not Norway be able to follow these Scandinavian models, even if they are only temporary positions? He recalled what England had managed to accomplish for the displaced, guessing that at least half a dozen mathematicians had found work there. Countries where science is still free should feel duty bound to act accordingly. As far as Dehn was concerned, Nielsen thought Norway was the ideal place for him:

> I can remember his enthusiasm for Norway since my youth. In 1913 he invited me to his house in Fiskeløs in Telemark for the summer vacation. Unfortunately, I didn't accept, but postponed the plan for a year, and then the war came and it was too late. In 1930, after he and I served as examiners for Ingebrigt Johansson's dissertation,[8] we traveled with our wives along the Hardangerfjord, where he knew every path. I also think that this bond he feels for Norway has been reciprocated on the Norwegian side, not only in the form of personal friendship, but also more officially.[9]

Toni Dehn later recalled: "Norway was always a favorite country of my husband's. He loved it. He had gone there from very young years, and he even owned some land in southern Norway, in that same area where that infamous Mr. Quisling came from."[10]

Vidkun Quisling, a former military figure, led the Norwegian fascist party, which had little support at this time. After war broke out in September 1939, Norway hoped to remain neutral, but its strategic locality gradually brought the country into the struggle between Great Britain and Germany for control of the high seas. During this time, Quisling colluded with the Nazis, and after the German invasion in April 1940, he was eventually installed as head of a puppet government. After the war he was convicted of treason and executed. The Telemark region, where Max Dehn owned a small house, lies in southern Norway about midway between Oslo to the northeast and Kristiansand to the southwest. Little is known about how often he stayed there or when he sold this property. It seems likely, though, that he would have sold it during his stay in Norway, as their financial

---

[8] Dehn received a letter of thanks from the rector of Oslo University, dated 8 June 1931, for his role in assessing Johansson's dissertation (Dehn Papers, Dolph Briscoe Center for American History, University of Texas at Austin).

[9] J. Nielsen to P. Heegaard on 2 December 1938, Dehn Papers, Dolph Briscoe Center for American History, University of Texas at Austin; translated by Reinhard Siegmund-Schultze.

[10] Toni Dehn, Interview with Mary Emma Harris.

situation was very precarious throughout that time. And it would remain so for many years to come.

In his lengthy letter to Heegaard, Nielsen strongly supported a plan to bring Dehn to Norway for a short period, allowing him time to prepare for settlement in the United States:

> Wouldn't that be a good way to draw the consequences and give him an existence for just one year until he can move on? Either with a modest state scholarship or with a temporary job? This would achieve two things: first, it would make the transition to America easier for him because he would then come from a recognized position in Europe, not from a modern ghetto in Germany. And secondly, and this is crucial, it would preserve his ability to work. Because there is no doubt that his present mental state is threatened. A man with his great abilities and with such an important contribution to the mathematics of our time should not perish because all states close their borders to him. I can confess to you that in the silence I had hoped for a miracle that Dehn would be allowed to be Skolem's successor in Bergen for a short time, maybe a year. Could anything better be achieved? And could Norway secure Dehn in a nicer way? Such support would mean so much more than all these words with which Norway protests against the persecution in Germany. And that would not block the way for Norway's own scientists. But this will gradually be counteracted from other sides, and so one has to be content with something less momentous.[11]

Poul Heegaard wrote to the head of the Chr. Michelsen Institute in Bergen on December 6, 1938. He included a copy of Nielsen's letter along with messages of support from five other distinguished foreign mathematicians: Élie Cartan (Paris), Heinz Hopf (Zurich), J.H.C. Whitehead (Oxford), David van Dantzig (Holland), and Karol Borsuk (Warsaw). Cartan had been a plenary speaker at the Oslo ICM two years earlier, when he chose a most appropriate topic: Sophus Lie's theory of groups in connection with developments in modern geometry [Hollings and Siegmund-Schultze, 2020, 165–168]. Heegaard began his letter by reporting that several of those who had attended the Scandinavian Congress of mathematicians, held in Helsingfors (Helsinki) the previous summer, had spoken with concern about Dehn's situation and wondered if he might not be able to succeed Skolem in Bergen, if only for a short period of time. He then went on to describe the situation:

> Prof. Dehn was arrested on his 60th birthday. He is at large again, but he and his wife want to leave Germany as soon as possible. He has a brother and son in the US and hopes to get a permit through them within a year or two. In the meantime, friends of his in Denmark, Norway, and England have promised to allow him and his wife to stay, and a few days ago the Norwegian Central Passport Office gave him a six-month residence permit. In the meantime I have received a number of letters from abroad and I feel obliged to send them to the management

---

[11] Nielsen to Heegaard, 2 December 1938, Dehn Papers, Dolph Briscoe Center for American History, University of Texas at Austin; translated by Reinhard Siegmund-Schultze.

of the Chr. Michelsen Institute. Incidentally, in some of the letters the matter is misunderstood, as if there was a vacancy at a "university."[12]

Toni Dehn's interview with Mary Emma Harris illuminates in more detail how they made arrangements to leave Germany while in Hamburg. Knowing they could stay with Nielsen – who lived with his wife and three children in Hellerup, just north of Copenhagen – the Dehns first applied for a 90-day visa to visit Denmark. They learned then that they would have to provide evidence they had another country they could visit once their time in Denmark had elapsed. In the wake of *Kristallnacht*, Jews and others targeted by the Nazis were clamoring to leave, which led bordering countries to adopt measures like this in an effort to stem the tide. Once Jakob Nielsen became aware of this roadblock, he contacted Poul Heegaard, Dehn's old friend in Oslo, who extended a personal invitation to the Dehns. In the meantime, Maria had inquired about lodging for her parents with an English family she knew. Max very likely had met them when he taught at Bunce Court, his daughters' school in Kent, during the winter of 1938.

With these invitations in hand, the Dehns received visas that allowed them to stay in Copenhagen and then Oslo for six months in all. During that time, they hoped to plan the next leg of their journey. As Toni recalled, "...we loved England, and we thought it would be fine. This was straightened out so we could leave...Germany by the 18th of January. Then we said goodbye to our relatives in Hamburg and my parents in Berlin and went on to Denmark."[13] She neglected to add that they departed with little more than the proverbial "shirts on their backs." Carl Siegel's account of what became of the 4,000 books in Dehn's library stands in sharp contrast with what Willy Hartner wrote about this in his obituary article:

> Before they left, Dehn's extensive library and many valuable pieces of furniture were bought by crafty Aryan merchants at ridiculously low prices: of course, many Germans took advantage of the Jews' plight. Used book dealers would pay 10 pfennigs on the average for their books; I've been told that mathematics students then bought books from Hellinger's and Dehn's former libraries for 50 to 100 times that amount from the dealers. Dehn was able to ship some of his furniture to London. He had hoped to be able to use some of the things he saved from Germany to start a new life elsewhere. But he lost even these last few possessions when they were confiscated and auctioned off as a result of his inability to pay the English storage charges. [Siegel, 1979, 228]

Some time before they left Denmark for Oslo in mid-April 1939, the Dehns wrote to Carl Siegel, though they had no news to impart about Max's chances of gaining a position in Norway. Siegel answered them in a letter written while he was vacationing in Nice.[14] Here, for the first time in their correspondence, he used the familiar "Du" form in addressing Max, a clear sign that despite the difference

---

[12] Poul Heegaard to Chr. Michelsens Institute, Bergen, 6 December 1938, Dehn Papers, Dolph Briscoe Center for American History, University of Texas at Austin; translated by Reinhard Siegmund-Schultze.

[13] Toni Dehn, "Interview with Mary Emma Harris."

[14] C.L. Siegel to Max and Toni Dehn, 13 April 1939, Siegel Nachlass, SUB Göttingen.

in age their friendship had grown more intimate in the wake of the events leading up to the Dehns' flight from Germany. Siegel was pleased, of course, to learn that the Dehns had enjoyed pleasant days at Nielsen's home. He made allusions to conflict in Göttingen, probably stemming from a nasty falling out with Helmut Hasse, although he was vague about this.

The general tenor of Siegel's letter was surprisingly upbeat, given the circumstances. He advised Dehn not to fret too much, since he thought that his Norwegian friends had good connections with the government in Oslo. Their main plan was still to find their way to England, but in the meantime they faced the problem that their money and Max's pension payments were lying in a Hamburg bank account that could not be tapped. So he desperately needed to find work of some kind. In the meantime, they had put some of their funds in Siegel's hands for safekeeping. He was planning to visit them at some point, and though he could not export money from Germany he promised to use this for clothing or whatever else they might need and bring these items to Norway then.

Siegel's longer term future plans were quite unsettled it seems, and since he had not burnt all his bridges to Princeton the thought of taking refuge there may well have been on his mind. In any case, he was happy to learn from the Dehns that Ernst Hellinger had found solid footing in the USA. Knowing his temperament and abilities, Siegel predicted he would manage well there: so long as he avoided the dangers of women and autos. In recalling his own experiences in Princeton, he merely noted local factors that had upset his equanimity; simply put, Mrs. Eisenhart and Solomon Lefschetz had gotten on his nerves.

Carl Siegel was hardly any more prescient about what was to come than anyone else. He had a deep affection for the Dehns, but at this point rather little appreciation of their desperate straits. The Hitler-Stalin Pact was still four months away, but the Dehns wanted to know his plans in the event the Nazis started beating the war drums. He answered that he would then flee to Scandinavia, but added that it was senseless to think of that now when he had no hope of finding employment there. Besides, he imagined it would take several years before the "culture beasts" went at each other. Siegel's pacifism was not so much a political conviction as a deep fear of losing his personal freedom and rather lavish lifestyle. He often behaved like an *enfant terrible*, and this erratic side of his character was on full display throughout the Nazi era. This should be borne in mind when reading what he had to say later in his oft-quoted Frankfurt speech from 1964, cited again immediately below.

Siegel grew particularly close with the Dehns, but he remained a loyal friend to all his former Jewish colleagues, including Paul Epstein. Max Dehn also remained in contact with Epstein, who was the oldest of the Frankfurt mathematicians. Epstein hoped to reach the United States, where his sister lived, but like the Dehns, he wanted to apply for a visa while in England. From Oslo, Dehn wrote on 27 April, 1939 to Louis Mordell in Manchester, hoping that funding for an interim stay could be arranged for him [Siegmund-Schultze, 2009, 206]. Epstein's health was already deteriorating by this time, though, as Siegel described in recalling his sad fate:

> Epstein had a chronic ailment and his condition had worsened in those last few days to the point where he could not be moved. Epstein thus escaped being dragged off to the concentration camp, but he knew that it was only a brief stay of execution.

Since he was already 68, he could not realistically expect to begin anew in another country. ...

I visited Epstein in August 1939 and we sat in the sunny garden of the house he was living in then in Dornbusch. He told me that he had a favorite cat put to sleep simply because it liked to chase birds and might have irritated the neighbors, but he didn't seem otherwise distressed. I still remember how he pointed to the trees and flowers in the garden and said, "Isn't it lovely here?" Eight days later he killed himself with a lethal dose of Veronal after receiving a summons from the secret police. Everyone knew that a summons from the Gestapo often meant torture and death, and with that in mind Epstein decided to initiate the irreversible process with his own hand. ... Emigration would probably have been impossible in any case with the outbreak of war only a few weeks away, and Hitler's so-called final solution to the Jewish problem following shortly thereafter. I really think that Epstein acted as wisely as he could. [Siegel, 1979, 228]

From Copenhagen and Oslo to Trondheim

The possibility of succeeding Skolem in Bergen was apparently still up in the air when the Dehns left Copenhagen for Oslo, but then an unexpected turn of events took place. "While we were still in Denmark," Toni explained, "my husband got a letter from a mathematician friend in Trondheim ... Viggo Brun was his name. He [had] always wanted to take a year's leave-of-absence to take a trip around the world, and he had never been able to find someone who could take his job for a year. Now there was Max."[15]

Viggo Brun had been teaching since 1923 at the Trondheim Institute of Technology. He had studied in Göttingen back in 1910, but it is unclear when and how he and Max Dehn first met; it might have even been at the ICM held in Oslo in 1936. Brun's work in additive number theory was well known to Carl Siegel and other experts, as he had managed to make significant progress on two famous classical problems: Goldbach's conjecture and the twin prime conjecture. The latter simply states that there are infinitely many twin primes, beginning with $3, 5; 5, 7; 11, 13; 17, 19; 29, 31; \ldots$. Euler famously proved that the infinite series given by the reciprocals of the prime numbers diverges, whereas Brun showed that this is not the case for the series $\left(\frac{1}{3} + \frac{1}{5}\right) + \left(\frac{1}{5} + \frac{1}{7}\right) + \left(\frac{1}{11} + \frac{1}{13}\right) + \left(\frac{1}{17} + \frac{1}{19}\right) + \left(\frac{1}{29} + \frac{1}{31}\right) + \ldots$, whose sum is less than 2. He obtained this result by using a refinement of the sieve of Eratosthenes, a method that is now called Brun's sieve.

The Dehns accepted this chance to extend their stay with alacrity. Both of them loved Norway and would have been happy to live there permanently, though they would have chosen a locality in the south over Trondheim, which lies very far to the north at 63°N latitude. Winters there were long, dark and dreary, and around the time of the winter solstice its inhabitants had to spend over 20 hours a day without sunlight. By contrast, the summers were short, cool, and often wet. Over the course of the year, the temperature typically ranged from 24°F to 65°F. Since filling in for Viggo Brun meant moving on to another locale afterward, Dehn's Norwegian

---

[15] Toni Dehn, interview with Mary Emma Harris.

friends and colleagues continued their search for a more permanent position for him in their country. In the meantime, Max taught alongside another German émigré mathematician, Ernst Jacobsthal, who came to Trondheim from Berlin in September 1939. Jacobsthal managed to remain until January 1943, at which time he fled to Sweden, though he returned after the war had ended.

Toni, who was an avid letter writer, did her best to keep family members in touch. Once she and Max were settled in Trondheim, her correspondence proved even more important as the international situation deteriorated. Those who had already left Germany or were about to leave found it increasingly difficult to write their relatives back in Hamburg. By this time, none of Max's brothers were still in Germany, but three of his four sisters still lived there.[16] The oldest of the eight children of Maximilian and Berta Dehn, Rudolf, had been a successful lawyer in Hamburg until the Nazis came to power. After he died in 1938, his wife and their two children settled in England. Max's oldest sister, Elisabeth, married the banker Eduard Goldstein, with whom she had seven children. All of them managed to escape and settle abroad; the youngest of the seven, Gertrude (called Gego), was the last to leave the family's home when she fled to Venezuela. Her parents left for England in 1939 and later lived with their daughter Hertha, the wife of Robert Solmitz, in Los Angeles.

Max's second oldest sister Hedwig was married to Heinrich Wohlwill, a chemist from the distinguished Hamburg family associated with his father, Emil Wohlwill (see Chapter 1). Their son Max, who was also a chemist, was working for a company in Frankfurt until 1938. Realizing that he and his family had no future in Germany, Max Wohlwill accepted a job in Australia, thanks to the intervention of a friend who lived in Sydney. He and his wife Erika had three young daughters, the beloved grandchildren of Hedwig and Heinrich Wohlwill. After resigning from his job in Frankfurt, they planned to stay with the grandparents in Hamburg while preparing for their long journey to Australia. As a first step, they filed to de-register as residents at the local office in Frankfurt so that they could register their address on arrival in Hamburg.

Soon before departing, though, they were awoken in the middle of the night by a Gestapo official who confiscated their passports, apparently without explanation. In Hamburg Max Wohlwill received a phone call from his brother-in-law, Max Dehn, warning him that the Gestapo was searching for them in Frankfurt (this took place some months before Max and Toni fled the city after *Kristallnacht*). Wohlwill went to register the family's new domicile, but was warned not to do so until they had recovered their passports. This task largely fell to Erika, since as a non-Jew she could travel about freely. After several months of pleas, the authorities finally relented around Christmastime and returned the passports. To use them, however, they had to list all their personal possessions, which were duly confiscated. In mid-January they finally boarded a ship in Rotterdam bound for Sydney. As conditions for Jewish families worsened dramatically throughout Germany, Heinrich and Hedwig Wohlwill dreamed of joining their son and his family there, but their chances of escaping their fate in Hamburg were few. On 29 August 1939, just two days before Hitler launched the Second World War by attacking Poland, Heinrich wrote to his son Max in Sydney:

---

[16] The information described below concerning Max Dehn's relatives can be found in [Brandis, 2020].

My dear ones,

> You can well imagine how much our thoughts are with you in these days. These lines only have the purpose of sending you a short greeting so that you know that all of us are doing well. We are all of good hope that everything will still be good. But because one can never know when normal transportation possibilities will become available, we urge all those who receive this message to write in short time intervals to Max Dehn at Strandeten 9, Trondheim to let him know how you are doing. He will forward your mail. You can best judge for yourselves at what point this procedure is no longer necessary. [Brandis, 2020, 58]

By this date, nearly all of the younger members of the extended Dehn family were living abroad. Maria and Eva, the two daughters of Max and Toni, were living in England, where they had attended the Bunce School in Kent (see Chapter 10). Maria, who had been one of the 66 children brought over from Germany when Anna Esslinger reopened the school in Kent, later taught there as well. The three children of Hertha and Robert Solmitz,[17] the grandchildren of Elisabeth Goldschmidt née Dehn, also attended the Bunce School. Toni Dehn's extended family included a number of relatives who lived in the United States, a connection that enabled their son Helmut to settle in Cleveland and study medicine at the University of Virginia. He later became a children's doctor in the town of Berea, Ohio, today a southwestern suburb of Cleveland. After his father's death in 1952, Helmut's mother Toni went to live with him in Berea. Afterward, she made several trips to Hamburg and led an active life up until the early 1990s. She died at age 103 in 1995.

Toni was the main link between the far-flung family members, not only with the Wohlwill family in Australia, but also with the three children of Max Dehn's younger sister Marie, the wife of Wilhelm Mayer. The Mayers, too, were stranded in Hamburg, whereas their eldest son Wilhelm had already emigrated to Peru in 1935. The younger son Reinhard left to study agriculture in Denmark and then worked on a farm in Scotland; his sister Franziska studied weaving in Sweden and later taught this craft at the International Grenfell Association in Newfoundland. After the war, the two younger Mayer children eventually joined their older brother Wilhelm in Peru. Max Dehn's youngest brother Georg lived with his wife in Munich, but managed to flee to Ecuador after *Kristallnacht*. They were later joined by Max's sister Bertha, a professional violinist; she managed to obtain an exit visa for the journey in October 1941, shortly before the German government cut off all Jewish emigration [Brandis, 2020, 153–163]. Max's other younger brother, Karl Arnold, had already emigrated to Japan in the 1920s. He was a successful businessman specializing in the salmon trade. Eventually, he and his wife moved to Seattle, where Max and Toni later visited them on a tour of the western United States.

Back in March 1940, when they were living in Trondheim, the Dehns still hoped they would be able to set down roots somewhere in Norway. Max was still receiving his pension from Frankfurt and held out hopes for an appointment, though no

---

[17]Walter Solmitz, Robert's cousin, studied under Aby Warburg in Hamburg, and so surely knew his daughter, Frede. She later taught at the Bunce school, where she met and married Adolf Prag (see Chapter 10).

concrete possibilities had by then arisen. Carl Siegel had long before made arrangements to visit Norway as part of a plan to escape from Nazi Germany. Although he clearly had no burning desire to return to the United States, he apparently had a standing invitation to return to Princeton's Institute for Advanced Study, a possibility he welcomed now that Hitler's war had begun. As it turned out, Siegel barely managed to get away safely, as he later recalled:

> When I visited [Dehn in Trondheim] in March, 1940, he seemed to have an air of renewed hope about him after the sad events of the previous years, and he was happy to be lecturing again. While walking together one day we noticed several seemingly deserted merchant ships in the harbor flying the German flag. Dehn told me that they had been there quite some time already, reportedly with engine trouble. They were called pirate ships by the locals on account of the somewhat frightening impression they made. Because I left a few days later for a self-imposed exile in America, I learned only later the reason for those mysterious ships' presence. They were filled with war material for the German soldiers who suddenly occupied Trondheim on the day of the invasion of Norway. They were followed by the Gestapo and the national-socialist party organisations. [Siegel, 1979, 228–229]

Toni Dehn must have kept personal notes or correspondence from this time, as she was able to give Mary Emma Harris a very precise account of what happened during Siegel's visit to Trondheim. During that time, she recalled, he was to give a lecture in Oslo, but he then learned that the Swedish vessel he had booked only allowed Jewish emigrants from Germany. There was a boat leaving on the 7th of April from Oslo, and so Max told him he should leave earlier so that he could get on that ship after his lecture. As it turned out, he wasn't able to board the boat in Oslo either, but he took a train over the mountains from Oslo to Bergen, where on April 8 he departed for the United States on the very last ship available. Toni recalled how they were awoken by an early morning phone call the next day:

> I heard [Max] say in Norwegian, "No, you must be crazy." That person said, "Well, you can just look out of the window. If the snow stops maybe for a moment, you will see that the harbor is full of warships – German warships." Later a Norwegian friend of ours and my husband walked down to the harbor, and there were Nazis and Norwegian girls already making friends with German soldiers.

Preparing to Flee yet Again

The Norwegian fascist party, the *Nasjonal Samling* (National Union), had never enjoyed much popular support in Norway, but with the help of the German Nazis Quisling readied himself to seize power. At first the Germans imagined that the king's government would simply step aside; instead, however, its leaders merely relocated to Elverum, about 150 kilometers from Oslo, and refused to yield. Quisling and his German counterparts then decided to stage a coup, which was to receive Hitler's personal approval. After taking control of radio communications in Oslo, Quisling announced that he had formed a new government. This coup attempt quickly fell on its face, though, because the military refused to carry out his orders to arrest the heads of the old government. Germany's ambassador to Norway and the Foreign Ministry then tried to find a diplomatic solution. The ambassador paid a visit to King Haakon and presented him with Hitler's demand that Quisling be appointed head of a new government, thereby paving the way for a peaceful transition of power. When the king rejected this out of hand, the Germans quickly dropped their support for Quisling, and on 24 April Hitler appointed a German, Josef Terboven, as the new Norwegian Reichskommissar. Quisling remained in the background and later made a comeback of sorts by throwing the full support of his *Nasjonal Samling* behind Hitler. In 1942, this became the only legal political party in Norway, but it never had more than about 40,000 members.

The Nazi's takeover in Norway clearly had not gone as planned, largely due to their misplaced confidence in Quisling's assurances. Thus, the ensuing weeks were highly chaotic and Norwegians found themselves cut off from the outer world with essentially no news or mail delivery. The Dehns fled Trondheim and took refuge with a farmer and his wife, who took them in for some three weeks. Max grew increasingly agitated and wanted to return as soon as possible to resume teaching his classes. Once he received word that the situation in Trondheim had stabilized, he and Toni left their gracious hosts and began planning the next phase in their lives. Soon after the semester in Trondheim ended, they took up residence again in Oslo, apparently without any great concern that they might be arrested.

Although the German war machine was now in high gear and readying its soldiers to overrun France, Max Dehn still held out hope that the Prussian Ministry of Finance would continue to pay his pension. Throughout this time abroad, Dehn continued to inform local authorities in Frankfurt of his whereabouts. They had left Germany legally and since he still enjoyed the status of professor emeritus, the government continued to pay his pension, even though he could not withdraw funds from his personal bank account. By July 1940, however, these payments had stopped, and despite months of efforts to reinstate them or at least clarify the situation, he received no official reply. On June 5, 1940, while still in Trondheim, he wrote to Frankfurt University requesting an extension of his leave of absence. Then, after moving to Hvalstad, near Oslo, he sent a message on August 29, 1940, to inform German authorities of his new address. In both instances, he signed his name as Max Israel Dehn, in accordance with the new Nazi regulations. All this may seem quite incredible, unless one keeps in mind that few could imagine at this time how far the Nazis would ultimately go in murdering millions of innocent people throughout Europe, very few of whom posed any threat to Germany at all. Moreover, as Sanford Segal documented over and again, German academics and government bureaucrats largely went about their business as before [Segal, 2003]. Some surely realized that the Nazi regime was headed by lawless thugs, but even so, given the risks involved in showing any form of overt resistance, decent people still hung on to their hopes that established routines of normal behavior would prevail in the end.

The United States was by now the only safe haven left to the Dehns, though they could no longer attempt to reach American shores via a trans-Atlantic crossing. Before they could do anything, though, Max needed to secure an academic position in America, even if only for one year [Parshall, 2022, 317–321]. Toni had stayed in touch with a Jewish physician from Frankfurt named Clare Haas, who had fled Germany after the Nazi takeover. Dr. Haas eventually found a position as a hospital psychiatrist in Pocatello, Idaho, home of Idaho University – Southern Branch (today Idaho State University). This was a two-year school, headed since 1927 by Dean John R. Nichols [Yandell, 2002, 129]. Haas approached Nichols, who then contacted Ernst Hellinger at Northwestern, mainly to explain that Dehn would be teaching students with very little background in mathematics. He wondered whether Dehn would be prepared to teach low-level courses at an American junior college. Hellinger assured him that Max Dehn had very wide experience in teaching, which included courses he taught in 1938 at an English high school, referring to the school his daughters had attended in Kent.

The first good news arrived on the 4th of July – a telegram from Nichols offering Dehn a position in Pocatello. This part of the story eventually had a happy ending (see Chapter 10), as the Dehns found their new natural surroundings very beautiful to behold. Dehn's appointment as associate professor of mathematics and philosophy lasted, however, only from February 1941 through the spring of 1942. Moreover, the whole ordeal of first obtaining an entry visa from the US Consulate and then traveling by plane, train, and then ship to reach the western shores of their future homeland would take another six months.

Max Dehn chronicled their journey in a talk he gave to the Sierra Club in Pocatello not long after his arrival.[18] Speaking to a group of nature lovers, like himself, he perhaps thought it more appropriate to emphasize the sights, sounds, and smells he experienced rather than the personal circumstances that had forced him and his wife to undertake such a difficult and dangerous trip. He told them at the start that this would be something like an objective travel report, but colored by personal impressions, such as the feeling Dehn had on the cold, dark morning of October 31st, the day their journey began. How gratified he and Toni must have felt to be greeted on the train platform by a small group of friends. These people had made the long trek there from their homes, either by foot or public transportation (due to the oil shortage in occupied Norway, no one was allowed to drive a private vehicle). The night sky was clear and impressively visible, with Saturn and Jupiter rising in the West, Venus and the dying Moon in the East. One of the friends at the Oslo station happened to be an astronomer, who saw that particular planetary constellation as a good omen for Max Dehn. The dying Moon, he later wrote him, representing the end of Dehn's European career, but "the rising of Venus meant that now there was coming for us an American period under the domination of love with Jupiter and Saturn, the stars of Bethlehem, as protectors."

What the stars had to say about the trip that lay ahead, though, was another matter, and love had little to do with it. Already at the border crossing from Norway to Sweden the Dehns got a taste of the kind of nasty treatment suffered the world over by refugees. They were dismayed by the way border guards ransacked their luggage and spoke to them in a malicious way, though Dehn recognized this was a reflection of "German policy." Still, he wondered how "young people could exult in [such] unkindness without any real profit for themselves or their community."[19] Norway, the country he had long loved, would soon enter a dark era under German occupation. Only around 2,000 Jews lived in Norway at this time, and arrests only began the following year. Up until then, only around 150 Jews had fled the country,

---

[18]See Chapter 10; the manuscript can be found in the Dehn Papers, Dolph Briscoe Center for American History, University of Texas at Austin; see also [Dawson, 2002]

[19]Toni Dehn, Interview with Mary Emma Harris.

probably because they saw themselves as part of the larger group of Norwegians who resented the occupation of their country by a foreign power, whatever their views regarding the Nazi's racist ideology may have been. Beginning in the fall of 1942, however, German soldiers and Norwegian police undertook measures that would lead to the eradication of the Jewish population.

Once they had settled in Idaho, Toni Dehn surely tried to connect again with their many friends and relatives, but obtaining news from the remaining family in Hamburg was no doubt next to impossible. She presumably still had regular contact with her parents, who fled Berlin for Switzerland; both managed to live out their twilight years in Lugano. Over the course of the year 1941, it became ever more difficult to leave Germany. Robert Solmitz headed the charitable organization founded in 1938 by Fritz Warburg, "Max M. Israel Warburg, Sekretariat," located in the family villa. After Fritz Warburg's release from a concentration camp in early 1939, he fled to Sweden as the the last member of his family to leave Germany. Solmitz ran this charity from October 1938 to June 1941, when he and his wife Hertha managed to escape. In looking back on this time, he recalled how the Jews in Hamburg who frequented the Warburg villa called it "The Oasis," a cultural center where they could gather amid the growing hostility outside. They attended lectures, readings, and concerts; one could hear the violinist Bertha Dehn perform here on occasion. Only a few months before she left to join her brother Georg in Ecuador. Hertha and Robert Solmitz reached safety in New York, after taking trains through occupied France, Spain, and then boarding a ship in Lisbon, Portugal.

Max Dehn's other two sisters, Hedwig and Marie, had no such luck. They and their husbands were forced to leave Hamburg in July 1942 on a transport train that took them to Theresienstadt. Marie's husband, Heinrich Mayer, died there before the year had ended; Heinrich Wohlwill's death came only shortly thereafter in January 1943. After losing her husband, Hedwig Wohlwill then had to say good-bye to her sister Marie on the 16th of May 1944. Marie Mayer then left Theresienstadt for some undisclosed location in Poland; only after the war did family members learn that she had been sent to Auschwitz, where the Nazis murdered more than a million innocent people. Hedwig only barely survived the Holocaust; by the time she returned to Hamburg, she was nearly blind and very weak. Her daughter Marianne, who had fled to England and worked there as a nurse, came for a time to care for her ailing mother. She returned in 1948 to attend Hedwig Wohlwill's funeral.

References

Brandis, Matthias: *Meines Großvaters Geige. Das Schicksal der Hamburger jüdischen Familien Wohlwill und Dehn*, Leipzig: Hentrich & Hentrich.

Braun, Hel: *Eine Frau und die Mathematik, 1933–1940*, Heidelberg: Springer.

Burde, Gerhard, Schwarz, Wolfgang and Wolfart, Jürgen: Max Dehn und das mathematische Seminar, preprint.

2 Dawson, John W.: Max Dehn, Kurt Gödel, and the Trans-Siberian Escape Route, *Notices of the AMS*, 49(9): 1068–1075.

Dehn, Max and Heegaard, Poul: Analysis situs, *Enzyklopädie der Mathematischen Wissenschaften* Ill, AB 3, Leipzig: Teubner, 153–220.

Hartner, Willy: *Judentum und Abendland*, Frankfurt am Main: Franz Jos. Henrich.

Hollings, Christopher D. and Siegmund-Schultze, Reinhard: *Meeting under the Integral Sign?: The Oslo Congress of Mathematicians on the Eve of the Second World War*, Providence, RI: American Mathematical Society.

Parshall, Karen Hunger: *The New Era in American Mathematics, 1920-1950*, Princeton: Princeton University Press.

Segal, Sanford L.: *Mathematicians under the Nazis*, Princeton: Princeton University Press.

Siegel, Carl Ludwig: On the History of the Frankfurt Mathematics Seminar, *Mathematical Intelligencer* 1(4): 223–230.

Siegmund-Schultze, Reinhard: Theodor Vahlen – zum Schuldanteil eines deutschen Mathematikers am faschistischen Mißbrauch der Wissenschaft, *NTM Schriftenreihe für Geschichte der Naturwissenschaften, Technik und Medizin*, 21(1): 17–32.

Siegmund-Schultze, Reinhard: *Mathematicians Fleeing from Nazi Germany: Individual Fates and Global Impact*, Princeton: Princeton University Press.

Yandell, Ben H.: *The Honors Class. Hilbert's Problems and their Solvers*, Natick, Mass.: AK Peters.

CHAPTER 10

Max Dehn's Long Journey

Marjorie Senechal

And so Max Dehn joined the thousands of scholars "of Jewish extraction" (his words) fleeing, or trying to flee, Hitlerian Europe during the Second World War. His journey was a long one, from Scandinavia, across Russia and Siberia to Vladivostok, and to America via Kobe, Japan.

1. From Oslo to San Francisco

We can read about Max Dehn's journey from Europe to America in his own words: the text of a lecture he gave, in English, to the curious residents of his first American home, Pocatello, Idaho. The typescript, which runs eight pages, reads more like a pleasure trip around the world than the harried flight of a desperate refugee. In its generosity and curiosity, the Pocatello lecture echoes Max's birthday celebration with the Hartner's. This portion of Max's talk is given verbatim, that is, the spelling and grammar are his.[1]

> To begin with the beginning: we left Oslo, the Norwegian capital, on an icy cold dark morning at the end of October. By the way, perhaps you don't realize that nowadays almost the whole of western and central Europe is totally dark during the night, in the cities as well as in the country ....
>
> There was a small group of friends at the platform, notwithstanding the cold and the darkness, coming to see us off from their homes far from the depot. Incidentally, no private person is allowed to use his car because of the shortage in oil ....
>
> In Stockholm we expected to be only three days, getting during this time the necessary visa for Russia and Japan. The visa for Japan was given at once. But the very kind Russian consul told us that our visas were granted – we had applied for them many weeks before from Oslo – but that they could not be given before two weeks. "What is the reason for this delay?" we asked. "Plague in Mandschukuo" he said. "But our way goes around Mandschukuo directly to Vladivostok." "Yes, but there is also plague in Vladivostok." "And you believe that the plague is over after two weeks?" "Hm, no. But the government might provide other ways of transport" a.s.o. He smiled because he

---

©2024 American Mathematical Society

[1]The full text of Max Dehn's Pocatello lecture is held by the Max Dehn papers, Briscoe Center for American History, the University of Texas at Austin.

knew we could not take in earnest his reasons. I don't know the real ones. Probably some obscure political ones.

But at all events this delay gave us a three week stay in Stockholm, which we spent very pleasingly. We were time and again impressed and delighted by seeing once more a splendidly illuminated city, – by the way: with rather queer looking cars. Almost all the cars are driven by generator gas, and this gas is produced by charcoal in a kind of oven attached at the back of the car or on a separate small car running behind. Once in a while you can see the driver getting down his immense bag with charcoals from the roof of his car, filling his oven and stirring it as if he was cooking his dinner....

Then, after repeated unexplained delays of the Russian planes, our trip began in earnest. We were forced to recognise that we were travelling around the world from west to east because the difference in time between Stockholm and San Francisco as we traveled was 15 hours.... So we had to advance our watches. And almost every day then came with the nuisance date line, and the fact that we had 2 days with the same date. Of course, I had to bear the heaviest burden of this incident, because I, as a mathematician, had to explain to most of the passengers this wonder – a hard work. And I had to solve the question of a law-abiding Russian Rabbi whether he had to count 8 days from Sabbath to Sabbath....

What did we see? Under our plane from Stockholm to Moscow the winterbrown soil of Sweden sprinkled with many shining little lakes, the gentle ripple of the Baltic sea. After that the endless Russian plain, for a week without any hill – we passed the Ural mountains during the night – the great streams, especially the giant rivers of Siberia....

But mankind has transformed the landscape.... On the streets and streetcars [of Moscow], in the huge department stores and shops at all times of the day were crowds of people, all dressed somewhat alike, a vizualization of communism. Russia is perhaps the only country in Europe, where nothing is rationed ... you can buy in Moscow all you are wanting, provided it is to be found in the shops or department stores. There is plenty of all sorts of food, but f.i. it might be impossible to buy a common wrist watch or silk stockings.... I had to consult a doctor. Of course he charged no fees, he also is an employee of the State ....

A special difficulty is the shortage of room and this is particularly true in the extremely fast-growing city of Moscow. A family has to be content with one room and with sharing the kitchen with perhaps 6 other families. A guide told us: "Yes, our standard of living is too low, it should be twice as high" – incidentally also the Russian have the somewhat curious fashion of expressing all things by numbers – "But" she added "we have

> concentrate all our energie on armaments. Since the whole world is rearming, we have to do the same." ...
>
> We saw a classical opera with splendid decorations and costume and the famous Russian Ballet with the Prima Ballerina and the other paraphernalia of the bygone times. This place too was crowded with a simply dressed, very bright audience. During the intermission the foyers were crowded with people enjoying caviar- and other sandwiches, sweets, beer, and Krim-Champain .... The subways stations present resplend in Kaukasian marble and aluminum ....
>
> In Siberia, all along the track were huge plants, apparantly in a quite modern style.... At Novosibirsk, where the grand railway track starts to Afghanistan, surmounting one of the highest mountain ranges, we saw a tremendous railway station ....
>
> We were forced to a six day's stay in Vladivostok, to await the first boat to Japan .... It has rather poor buildings, but some of them are looking fairly well from the outside but inside somewhat disappointing. We entered the most conspicuous of these buildings as it showed over its entrance the inscription: Padagological Institute. It proved to be a Teacher's College ... we were shown around very kindly by the headmaster ....

From Vladivostok the Dehns sailed to Kobe, Japan, and then to San Francisco, arriving on January 1, 1941.

> In glorious sunshine we passed under the Golden Gate Bridge and sailed into the peaceful San Francisco Bay. I hope that sunshine and peace may be the meaning of this earthly constellation on New Year's Day and at the last moment of our long trip.

But it was not the last moment, not even the middle of their very long trip. Their journey continued across America, connecting the dots in the landscape of American higher education at that time (outside the major private universities):

- Idaho State University, a two-year state college in Pocatello, Idaho
- The Illinois Institute of Technology, a new hybrid engineering institute, in Chicago, Illinois
- St. John's College, a Great Books curriculum private college for men in Annapolis, Maryland
- Black Mountain College, an arts-centered experiment near Asheville, North Carolina

and by-passing two might-have-been havens: a small pioneering mountain college in Kentucky (Berea College), and a large western state university (the University of Wyoming at Laramie).

2. The Emergency Committee Lends a Hand

We have seen, in Chapter 9, how the Dehns were assisted on their way by friends and family. The Emergency Committee in Aid of Displaced Foreign Scholars assisted and coordinated their efforts.

The Emergency Committee, as we will call it, was founded in 1933 by a group of influential educators in New York City [Duggan and Drury, 1948]. Over 6,000 displaced scholars and professionals appealed to it for aid; 335 received grants.[2]

The Emergency Committee faced a delicate task: American universities, hard-pressed by the Great Depression, were wary of an influx of foreign teachers and researchers competing with their own faculties. Like most of the German scholars they were trying to help, the Emergency Committee did not foresee the war and hoped that Hitler's reign would be short-lived. To mitigate tensions, it sought short-term appointments for the refugees, and raised money to cover their salaries and other necessary expenses. The Emergency Committee also decided, as Duggan and Drury explained [Duggan and Drury, 1948], to confine "its efforts to displaced scholars of such eminence in their fields that there would be no thought of competition with young American scholars." Thus it would "not, except in exceptional cases, make grants to refugee scholars under thirty years of age or over sixty." Furthermore, it would "make no direct grant to an individual scholar. The request for a grant must come from a college or university, which would indicate the scholar it wanted and express willingness to disburse our stipend in his behalf in the form of salary payments."

The Emergency Committee also had an unannounced operating policy: it would spread the refugees around. Clustering them would stir extra resentment at the universities in which they were placed, and could fan discomfort among the refugees themselves. "Anything that could be done to keep scholars from living and brooding among members of their own language groups was a forward step in orientation."[3]

The refugee scholars had a lot to brood about. Concern for families and friends still in Europe were first and foremost. But in America they faced language difficulties, xenophobia, culture shock (in academia as well as on Main Street), and low salaries. And job insecurity: many were filling in temporarily for faculty serving in the armed forces.

To ask why Harvard, Yale, Chicago, and other leading American universities didn't vie to host a mathematician of Max Dehn's stature, even under the difficult circumstances of that time, is betrayal by hindsight. Though his contemporaries certainly knew Max Dehn as a leading figure, "Dehn" was not yet the adjective known to all students that it is today. "His name is probably more widely known – and certainly more widely used – today than it was at the time of his death in 1952," his former student Wilhelm Magnus remarked at Dehn's centennial in 1978 [Dehn and Magnus, 1978]. And, in the early '30s, neither Max Dehn nor anyone else thought he was in danger. In 1933 the eminent German Jewish refugee mathematician Richard Brauer wrote to the Emergency Committee from the University of Kentucky in Lexington. The Emergency Committee had placed Brauer in Lexington in 1933.

> The rumor that Professor Dehn is supposed to be in great danger, has no background, whatsoever, on the contrary, one supposes that he will stay in his sphere of duty since he is being protected by the official law (Beamtengesetz). None of my acquain-

---

[2] http://www.archives.nypl.org/mss/922.

[3] Emergency Committee to the American Friends Service Committee in *Resettling the Refugee Scholars*, excerpt from the Trustees' Confidential Monthly Report, Oct. 1, 1941, Rockefeller Foundation Archives, RG 1.1, Series 200, Box 47, Folder 538.

tances in Germany ever heard of any measures taken against Professor Dehn.[4]

The picture worsened quickly. Yet, except for the first four months of 1938, Max and Toni remained in Frankfurt. Those four months both puzzle and instruct us. On the one hand, they make the Dehns' decision to stay in Germany even more difficult to understand. On the other hand, they suggest why, once he reached the North Carolina mountains, Max was content, even happy there.

3. Flashback: England, January–April 1938

Line 12 of Max Dehn's CV in the records of the Emergency Committee reads: "1938 January–April Teacher of mathematics at the New-Herrlingen School near Faversham, Kent (England)." Had the Dehns remained in England, there would have been no Long Journey. They didn't stay, but they planned to return to England, should emigration become necessary. But when it did, they couldn't.

The Dehns' connection with the school that became the New-Herrlingen began in 1931. Their older daughter, Maria, recalled, "I was a teenager and refused to eat adequately. Ado Prag suggested my parents send me to the Landschulheim Herrlingen, near Ulm. This they did."[5]

Adolf Prag, a teacher Latin, Greek, and mathematics at that school, had been Max Dehn's student at Frankfurt University and an active participant in the history of mathematics seminar. [Scriba, 2004], translating "Latin and Greek texts fluently into the German language." Though he chose a teaching rather than a university career, he had remained close to the Dehns and to his fellow mathematics students Ruth Moufang and Wilhelm Magnus.[6]

The school was five years old when Maria Dehn enrolled. Its director, the remarkable German-born Anna Essinger (1879–1960), had lived in the United States for twenty years. Influenced by the Quakers and new currents in pedagogy, she returned to Germany after World War I to work for humanitarian relief. Her sister had founded an orphanage in Ulm; together they turned it into the Landschulheim Herrlingen, a boarding school for boys and girls of all religions. It opened in 1926, with Anna as headmistress.

"She ran Landschulheim Herrlingen like a Montessori program," Wikipedia explains,[7] "placing high value on communal living, mutual respect, and a shared sense of responsibility for the school.... Learning was accomplished through living, whether from daily walks in the woods, from the tasks required of the children in and around the building, or at meal time, where there were 'English' and 'French' tables and those sitting at them would speak in those languages during the meal. The arts were also offered. In addition to painting, drawing, singing, and drama, the children learned to play music." Children's work was assessed, not graded. Staff and pupils were on a first-name basis.

Daily walks in the woods; mutual respect; foreign languages; no grades; playing music, first names: Landschulheim Herrlingen was an unwitting prototype for Black Mountain College.

---

[4]Richard Brauer to the Emergency Committee, November 9, 1933.

[5]Maria Dehn Peters to Constance Reid, Max Dehn papers, Briscoe Center for American History, the University of Texas at Austin.

[6]See Chapter 6 in this volume.)

[7]See "Anna Essinger," Wikipedia.

As Nazi influence grew in Germany, Anna Essinger, who was Jewish, read the writing on the wall and decided to move the school to England.[8] She bought an old manor house near Faversham, in the County of Kent, and with Adolf Prag, brought sixty six children there on the pretext of a school outing. "In September, 1933, after much exploring and planning in which Father was deeply involved, the school moved to England and I moved with it," Maria recalled. The transplanted school was first called New-Herrlingen, then Bunce Court. Eva Dehn joined her sister there in 1936; that same year, their older brother Helmut left to study medicine in America [Yandell, 2002].

Max Dehn was a frequent visitor at Bunce Court.[9]

"Father's visits to England bore little relation to his professional life," Maria wrote. "My brother, Helmut, and I agree that he went to Bunce Court, the New Herrlingen School, largely to, as my brother puts it, hold Anna Essinger's hand .... She admired my father and valued his judgment. Thus she liked to consult with him. Indeed, this continued after the Dehn girls had graduated. When Max taught at the school, in 1938, Maria was finishing college and Eva was in training as a nurse.[10]

At Bunce Court, Max Dehn gained first-hand experience teaching in an experimental setting, in which educating the "whole student" was as important as teaching any particular subject matter. Other refugee Herr Professor Doktors would balk at this, but Max found it to his liking. At Black Mountain College, the last stop on his long journey, Bunce Court echoed daily.

4. Across America

The Dehns left Germany for Copenhagen on January 18, 1939, unaware that they were headed for America, not England, and that it would take them nearly two years to get there. Travel to England became impossible after World War II began, on August 1, 1939, a few weeks after the Dehns arrived in Trondheim. And staying in Norway became impossible after April 9, 1940, when Germany invaded the country.

4.1. Pocatello, Idaho (January 1941–July 1942). A friend of Toni's, Clara Haas, a Jewish refugee physician working in a hospital in Pocatello, Idaho, persuaded John R. Nichols, the dean of the branch of Idaho State University in Pocatello, to offer Max a position there if his friends raised the funds for his salary.[11] Dean Nichols thought it best to warn them:

> May 31, 1940 John Nichols (Dean of ISU) to Ernst Hellenger (sic),

---

[8] Anna Essinger did not close the Herrlingen School in Ulm; Hugo Rosethal, a Zionist and educator, ran it as a school for Jewish children until 1939 [Schachne, 1988]. The mathematician Marion Walter was a student there (http://www-history.mcs.st-andrews.ac.uk/Biographies/Walter.html)

[9] On one of those visits, Max introduced Adolf Prag to Prag's future wife, Frede Warburg, who had applied for a position at the school. When Bunce Court closed, in 1948, the Prags stayed in England. Adolf Prag taught in various English schools and contributed to the history of mathematics by translating and editing classical texts.

[10] Maria Dehn Peters to Constance Reid, Max Dehn papers, Briscoe Center for American History, the University of Texas at Austin.

[11] All of the correspondence with Nichols arranging Dehn's position at Idaho is held in the Max Dehn papers at the Briscoe Center for American History, the University of Texas at Austin.

> ... Of course we could always use another good man on the faculty especially when it does not cost us anything .... We are teaching only freshman and sophomore mathematics which would be about the equivalent of the first year in the German Gymnasium. He would find the students woefully unprepared, generally disinterested in mathematics, and possibly, from his point of view, exceedingly lazy and slipshod in all their work and thinking. On the other hand such a person might arise to the challenge and see what he could do with the type of students who come to an American junior college.

Hellinger replied promptly:

> June 7, 1940 Ernst Hellinger to John Nichols
> Knowing Max Dehn as I do, I am quite sure that he would enjoy thoroughly working with that type of students which you describe, and that he would get along extremely well with them. I suppose I mentioned already that Dehn is remarkably vivacious and adaptable and that he taught Mathematics at an English high school in 1938. Incidentally, as a kind of hobby he liked to interpret Mathematics to young people of high school age.

On June 22, 1940, Nichols offered Max Dehn a temporary position at Idaho State. A week later, Herman Weyl wrote to Dean Nichols: "I am delighted that Professor Dehn is to be at your University, and I congratulate you on this appointment,"

But it took the Dehns four months to arrange their travel. Meanwhile the Emergency Committee gave his case low priority (Figure 1).

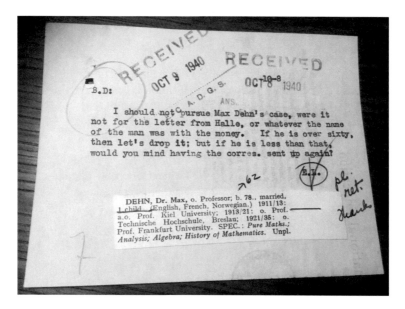

FIGURE 1. A note to Betty Drury of the Emergency Committee, concerning Max Dehn (in the Committee's papers, New York Public Library).

"I am very glad to tell you that after all, Professor and Mrs. Dehn were able to start their journey to this country," Hellinger wrote to John Nichols on November 20, 1940.[12] "Although they were anxiously trying through all these months to find a direct passage, the shortness of boats and the different new regulations which happen to be made in the Nazi ruled countries prevented a success again and again. Eventually they were compelled to choose the long and expensive route through Siberia - Japan .... They are hoping to arrive in San Francisco about New Year, hoping there is no further delay. From San Francisco they will come at once to Pocatello .... Following your suggestion, I shall send the money for Dehn's salary by the middle of December to your bursar."

The Pocatello community followed the Dehns' journey, through letters and press reports, as it progressed across Siberia and the Pacific Ocean, and were eager to hear all about it. Soon after arriving in Pocatello, Max Dehn satisfied their curiosity with the lengthy public lecture, in English, excerpted above.

Idaho State University was founded in 1901 as the Academy of Idaho, renamed the Idaho Technical Institute in 1915, and given the name by which Max Dehn would know it – the University of Idaho, Southern Branch – in 1927. John R. Nichols had been the Branch's chief executive since the last renaming. The Southern Branch was a two-year college when the Dehns were there; Dean Nichols worked to strengthen and expand it. By 1947, when he left,[13] the college was the four-year Idaho State College. It was renamed a fifth and final time in 1963.

The local press rushed to interview Max and Toni when they finally arrived. In an article titled "Dr. Max Dehn Arrives Here After Crossing Siberia," the Pocatello journalist Joe Ruffner noted,[14] among other things, that

> The trials of a refugee scholar are many. At any rate, that's what this reporter learned Friday when he interviewed Dr. Max Dehn, German-Jewish refugee scholar about an hour after he arrived in Pocatello, where, as associate professor of mathematics and philosophy at Idaho Southern University, he will make his home for an indefinite period.
>
> Mrs. Dehn surprised us by her fluent use of the English language. The question of the Dehns' ability to speak English had been a large one at Idaho Southern for some months.
>
> His salary is not being paid by the state of Idaho, but by a non-government fund which numbers leading American business and academic figures among its subscribers.

Pocatello nestles between Idaho's welcoming (to hikers like the Dehns) Bannock Range and Samaria mountains. The academic community, faculty and students alike, welcomed them warmly too. By all accounts, both Dehns returned the affection.

But, outside academia, some Pocatellians harbored suspicions. A member of the American Legion wrote to Dean Nichols in alarm.

---

[12] Ernst Hellinger to John Nichols, Max Dehn papers, Briscoe Center for American History, the University of Texas at Austin.

[13] John Nichols left Idaho in 1947 to become president of the New Mexico College of Agriculture and Mechanic Arts and, subsequently, the U.S. Commissioner of Indian Affairs.

[14] J. Ruffner, Idaho State Journal, January 19, 1941.

Dear Mr. Nichols,

The Salt Lake Tribune of recent date carried an item, or rather a statement by J. W. Condie relative to a German proffessor (sic) being assigned to the So. branch to teach math and philosopy (sic). It stated that the professor was assigned by some organization to the university without cost to the school.

The Hazleton Post of the Am. Legion has appointed a committee, myself as Ch. to contact you for information pertaining to the facts of the matter. We are interested in knowing where this money comes from, who is sponsoring him and who is paying his salary.

How does he happen to be assigned to the So. Branch.

Our next meeting is on Mon. night Oct. 7th and I would appreciate hearing from you so that I may report at that meeting.

Very Truly Yours, W. H. Detweiler (undated)

At first Dean Nichols laughed off such objections. "My only problem has been to assure legislators, American Legionnaires and other super-patriots that he was not a Communist!"[15]

But soon Nichols concluded, reluctantly, that Max Dehn's appointment could not be continued. "The Dehns are popular on campus," he wrote. "The students of a dormitory invited Toni Dehn to be its manager; she was pleased to do it, and the Dehns received free room and board.[16] But we have no money in our budget for a permanent position and have some state-wide opposition to 'hiring' an alien German Jewish refugee."

Dean Nichols reappointed Max Dehn for the 1942–43 academic year nevertheless, and the Emergency Committee came through with $250 to help support him. But, they told him, this was all they could do, and one time only. Meanwhile they tried to help him find a position elsewhere. The Emergency Committee charged its field secretary, Laurens Hickok Seelye, with visiting "colleges and universities in various parts of the United States in order to stimulate interest in the program of the Emergency Committee."[17] In other words, Seelye's job was to spread the refugees around.

4.2. Bypassing Berea and Laramie. Laurens Seelye thought Max Dehn might be a good fit for Berea College in Kentucky, and vice versa. Berea College nestles in The Glade, where the foothills of the Cumberland Mountains slope down to the central plains of Kentucky's bluegrass. The Dehns would have loved the setting, if they had seen it.

The college was founded in 1855 by an abolitionist, the Reverend John G. Fee, From its start, Berea was co-educational, interracial, and free; even today, its students pay no tuition. Indeed, the college only accepts students of limited financial means.

---

[15]18 John Nichols to Kathleen Hanstein, American Friends Service Committee, March 6, 1941, NYPL; see also Nichols to the Emergency Committee, March 3, 1942.

[16]John Nichols to Francis Hutchins, Nov. 4, 1941, Emergency Committee.

[17]Seelye was well-connected in the American academic world. The grandson of a president of Amherst College and grand-nephew of a president of Smith College, he had taught at the American University in Beirut and had been president of St. Lawrence University in Canton, NY.

After Kentucky's legislature passed a law in 1904 prohibiting "White and Colored Persons from Attending the Same School," the college served primarily Appalachia's poor whites. But its biblical motto, *God has made of one blood all the people of the earth*, remained its Christian mission, and Berea was the first Kentucky college to integrate (in its case, reintegrate) when the law was amended in 1950.

Prompted by Laurens Seelye, Berea's President Francis Hutchins told John Nichols and Max Dehn that he had an opening for a math teacher for the spring semester. Max replied cautiously, "I have read about the college and I must confess that I am enthusiastic to know that there have been and still are people devoting their whole force and their whole abilities to purposes of such high social value. I am a little doubtful whether, with my age, I would wholly satisfy the requirements for a guide, philosopher, and friend of these young people."[18]

Toni Dehn recalled this round-the-clock availability as her husband's principal objection to Berea, but he had others. He was reluctant to move across the country to teach for one semester, at a very low salary. And then there was "the religious basis of your institution." He and Toni would not miss alcohol or tobacco, which the college forbade, he told Hutchins, but "we hesitate to guarantee our regular attendance to church service."

President Hutchins surely chuckled on reading this. When he was offered the presidency of the college, in 1939, he too had objected that he was not religious. But William Hutchins, Berea's retiring president, refused to take no for an answer, and Francis gave in. William Hutchins was Francis Hutchins' father.

Seelye, astutely reading Dehn's demurral as a polite refusal, urged him to change his mind.[19] Max could go to Berea for the semester alone, while Toni ran the dorm in Idaho. "I think it would furnish you with an excellent opportunity to come in touch with another region and with other colleges which might lead to further connections. It seems to me very important that a displaced scholar should keep 'in circulation.' I think that in your position I would even be willing to promise to go to church every Sunday for half a year, if that is the idiosyncrasy of the college!"

The Dehns would have enjoyed the company of Francis Hutchins and his Yale M.D. wife. Both had taught in China for many years; in 1942, in Berea, Dr. Louise Hutchins directed the Mountain Maternal Health League, which brought health care to women in isolated communities. And my parents, in nearby Lexington, would have heard of them through medical circles, and warmly welcomed them into our lives.

By 'further connections,' Seelye may have had Francis Hutchins' brother in mind. Robert Maynard Hutchins was a member of the Emergency Committee and president of the University of Chicago.

Would Max have been happy at the University of Chicago? R. M. Hutchins took pride in the undergraduate Great Books program he'd created in close consultation with Scott Buchanan and Stringfellow Barr at St. John's College, a program

---

[18] Nov. 6, 1941, copy to Seelye. Emergency Committee.

[19] Laurens Seelye to Max Dehn, Nov. 18, 1941, Max Dehn papers, Briscoe Center for American History, the University of Texas at Austin.

that Max, as we will see, very much disliked. But he would have been in his mathematical element in the graduate program: the University of Chicago's mathematics department was one of the best in the world.

But in the end, Max Dehn did not visit Berea, and Francis Hutchins did not offer him the job.

Laurens Seelye also suggested the University of Wyoming in Laramie, a land-grant state university founded in 1886 on a high plain between two snowy mountain ranges. Max visited Laramie and received an offer. But, he told Seelye, the salary was low and the teaching load heavy.[20]

Just then, Ernst Hellinger called Max with the offer of a one-year position at the Illinois Institute of Technology in Chicago, with a probable one-year reappointment. The two long-time friends would be practically neighbors: Northwestern University was only fifteen miles away. John Nichols urged him to accept the offer. "They hated to leave here," he told the Emergency Committee.[21] "However I thought the other proposition was better than anything we could do for them permanently and I believe I was right."

Dehn accepted the offer. The Emergency Committee transferred the $250 grant from Idaho to IIT – emphasizing again that it would not be renewed.

4.3. Illinois Institute of Technology in Chicago (1942–43). Chicago's Lewis Institute and Armour Institute of Technology were merged in 1940 to create the Illinois Institute of Technology (IIT). When Max Dehn arrived in 1942, IIT did not yet have a graduate program, and its undergraduate offerings were depleted by the exigencies of wartime service. But the Institute's leaders envisioned a brilliant future: IIT, they hoped, would be the midwest's node in a three-point arc of engineering excellence from Cal Tech on the west coast to MIT on the east. They had hired Mies van der Rohe, the Bauhaus's last director, to design a new campus in the center of the city; construction had not begun, but excitement was in the air. Laszlo Moholy-Nagy's Institute of Design joined the merger in 1949. Van der Rohe's Crown Hall, built in 1956, is the main campus's crown jewel to this day, and a national landmark.

Lester R. Ford (1886–1967), the chairman of the mathematics department at the Armour Institute at the time of the merger, created a new one in and for the hybrid institution. He had to start from scratch. "The mathematics syllabus was like that of a pretty good high school course of today, with algebra, calculus, and differential equations," says the institute's website.[22] Max Dehn found himself not on a continent-spanning arc, but on a straight line from Idaho.

Besides Lester Ford, Max's mathematics colleagues at IIT in 1942-43 included Herbert Busemann, also a German refugee mathematician, and the young George Mackey, a new Ph.D. from Harvard. (Karl Menger, Gordon Pall, and Josephine Mehlberg joined the IIT department a few years after Max left; they were its linchpins when I was a graduate student there, in the '60s.)

Did the Dehns meet the Russian-French refugee chemist-dancer, Natasha Goldowsky, at IIT? She taught there the year that Max did. Nothing suggests that they met then, or that they met Mies van der Rohe either. But the Dehns would meet

---

[20]Max Dehn to Laurens Seelye, August 23, 1942, Max Dehn papers, Briscoe Center for American History, the University of Texas at Austin.

[21]Dec 15, 1942. John Nichols to Betty Drury, secretary of the Emergency Committee.

[22]https://science.iit.edu/science-iit-past-present-future

other Bauhuas leaders, and Natasha Goldowsky, at Black Mountain College a few years later.

According to his biographer, Herbert Busemann was highly critical of IIT.[23]

> His first permanent position was at the Illinois Institute of Technology in Chicago, a position which he described as a "horrible permanent job." He recalls that this was a period where "everybody was looking for jobs, and one had to take whatever." He spent "five miserable years" in Chicago, from 1940 to 1945. "The head of the department made it difficult. He did not like foreigners in the first place. He belonged to those people who had done a couple of good things when they were quite young and he was against anyone who was too active mathematically. On the other hand the administration forced him to take good people, and he resented them."

Max's relationship with Lester Ford seems to have been happier. Ford encouraged him to write articles on the history of mathematics for the American Mathematical Monthly, which he edited at that time, and encouraged him to meet American mathematicians at meetings of the MAA, of which Ford would later be president. Dehn would write five articles (one of them with Ernst Hellinger) and two book reviews for the Monthly.

The first paper in the series, Mathematics, 600 B.C. – 400 B.C., appeared in the June-July issue of the Monthly in 1943, while Max was still at IIT [Dehn, n.d.]. A footnote read, "This is the first of a series of articles by Professor Dehn which, it is hoped, will cover the whole history of mathematics in compact form. An important feature of the series is the collection of maps, which are the work of Mrs. Dehn. Professor M. H. Heins, of Illinois Institute of Technology, is doing some of the editorial work connected with the articles." There are no acknowledgements in subsequent articles.

But despite Ford's encouragement and the proximity of his close friend Ernst Hellinger, Max Dehn too was unhappy at IIT. His friend Carl Ludwig Siegel blamed the noisy city and a remedial class Max had to teach. Toni Dehn recalled that "Max didn't like to use public transportation and it was too far to walk." That it was: the Dehns lived on Blackstone Avenue, in the Kenwood neighborhood near the University of Chicago; by foot, IIT was an hour and a half away (see Figure 2).

Perhaps the Dehns chose Kenwood for its proximity to beautiful Lake Michigan. But Chicago isn't called "the Windy City" for nothing. All winter, a bitterly chill wind roars across the lake and into the city. And there are no mountains to climb, not even gentle hills: Chicago is flat. The highest point in the state of Illinois is Charles Mound in Galena, a meager 375 meters high (and 156 long miles from Chicago).

Max could not have stayed much longer at IIT even if he had liked it there. "We are very sorry to lose him," Lester Ford wrote to Dehn's next employer, but "he is very near our retiring age, and there seemed nothing for us to do except to view the situation from Professor Dehn's point of view and let him go to an institution where he could look forward to additional years of service." But, as we will see, he couldn't.[24]

---

[23] In this brief biography, Busemann does not mention Max Dehn.

[24] All the correspondence concerning this appointment is from the St. John's College archives.

FIGURE 2. From Hyde Park to IIT.

"Max Dehn Terminates 11 Month Lecturing Visit," an IIT newsletter announced on August 2, 1943. "Dr. Dehn accepted a position as head of the mathematics department at St. John's College, Annapolis, Maryland." That's not quite accurate: St. John's has no departments and all faculty members hold the rank of tutor or fellow. But he did accept a position there.

4.4. St. John's College (1943–44). Herman Weyl had suggested Dehn to St. John's Dean, Scott Buchanan. Dean Buchanan was delighted with Weyl's letter and offered Max a generous-for-the-time salary of $3000 for the year. But, he warned Max, his long-term prospects at St. John's were not good: "in discussing the matter with President Barr we were both reminded of the problems that may face us after this year. We have a good many valuable teachers in the armed forces. They will want to return and we shall want to take them back .... I am afraid that I did not emphasize this enough in our conversations while you were here and therefore did not emphasize the extra sharpness of the policy during the war of appointing tutors for one year only."[25]

Max accepted the appointment in spite of, and aware of, this. Prudently, Toni stayed in Chicago, working as an advertising designer for the mail-order retailer and department store chain Montgomery Ward.

St. John's College was founded in 1696 as King William's. It was renamed St. John's in 1784, after the Revolutionary War. In 1937, the college adopted a Great Books program, in which students read primary sources (in translation) cover to cover, instead of excerpts and textbooks. In 1943, when Max Dehn arrived there, the students were still all men, and most of them were on leave, serving in the war. Still, the college persisted on its chosen course.

---

[25] Scott Buchanan to Max Dehn, July 12, 1943.

"The real ultimate and original teachers at St. John's are the authors of some hundred greatest books of European and American thought," says its 1943-44 catalogue. In each year, all St. John's students read exactly the same works at exactly the same time.

The twenty-person St. John's faculty included that year the classicist and historian of mathematics Jacob Klein, also an Emergency Committee grantee, and the emigré musician Nicolas Nabokov. Max Dehn and Jacob Klein shared a deep interest in Greek mathematics, which they both studied in the original Greek; surely they would have discussed Klein's book, Greek Mathematical Thought and the Origins of Algebra, and its significance, or lack thereof, for the St. John's curriculum. But, perhaps because they saw each other every day, there is no record of their conversations.

The official correspondence between Dehn and St. John's leadership is cordial (see Figure 3).[26]

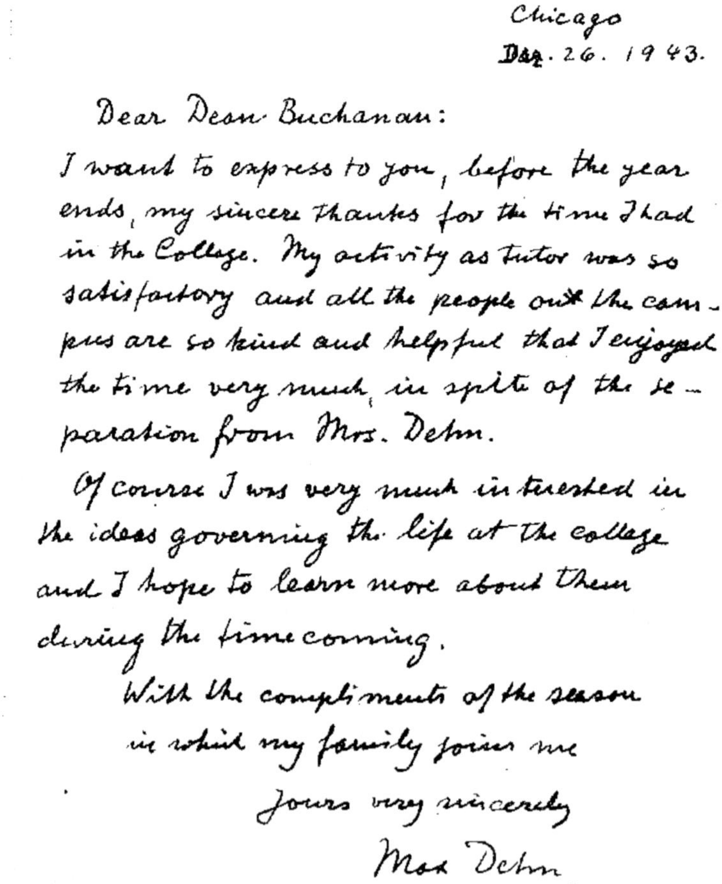

> Chicago
> Dec. 26. 1943.
>
> Dear Dean Buchanan:
>
> I want to express to you, before the year ends, my sincere thanks for the time I had in the College. My activity as tutor was so satisfactory and all the people on the campus are so kind and helpful that I enjoyed the time very much, in spite of the separation from Mrs. Dehn.
>
> Of course I was very much interested in the ideas governing the life at the college and I hope to learn more about them during the time coming.
>
> With the compliments of the season in which my family joins me
>
> Yours very sincerely
> Max Dehn

FIGURE 3. Max Dehn to Scott Buchanan. Used with permission from the Dean's Office, St. John's College, Annapolis, MD.

---

[26]Max Dehn to Scott Buchanan, Dec. 26, 1943.

But Carl Ludwig Siegel remembers otherwise [Dehn and Magnus, 1978]:

> He was particularly unhappy. It was wartime then, and the College was attended only by very young students ranging in age from 15 to 18 .... You must keep in mind that, in general, American schools cover far less than German schools even today, and that many 15 year-olds had not even mastered English properly. But St. John's pretentious plan of study would have these immature youths reading Homer, Dante, Descartes and Goethe in the original.[27] The whole thing seemed a malicious parody of the goals of the history of mathematics seminar in Frankfurt. Dehn, of course, saw the insanity of such an undertaking from the start and presented his criticism to the responsible authorities. The result was a complete falling out followed by a long period of friction which he had to endure ....

As Max had been warned, his position at St. John's was not continued (see Figure 4).

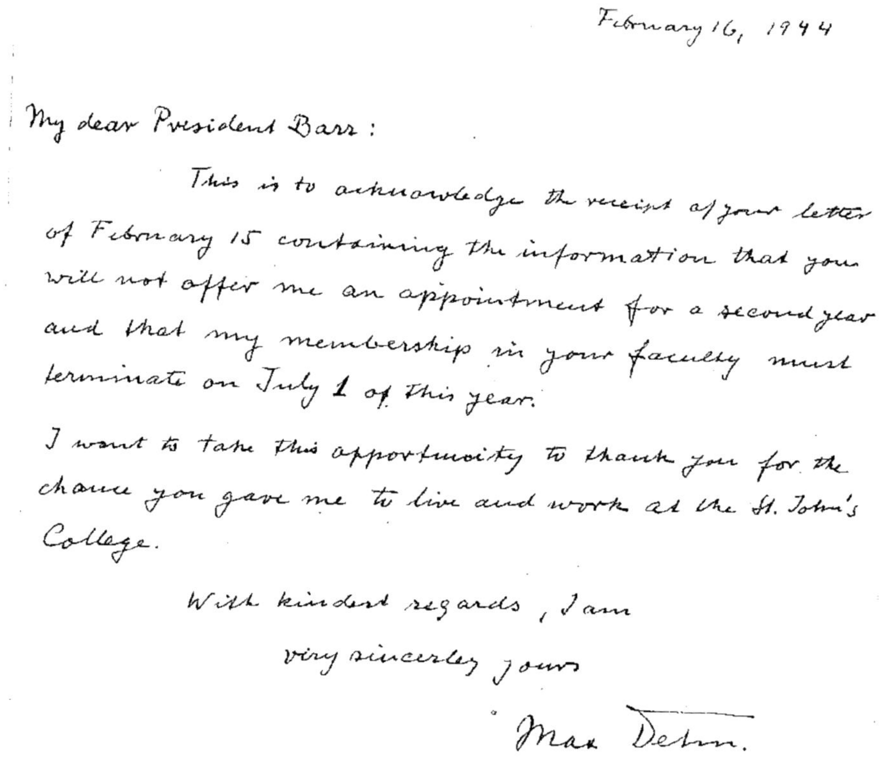

> February 16, 1944
>
> My dear President Barr:
>
> This is to acknowledge the receipt of your letter of February 15 containing the information that you will not offer me an appointment for a second year and that my membership in your faculty must terminate on July 1 of this year.
>
> I want to take this opportunity to thank you for the chance you gave me to live and work at the St. John's College.
>
> With kindest regards, I am
>
> very sincerely yours
>
> Max Dehn.

FIGURE 4. Max Dehn to Stringfellow Barr. Used with permission from the Dean's Office, St. John's College, Annapolis, MD.

---

[27] Siegel is mistaken. Most of the Great Books were read in English translation.

4.5. On to Black Mountain. But plans were already underway to bring the Dehns to Black Mountain College.

The prime mover was Black Mountain College's philosophy professor Erwin Straus, a German refugee psychiatrist. Dr. Straus and Jacob Klein had known each other well in Germany and, finding themselves building new lives on the same strange shore, they visited frequently and gave lectures at each other's institutions.

TABLE 1. Visiting Speakers Provoke Discussions (adapted from the BMC Newsletter, March, 1941)

St. John's	Black Mountain
modern education is on the wrong track	modern education is on the wrong track
man is a rational being	man is a thinking being
education should train the mind	education also concerns character, etc.
great books, seminars, lectures	community, labor, work in the arts
fixed curriculum, math at the center	elective curriculum, arts at the center

On one of his trips to St. John's, Erwin Straus met Max Dehn, learned he was unhappy there, and proposed a visit.[28]

Max Dehn liked what he saw and Black Mountain's faculty and students did too. He was offered a position and happily accepted. Toni kept her Chicago job until they were sure Black Mountain would work out. She joined him in 1947.

Erwin Straus was not at Black Mountain College to greet Max when he arrived for good; they had switched places, approximately. In 1946, after a two-year research appointment at Johns Hopkins University in Maryland, Straus became director of research at the Veterans' Hospital in Lexington, Kentucky; the Strauses became our close family friends, but that's a story for another book.

The next chapters of this book explore who and what Max did find at Black Mountain College, and what he brought to them and to it.

Acknowledgments

It is a great pleasure to thank Carol Mead, head archivist for the Archives of American Mathematics, Dolph Briscoe Center for American History, The University of Texas at Austin; Heather South, archivist at the Western Regional Archives in Asheville, North Carolina; and the staff of the Emergency Committee papers at the New York Public Library, for their unstinting help in documenting the Dehn's long journey.

I am also grateful to Donald Crowe, John Dawson, Cameron Gordon, and Karen Parshall, and to David Rowe, John McCleary, Jemma Lorenat, and Volker Remmert, my co-editors for this book, for their generous advice and for sharing materials.

And warm thanks to Mirela Ciperjani, Tim Perutz, and Francis Perutz for their warm hospitality in Austin. (Arianna Perutz arrived a few months after I left.)

---

[28] All correspondence concerning Max Dehn's appointment to Black Mountain College is held in the Black Mountain collection at the Western Regional Archives (NC).

Archival Sources

Archives of American Mathematics, Dolph Briscoe Center for American History, The University of Texas at Austin (UT)

University of Idaho, Pocatello, Idaho

Illinois Institute of Technology, Chicago, Illinois

St. John's College, Annapolis, Maryland

Emergency Committee in Aid of Displaced Foreign Scholars records, Box 6, Dehn 1941-44, Manuscripts and Archives Division, The New York Public Library, Astor, Lenox, and Tilden Foundations.

Western Regional Archives in Asheville, North Carolina

References

Stephan Duggan and Betty Drury, *The Rescue of Science and Learning*, MacMillan, New York, 1948.

Max Dehn, Wilhelm Magnus, *The Mathematical Intelligencer*, vol. 1., no.3, Nov. 18. 1978.

C. Scriba, In Memoriam, Adolf Prag (1906 - 2004), *Historia Mathematica*, 31 (2004) 409 - 413.

Luce Schachne, *Education Towards Spiritual Resistance: the Jewish Landschulheim Herrlingen, 1933 to 1939*, Frankfurt, 1988

Ben Yandell, The Honors Class, A. K. Peters, 2002.

A. Papadopoulos, *Herbert Busemann (1905 - 1994). A biography for his Selected Works edition*, https://www.ams.org/journals/notices/201803/rnoti-p341.pdf.

Max Dehn, Mathematics, 600 B.C.-400 B.C., Amer. Math. Monthly, Vol. 50, No. 6 (Jun. - Jul., 1943), 357-360

Max Dehn, Mathematics, 400 B.C.-300 B.C., Amer. Math. Monthly, Vol. 50, No. 7 (Aug. - Sep., 1943), 411-414

Max Dehn, Mathematics, 300 B.C.–200 B.C., Amer. Math. Monthly, Vol. 51, No. 1 (Jan., 1944), 25-31

Max Dehn, Mathematics, 200 B.C.-600 A.D., Amer. Math. Monthly, Vol. 51, No. 3 (Mar., 1944), 149-157

Max Dehn and E. D. Hellinger, Certain Mathematical Achievements of James Gregory, Amer. Math. Monthly, Vol. 50, No. 3 (Mar., 1943), 149-163

Die Mathematische Denkweise by Andreas Speiser, reviewed by Max Dehn, Amer. Math. Monthly, Vol. 54, No. 7, Part 1(Aug. - Sep., 1947), 424-426

A Concise History of Mathematics by D. J. Struik, reviewed by Max Dehn, Amer. Math. Monthly, Vol. 56, No. 9 (Nov., 1949), 643-644

CHAPTER 11

Max Dehn's American Students

Marjorie Senechal

> Knowing that Max had only one student – a bright-eyed, blond, clean, and courteous young man with whom I'd seen him walking the mountain paths day after day, talking, talking – I asked him if it wasn't frustrating to give all that time to just one pupil. Not at all, he assured me, for Trueman was "a real student, a real mathematician"; it was a privilege to be his teacher. "In fact," he went on, "I have been very fortunate. In my sixty years of teaching, I have had at least fifteen real students."
>
> *Mildred Harding*[1]

Max Dehn was a popular teacher at Black Mountain College and influenced many students, including several later-famous artists, Dorothea Rockburne and Ruth Asawa among them. But in this remark to Mildred Harding, Dehn was speaking as a German Herr Professor Doktor: a real student was a promising mathematician, able and eager to explore the subject's most difficult reaches.

Black Mountain College, miniscule, out-of-the-way, unconventional and unaccredited, was hardly a magnet for budding academic specialists. Yet, thanks in part to Max Dehn's international reputation, it drew not one but two "real" mathematics students in the decade he taught there. A third real American student earned his Ph.D. with Dehn at the University of Wisconsin.

If Max Dehn's students in pre-war Germany were representative of serious mathematics students in that time and place, his three real students in America are a study in cultural contrasts, not only with their German counterparts but also with each other.

Note: Unless otherwise stated, Regional Western Archives, Asheville, N.C. is the source of all correspondence with Black Mountain College in this paper.

Joseph Engel

Joseph Henry Engel (1922–2011) was born in the Bronx, in New York City, a grandchild of Russian Jewish immigrants. When Joe was small, his father earned his living picking up laundry at customers' homes in a horse-drawn wagon and returning it the same way. Sometimes he took his son along; Joe remembers he sat "up high on the seat and watched the road and saw my father very professionally

---

©2024 American Mathematical Society

[1] In Black Mountain College: Sprouted Seeds. An anthology of Personal Accounts. Edited by Mervin Lane. The University of Tennessee Press, 1990, p. 298. Mildred Harding's spouse was a BMC faculty member in 1951-52.

pulling on the reins and had the fun of feeding sugar to the horse."[2] Nevertheless, the boy chose not to follow his father's career path. In kindergarten, which he entered at the age of 4 1/2, Joe discovered science, engineering, and counting. The next year he discovered the public library and his interests expanded to journalism. By high school he was determined to become a mathematician. He enrolled in City College in New York in 1938, and also joined the Reserve Officers Training Corps (ROTC). When World War II began, he was assigned to the Third Army Corps; later, as a B-29 navigator, he flew 39 missions over the Pacific and Japan. After the war, Joe entered graduate school with the aid of the G. I. Bill of Rights.[3] "I first heard about Professor Max Dehn in the Summer of 1946, just before my second year of graduate work in Mathematics at the University of Wisconsin," he told John Dawson half a century later.[4] "I had been told that he and his wife, who were then in their late sixties, had escaped from the Nazis in his German homeland a few years earlier. Then, after a harrowing journey through Siberia and China they were able to come to a safe haven in the United States.

"I learned that he was a very well known mathematician before coming here, as attested to by the fact that, near the beginning of this century, he had founded two specialized fields of mathematics. An indication of his eminence in the eyes of other mathematicians was the fact that the world-renowned 1911 edition of the Encyclopaedie der Mathematischen Wissenschaften contained articles by Professor Dehn on those two subjects. I also learned that, on the basis of his great reputation, he had been given a position at Black Mountain College in North Carolina (which provided a refuge for many intellectuals and artists who were fleeing from the war). Then he'd been offered a position as a Visiting Professor at Wisconsin starting in the fall of 1946. Knowing this much about him, I decided that I would like to take some of his classes."

How had the invitation to teach at Wisconsin come about? The written record is blank, but we can surmise that L. R. Ford, the chair of the mathematics department at the Illinois Institute of Technology the year that Dehn taught there, had a hand in it, if only indirectly. Ford had introduced Dehn to the wider American mathematical community through the publications and meetings of the Mathematical Association of America or MAA. Two members of the University of Wisconsin mathematics department, C. C. MacDuffee and R. L. Langer, were active in the MAA too: MacDuffee would (directly) precede Ford as MAA president (1945–1946), and Langer would directly succeed him. Langer chaired the Wisconsin math department from 1942 to 1952. In 1946 he invited Dehn to fill a one-semester vacancy.

A semester in Madison, Wisconsin held attractions for Max Dehn beyond its hills and lakes. "Madison offers possibility to direct advanced students to a deeper understanding of mathematics stop.", he telegraphed BMC. He assured the college that he would return at the semester's end. The leave was granted.

"When I met him a few weeks before I registered for the fall semester," Joe Engel continued, "I was amazed at how small and frail he seemed, and found it hard to understand how he and his wife had been able to survive their horrible journey to

---

[2]Calhoun: the NPS (Naval Postgraduate School) Institutional Archive, Military Operations Research Society (MORS) Oral History Interviews. http://hdl.handle.net/10945/49232

[3]Joseph Engel to John Dawson, June 25, 2001.

[4]Joseph Engel, unpublished memoir, 1997, quoted here courtesy of John Dawson.

safety. I was very impressed by his inner peace, his good humor and his innocence. Despite the hardships he had endured, he remained an idealistic man who loved goodness and truth, and the elegance of a well developed mathematical argument." One of the courses Joe took from Max Dehn that year was entitled 'Fundamental Concepts of Modern Mathematics'. Some of his friends in the Department thought, from seeing the title, that this would be a "crap course" that they couldn't afford to waste time on. "I'm so glad I didn't listen to them. I enjoyed the course thoroughly, as much for its intellectual content as the wonderful way he taught the course. Speaking softly (in his mild German accent) as he drew illuminating diagrams on the blackboard, he had a way of revealing the profundity of some of the simplest ideas that most of us take for granted.

"Once he challenged us all to go home and construct a set of blocks that would solve a brain teaser he'd given us. I was delighted to be able to build the blocks in clay and bring them to class the next day. My solution was correct and Professor Dehn was delighted. On another occasion he showed us a remarkably simple proof of the Pythagorean Theorem (that the sum of the squares of the arms of a right triangle equaled the square of the hypotenuse). His proof was so simple that we could visualize it in our heads, and go through the entire proof mentally. This was in tremendous contrast to the highly involved and most tedious way we had been taught how to prove this theorem in high school."

The course described below (Figure 1), which Max Dehn taught at the University of Wisconsin that fall, may be the course Engel found so compelling. (But what was the brain teaser?)

Max Dehn already had a relaxed view of exams and grading. As Engel recounted, "At the end of the semester, Professor Dehn told us he would give us our final exam in the course at the Rathskeller," the University of Wisconsin's still-extant version of a German beer hall.[5]

"We could quaff a brew, have a snack in a pleasant and relaxing atmosphere, talk, enjoy the view of the lake, and, it appeared, take our final exam, all at the same time. About a half dozen of the class members showed up at the Rathskeller, and we sat around and chatted and looked at the ice on the lake (this was the end of the fall semester, and a typical freezing Wisconsin mid-winter season). Eventually it dawned on us that two of our classmates had not shown up for the "exam," and we felt sorry for Professor Dehn and the slur that their disrespectful behavior implied. But he wasn't upset. He just said, 'Vell, I vill only gif dem a B in der course.'

"With that the exam was over, and Professor Dehn said, 'Undt now ve go for a valk on der ice.' And before we could stop him, there we were, scampering across Lake Mendota, enjoying a childlike adventure with this innocent but wise old man.

"As we approached the Vilus Park peninsula jutting out from a point a bit further up the shore, I noticed that the wind had built up a small ice barrier bordering the shoreline. Knowing that often on the protected side of such barriers the ice might be thin and cracked, I shouted to Professor Dehn, who was about ten feet ahead of me, "Be careful crossing that ice." But he ignored me, scrambled over the barrier, and fell through the ice on the lee side. He was in water up to his

---

[5]https://union.wisc.edu/visit/memorial-union/der-rathskeller/der-rathskeller-history/

```
MATHEMATICS 282.    CONCEPTS, PROBLEMS AND METHODS.

3 Class Meetings per week throughout the semester.  Sem.  3 Cr.  Mr. Dehn.

Catalog Description:   Basic operations, the concept of numbers, and
                       projective and metric geometry.  Axiomatics.
                       General methods, the non-logical elements in
                       Mathematics.

Prerequisite:          Consent of instructor.

Course Content:

    Concepts:          Couples, integers, basic operations, groups, the
                       inverse operations, and the successive generaliza-
                       tion of the concept of numbers.  Concepts of pro-
                       jective and metric geometry.  Continuum and
                       dimension, Measure.  Different approaches to the
                       basic concepts of Analysis.

    Problems:          Identities.  Generation and decomposition.  Asymptotic
                       approximation.  The continuum in Geometry.  Axiomatics.
                       Analysis of the different means and ways to the same
                       ends.

    Methods:           General methods:  observation and experiment.
                       Mathematical methods: the mathematical induction.
                       Transformation and reduction of problems.  Use of
                       geometric representation and physical intuition.
                       The non-logical elements in Mathematics.
```

FIGURE 1. Outline for a course Max Dehn taught at the University of Wisconsin.

waist, and stuck on the ice that he'd broken through. We grabbed him under the armpits and yanked him out easily (because he was so frail).

"We were far from any shelter, and although we were all dressed warmly, we became aware of how bitterly cold it was. We were very concerned over the health of this gentle man who, we suddenly realized, made another visit was so dear to us. Soaked as he was, we made him walk briskly back to the nearest building we could find, to keep him from freezing, and he continued to chat with us in his usual cheery and benevolent manner. He survived this little adventure, as he'd survived many worse ones in his past."[6]

Joe Engel was still at the University of Wisconsin when Dehn made another visit, and he was delighted to see his beloved professor Dehn again. "By then I'd finished my course work for the doctorate, and was ready to start on my dissertation. When I met him in the corridor on the first day of the fall term, on impulse I blurted out, "I'm so glad you're back. Would you like to be my thesis advisor?" He said, "Yah sure," and immediately told me that the topic I would work on was an

---

[6] Engel adds, "and I was delighted that I was able to take another course from him in the Spring semester." But Max Dehn returned to Black Mountain College for the spring semester of 1947. Did he teach Engel by correspondence? Or did Engel misremember?

unsolved problem he'd mentioned during the class that had ended with our walk across the frozen lake.

"Working under his kind and understanding guidance was a joy and a privilege. I was able to finish my dissertation by the end of the term, and relax during the Spring semester while the remaining formalities for granting my degree (including my being examined on my thesis) were accomplished.

"Looking back at that wondrous time," Engel concluded, "I realize that I am older now than Professor Dehn was then, and I still love him, and am in awe of his wisdom and humanity and humor and compassion."

Joe Engel presented his thesis to the American Mathematical Society's meeting at the University of Kansas in April, 1949. The abstract was published in the AMS Notices (Figure 2):

> **400. J. H. Engel:** *Contributions to the solution of the word problem for groups.*
>
> The word problem for the lower central series of the free group (called hereafter hypo-abelian groups) is solved by a method (distinct from the methods used by Magnus, Fouxe-Rabinovitch, or R. H. Fox) which utilizes invariants connected with a lattice representation of these groups. There is demonstrated for each word in each hypo-abelian group the existence of a unique canonical form, expressed as a product, in canonical order, of powers of generators and powers of commutators. An expansion theorem is developed, proving the known result that elements equalling the identity in all hypo-abelian groups of a free group equal the identity in the given free group. The word problem is solved for hypo-abelian groups of the first order with any number of generators and one additional defining relation, and for hypo-abelian groups of the first or second order with two generators and any number of additional defining relations. These results yield new necessary conditions for the equality of two elements of a group. (Received March 16, 1949.)

FIGURE 2. Abstract for Joe Engel's thesis.(From the AMS Notices, July 1949.)

Peter Nemenyi

Just at the time that Max Dehn returned to Black Mountain College for the spring semester of 1947, its Registrar received a letter from a young serviceman hoping to study with him there. Peter Nemenyi (1927–2002) had been interested in Black Mountain College for several years, he reminded the Registrar.

"In the fall of 1945 I made enquiries as to the possibility of my entering Blackmountain (sic) College, but I was drafted for military service before the beginning of the Spring Semester 1946. Now that my prospects of being discharged in less than 3 months are very good, I am again considering Blackmountain among the colleges of my first choice."

Peter Bjorn Nemenyi's background was unusual even for Black Mountain's highly eclectic student body. He was born to Hungarian parents who, ironically, had fled from its fascism to enlightened Berlin. They parked Peter in a socialist, vegetarian boarding school when he was three and then separated, his mother making her way to Paris, and his father (eventually) to the United States. As the Nazi threat grew, the school relocated to different European countries; Peter also lived in several foster homes. He completed his secondary education in England and then joined his father. Paul Nemenyi, a mathematician and physicist well known for his work on fluid dynamics, was teaching at the State College of Washington in Pullman, Washington at that time. Peter studied at State College from June 1945 to January 1946, when he was drafted into the army.

"The nature of the educational needs I feel most strongly has changed somewhat during this year in the army," Peter explained. "Previously my strongest need was for human contacts and community life. Though this is still an important factor in my plans, I feel that it is most important for me right now to counteract the stultifying influence of army life and to train myself in mathematical and scientific thinking while I still have some of the flexibility of youth. I am therefore most strongly interested in the Mathematics Department. I hope to take an intensive study course in Mathematics together with some Music and practical work, for at least the first term after my discharge from the army. I believe that your college would be very suitable for this, especially if Professor Dehn is still teaching Mathematics there."

The Registrar sent Peter the college's narrative-style application form. What have your interests been during the past two or three years? was one of several open-ended questions. "I'm not sure where to begin," Peter wrote. Indeed, his preparation for intensive study in mathematics was unusual for that era.

"Mathematical games & various related problems have always been one of my prime hobbies," he wrote. "I awakened to music only a year ago: before that I had vaguely enjoyed it but didn't find time for it because I didn't know WHY. Now I can hardly live without music and I'm interested in learning to play an instrument as a means of self-expression. So far I've made very little progress on the piano, having only started a few months ago. I'm interested in poetry and in literature, though particularly ignorant in the latter. Languages fascinate me. So do the people of different backgrounds and walks of life, their ways of thinking and living. In connection with this I like to read biographical novels, especially Tolstoy and Dostoyevsky; though in this field too I haven't penetrated far. I have too many interests to pursue any one consistently and effectively. To sum up: my chief interests have been: 1) Reading – mainly biographical, historical, political & scientific material, mathematics & related pursuits, & getting acquainted with different people & philosophies, and, later, music (now perhaps the strongest) and art and 'nature'."

Peter Nemenyi enrolled at Black Mountain College in the summer of 1947 and studied calculus with Max Dehn. Dehn's handwritten note on Peter's course card says "Very eager, gifted for Math., seriously concerned with understanding."

He continued to do "very satisfactory" work in his tutorial classes with Max Dehn through the 1947–1948 academic year.

Then Langer invited Dehn to return to the University of Wisconsin for a year, and Dehn again asked Black Mountain College for leave. Not only could he work with advanced students again, having professional mathematical colleagues must have been an attraction too. (Engel recalled that Marshall Hall, and Steven Keene were members of the mathematics department at that time and R. H. Bing also visited.) Black Mountain College agreed, on the difficult condition that Dehn make the then-onerous trip to Black Mountain several times a month to give seminars. Fortunately, as it turned out, the college could not afford to pay for all that travel.

Peter Nemenyi wanted to follow his mentor. He too requested, and was granted, a leave of absence from Black Mountain, and transferred to the University of Wisconsin for the year.

Trueman MacHenry

Max Dehn and Peter Nemenyi returned to Black Mountain College in the fall of 1949 just as Trueman MacHenry was arriving. Thus, for the 1949–1950 academic year, Max Dehn had two real students at Black Mountain College! Peter wrote a thesis on the gamma function and graduated in the spring. His outside examiner was the eminent mathematician Emil Artin, who was then teaching at Princeton University (Figure 3).

```
BLACK MOUNTAIN COLLEGE                    BLACK MOUNTAIN, N. C.

                                          April 29th, 1950




          I came here to examine Mr. Nemenyi in Mathematics.

          He had sent me his thesis on the Gamma Function,
       which I have read.  It is an exhaustive study of axioma-
       tic introductions of the Gamma Function.  It is quite
       worthy of a thesis and reveals a thorough knowledge of
       the topic.

          In the oral examination I asked Mr. Nemenyi ques-
       tions concerning the foundation of real numbers, various
       questions on real variables and finally some on complex
       variables.  He showed a very good understanding of these
       topics.

          In the end Dr. Dehn asked him to give an account of
       the development of non-Euclidean Geometry.

          To sum up, I am quite satisfied with the achieve-
       ments of Mr. Nemenyi and propose to give him his degree.
       As grade of the examination I would propose "excellent".



                                          E. Artin
```

FIGURE 3. Emil Artin's report on Peter Nemenyi's examination.

And so we come to Trueman MacHenry. The following paragraphs are adapted from the eloquent essay he wrote for this book.

"During my time at Black Mountain, he had two students who did a full time degree course in mathematics, Peter Nemenyi and myself. Each of us had private meetings with Dehn in his study in the Dehn apartment. I remember sitting with him side by side at his desk in front of a window that looked out on the woods. Above me to my right was a shelf of books that included what was left of Dehn's mathematical library, most of which had been carted off in gunnysacks by the Gestapo while Dehn was still living in Germany. On that shelf were the collected works of Riemann and the works of Grassmann. And it was at this desk that I first learned about the ring of symmetric polynomials, a topic which has constituted the last several years of my research and publications.

"My biweekly meetings with Dehn were always very pleasant and very interesting. There was a lot of conversation about the background of the mathematics I was studying, as well as about the people whom Dehn had known as a student and teacher in Germany, for example, his teachers, David Hilbert and Felix Klein. Dehn, along with his very close friend, Carl Ludwig Siegel, conducted a well-known seminar in the History of Mathematics when they were both at Frankfurt University in Germany. When Dehn discussed mathematics, there was always a large element of its history involved. Dehn also had a gentle and tolerant disposition, and so his very wide and deep knowledge, which went well-outside of mathematics, did not intrude as a barrier to his students. However, when accuracy and clarity were at stake, he could be very demanding. He went over my thesis with me in the house of Hazel Larsen, a well-known photographer at Black Mountain, who knew him as this very kind and gentle man. Hazel was completely surprised by how demanding and unrelenting Dehn could be when truth and accuracy were at stake, and I too saw a side of Dehn that had not appeared before.

"The class with Dehn almost always ended with a cup of tea with Toni, then a walk in the woods with interesting conversation, sometimes about areas of philosophy that I was interested in. Along the way Dehn would often point out a rare orchid hiding below the foliage. Botany in general and orchids in particular were areas in which he was an expert. These conversational walks together were as valuable and interesting as our classes.

"It is perhaps of interest how I came to be at Black Mountain in the first place. I graduated from Natrona County High School in Casper, Wyoming in 1945, and went to the University of Wyoming in Laramie for two years where I studied French, German, Philosophy and Mathematics. Here I became friends with Andy Fischer, an interesting young poet, who had a talent for meeting people who were also interesting and who had unusual backgrounds. When Black Mountain had its 'time of the troubles' when the Albers left, Bolotovski and Wooden Day came to the University of Wyoming, and my friend, Andy, with his sixth sense for spotting interesting people immediately made friends with them. I had to left the University just a short time before because my money had run out, and I got a job with the United States Coast and Geodetic Survey. I worked first in Wyoming, then in Arizona and Southern California. Mostly the job entailed working in the desert and climbing mountains. My surveying party was later sent to Alaska, where we were set to work surveying the Kanai Peninsula, again climbing mountains. At my last camp site there, I was alone in a tent three-quarters of the way up the side of Mount Veniaminoff near Mount Aniakchak, both active volcanos. I was subsequently 'rescued' by a bush pilot in small plane from the side of this mountain, which, in the meantime had become engulfed in snow. My friend Andy, who by now was enthusiastic about Black Mountain College after hearing accounts about from it from Bolotovski, wrote me a letter together with a Black Mountain brochure, telling me that I must go to Black Mountain. So, when the Alaska project was finished, I left the Coast Survey, and did what Andy told me to do. I came to

Black Mountain. Natasha Goldowski's mother, Madame Goldowski, and who as a child had known Tolstoy, and now in her mid- eighties, renamed me "Alaska Boy".

"In addition to studying mathematics with Max Dehn, I also studied physics with Natasha Goldowski, French and Linguistics with Flola Shepard, Russian with Madame Goldowski, and, as mentioned above, German with Toni Dehn. I was also very interested in Anthropology at the time, and while I had no formal courses, I was good friends with both Paul Laser and with John Adams, and had extensive conversations with both of them concerning my particular anthropological interests. I was also interested in theatre: Mary Fitton and I were the Phonographs in MC's translation of the surrealistic Cocteau play, 'Marriage on the Eiffel Tower', for which I also made some (surrealistic) props. I also acted in a play produced by West Huss and in one written and produced by Mark Hedden. I had a course in counterpoint from Lou Harrison. I also had a literature course with MC Richards.

"A student who expected to finish with a degree or a certificate at BMC first had to pass a qualifying examination conducted by the entire faculty, which was a test of what the faculty deemed to be a reasonable grasp of general subject matter. They then selected an area of specialization and were assigned a faculty advisor with whom they wrote a thesis or did a project in their field, and were again examined before an assembly of the faculty, this time by an outside specialist, who gave a critical judgment of their project. If all of this was successful, the student was awarded a degree or certificate. This choice was a result of the underlying philosophy at Black Mountain College of breaking away from the usual way that marking and judging were handled in other educational institutions. Some of the more educationally adventurous students preferred to settle for a certificate of graduation, and avoid the more prosaic formal Degree. Dehn came down on the realistic side, and insisted that I graduate with a Bachelor of Science degree. Peter Nemenyi, on the other hand, was more idealistic; he went on to Princeton with a Certificate of Graduation from Black Mountain College.

"My thesis work was in Foundations of Geometry, an area in which Dehn earned his initial fame. My background reading for my thesis included Euclid's Elements. The only copy of the Elements available at Black Mountain belonged to Max Dehn. It was a beautiful edition (Elements of Euclid, 2 vols, Berlin, 1824) facing pages in Greek and Latin. I had no Greek, but I was able to read it in Latin. This book, as well as two others, were left to me after Dehn's death. My outside examiner at Black Mountain was Alfred Brauer, who was then teaching at the University of North Carolina in Chapel Hill."

In contrast to Joe Engel, Trueman MacHenry had no idea of Max Dehn's towering stature.

"When I entered Black Mountain College, I did not know that Max Dehn occupied such an important spot in the mathematical world. In fact, the faculty and students at Black Mountain in general did not understand Dehn's outstanding position as a mathematician. He could have been an obscure high school teacher, as far as the people at the College were concerned. I learned differently when I came to know Wilhelm Magnus, a professor in the Courant Institute at New York University, one of Dehn's former German Ph.D. students, and a well-known mathematician in his own right. I think that I then helped to spread the word at Black Mountain about Dehn's mathematical reputation, letting it be known the extent of his regard in the mathematical world, and that Max Dehn was not a simple high school mathematics teacher, but a famous professor and mathematician at Frankfort University in Germany, and his name then began being mentioned along with the other luminaries at the College."

Postscripts

The subsequent careers of Dehn's three American students were as varied as their backgrounds.

In his Ph. D. thesis Joe Engel acknowledged "with sincere gratitude, the assistance of Professor Max Dehn, who suggested this problem and method of attack to him, and gave him frequent advice." But afterwards their paths diverged. "I took a position on completion of my stay at the University of Wisconsin with M.I.T.'s Operations Evaluation Group which served as the U. S. Navy's in house Operations Research group," Engel told John Dawson. "When I told Professor Dehn where I was going he shook his head sadly and said, "Ach – ya – you will never return to mathematics." And he was almost right. I stuck with operations research in various guises for 41 years, and had a pretty good run for my money, including a tour as President of the Operations Research Society of America."

After graduating from Black Mountain in the spring of 1950, Peter Nemenyi enrolled in Princeton University for graduate study. A man of wide interests which he continued to pursue, he must have expected earning a Ph. D. to take some time. But he could not have expected the detours his life would soon take. By June, 1952, Peter had lost not one but two fathers: his father-like mentor Max Dehn (on June 27), and Paul Nemenyi, killed at 56 by a sudden heart attack just a few months earlier.

Paul Nemenyi left not one but two sons.[7] Peter had met his nine-year-old half-brother Bobby Fischer only twice. After Paul's death, the boy's mother, Regina Fischer, asked Peter to tell her son that Paul was in fact his father. (Bobby knew Paul well, but thought he was as a family friend.)[8] Complicating the situation, Regina was penniless and Paul Nemenyi left no will. Peter later assumed some responsibility for the budding chess prodigy's care.

Royalties from *Geometry and the Imagination* might have helped cover those expenses – if Peter had received them. That same year, 1952, the Chelsea Publishing Company released an English translation of the German classic *Anschauliche Geometrie* by Max Dehn's mentor David Hilbert and his assistant Stephan Cohn-Vossen; *Geometry and the Imagination* is still in print today. The translator, listed only as P. Nemenyi, is usually assumed to be Paul. Dear Reader, the translator is almost surely Peter. But it's less obvious how he came to do this translation. Did Dehn arrange for it? Or Artin? We hope to tell that story another time, and to clarify another mystery. In *Anschauliche Geometrie*, the authors discuss the single-sided surface that all mathematicians know as a Klein bottle, named for the German mathematician Felix Klein who first described it in 1882. For its first seventy years, this remarkable surface was known as a Klein surface (Klein'sche Fläsche). Then, in a casual aside in *Geometry and the Imagination*, it was redubbed a Klein bottle. This took, not only in English but in German too (Klein'sche Flasche).[9] Was

---

[7] Peter's mother had died in 1939.
[8] See https://www.chess.com/article/view/who-was-fischers-father
[9] "What Color Was George Washington's White Horse?" The Mathematical Intelligencer, 39, 9-17 (2017).

"bottle" Peter Nemenyi's stroke of genius? And who received the royalties for the translated book?

But back to known facts. Peter Nemenyi received his Ph. D. from Princeton University in 1963, under the direction of the eminent statistician John W. Tukey. He taught at several American colleges and universities, and was prominent in civil rights movements in the United States and Latin America. He is remembered by statisticians today for the several statistical tests named for him.

Trueman MacHenry continues: "Thanks to my work with Dehn, I was given a teaching fellowship at Chapel Hill, where I went to do my graduate work. There I taught a beginning course in Algebra. Unfortunately, my studies (and teaching) were interrupted by my draft board in Casper, Wyoming, which thought that I would be better at fighting in the Korean War. Rather than go into the Army, I chose to change my Naval Reserve status to active duty, so I spent two and one-half years in the U.S. Navy. After Boot Camp I married Selma Weisberger. My wife and I went to San Francisco where I attended a Navy Electronics School on Treasure Island near San Francisco for training in electronics. There my marks were high enough that I had a choice of berths when I graduated, so I chose to go to an air force base in southern California, in the same desert where I had worked as a surveyor earlier on. When my tour of duty was over I went to the University of Manchester in England to continue my graduate work. There I received an MSc. degree. My thesis was entitled "Abelian Groups and Tensor Products". My supervisors were Peter Hilton and B.H. Neumann. Later I received a Ph. D. at Adelphi University, where my thesis was on Roots of D-Pi groups. My teaching career included a course in mathematics taught at Black Mountain College, then at the University of North Carolina, a beginning algebra course, as mentioned above; after Manchester, I taught for several years at Rutgers University in New Jersey, then at Adelphi University on Long Island. Finally, I went to Canada and taught for 46 years at York University in Toronto, until retirement. My special areas of interest in mathematics are geometry, group theory and algebra in which I have published a number of papers."

The Other Side of the Mountain

"Of course, there never was any serious science at BMC, although there were serious scientists," Trueman MacHenry remarked. Today Max Dehn is perhaps the most renowned of the BMC scientists, but he was not the only one. In the course of the college's 24 years, "the other side of the mountain" was home to, among others, engineer Theodore Dreier, chemist Natasha Goldowski, electrochemist and metallurgist Fritz Hansgirg, psychoanalysts Fritz and Anna Moellenhoff, physicist Nathan Rosen, and neurologist Erwin Straus. A full account of the fascinating story of science, and the scientists and their students, at Black Mountain College awaits its author.

CHAPTER 12

Toward a Happy Life:
Max Dehn at Black Mountain College

Brenda Danilowitz and Philip Ording

For the historian, the purest joy is to relish the contemplation of the ups and downs of the development, of the connections, of the breaks and transitions, to try to see the divine spark in each of the creators and to relive their productive moments.
– Max Dehn [1983, p. 19].

Time decorates the grave of Max Dehn. Fallen twigs and leaves of wild rhododendron wreathe the site, bright green moss outlines the two terracotta rectangles that form his headstone, and pale blooms of lichen dapple the unglazed surface and shallow impressions that bear his name and dates. The modern type, composed of simple arcs and lines unburdened by serifs, and the well-proportioned ceramic tiles easily accommodate these natural ornaments. So does the stoneware itself, fired from found clay, a reddish-brown sediment deposited across the Appalachian landscape by thousands of years of weather and erosion.

The story of Dehn's time at Black Mountain College, to which he dedicated nearly the last eight years of his life, has been told many times. Following a series of unsatisfying appointments, the German WWII refugee mathematician found a home in the small progressive liberal arts college. Its communal form of education, European faculty – including a number of esteemed artists, musicians, and writers – and forested mountain landscape attracted him. For little more than room and board, he taught mathematics and philosophy, tutored ancient Greek and Latin, and inspired an intuitive appreciation for mathematics, often while walking nearby trails in the wooded foothills of North Carolina. Under his guidance two gifted students went on to become mathematicians, and many more benefited from his humane wisdom as well as the generous spirit that he and his wife, Antonie (Toni) Dehn, shared. One summer day, only weeks after the college granted him emeritus status, Dehn died abruptly of a pulmonary embolism probably released when he rushed up a trail to preserve a stand of flowering trees from being wrongfully cut down.

This narrative of Dehn's Black Mountain years, as spare and inviting to elaboration as his grave marker, has accommodated what evidence has surfaced since it was first presented in 1967 by his friend and former colleague Carl Siegel [Siegel, 1979]. By contrast, accounts of the college in the intervening decades have evolved dramatically. For the historian of visual, literary, or performing arts, Black Mountain College (1933–1957) signifies one of the most "productive moments," to borrow

---

©2024 by the authors

Dehn's words, in the cultural history of America. Over time, it has come to represent an experimental and interdisciplinary avant-garde that overtook and surpassed modernism. This chapter aims to reconsider Dehn's engagement with Black Mountain College in relation to the creativity and productivity that marked his life as a whole and to the pedagogical ideals and avant-garde imagination that have become the college's legacy.

Our portrait of the mathematician of Black Mountain College draws from his 1930s publications on the nature and origin of mathematics and the arts; course notes made separately by Dehn and by three students from his 1948 class Geometry for Artists; his undated manuscript entitled Psychology of Mathematical Activity; and college records that express his philosophy of education. Like many histories of the college, ours relies also on interviews and remembrances of faculty and students documented by others.[1]

"Max Dehn was a sort of mental 'Windex'," according to student Harry Weitzer. "He encouraged you to look through windows previously opaque, to think about dense subjects like Death, what is a Number(?), and Love. Few subjects failed to interest him, and through him, you" [Weitzer, n.d., p. 2]. The metaphor puts Dehn shoulder-to-shoulder with his students, looking out from the same side of their mental windows. In this way he shared with Josef Albers, the artist-teacher responsible for the college's educational structure, the goal "to open eyes." As Albers emphasized the perceptual rift between "physical fact and psychic effect," [Albers, 1963] Dehn was developing a psychological foundation of elementary mathematics that incorporated non-logical elements. Archival materials suggest that Dehn and the college further absorbed and clarified each other's instructional aims and creative possibilities in the 1949–1952 period between Albers's departure and the appointment of poet Charles Olson as college rector, the beginning of what was arguably Black Mountain College's most experimental final chapter before it dissolved in 1957.

A visit to Black Mountain College

Black Mountain College was established in the foothills of the Blue Ridge Mountains of North Carolina (Figure 1) by a group of disgruntled academics who, led by John Andrew Rice, a classicist and follower of philosopher John Dewey, had resigned from Rollins College in Florida. Accompanied by a small coterie of devoted students, they were in search of greater academic freedom, especially freedom from bureaucratic and high-handed administrators. As an acolyte of Dewey, whose theory of education was frequently broadly summarized as "learning by doing," Rice set out to make hands-on art education central to the general liberal arts curriculum then customary at American undergraduate colleges.

These events, which coincided with the Nazis' closing of the by then internationally known Bauhaus in Germany, resulted in the Bauhaus teacher-artists Josef Albers and his wife Anni Albers being invited to form the core of the educational

---

[1] The primary archives we consulted are: the Max Dehn Papers ("Dehn Papers") in the Archives of American Mathematics, The Dolph Briscoe Center for American History, The University of Texas at Austin; the Black Mountain College Project Collection and BMC Archives in the Western Regional Archives, North Carolina ("BMC Archives"), and the Ruth Asawa papers, Dept. of Special Collections and University Archives, Stanford University Libraries, Stanford, California ("Asawa Papers").

FIGURE 1. Studies Building across Lake Eden, Black Mountain College, ca. 1940s. Photograph by Claude Stoller.

framework for Black Mountain College. As Josef expressed it in the college Bulletin: "art is a province in which one finds all the problems of life reflected – not only the problems of form (e.g. proportion and balance) but also spiritual problems (e.g. of philosophy, of religion, of sociology, of economy). For this reason art is an important and rich medium for general education and development" [Albers, 1934, p. 2]. After 1934, as events in Europe escalated towards war, several scholars from the continent, drawn by the strong presence of the Alberses, began arriving. Among them were psychiatrists Fritz and Anno Moellenhoff, Swiss artist and theater designers Xanti and Irene Schawinsky, Viennese conductor Heinrich Jalowetz, and psychiatrist Erwin Straus. The Dehns were among the later arrivals.

The college invited Dehn as a guest speaker early in 1944. At the time, he was a tutor at St. John's College in Annapolis, Maryland on a temporary appointment, and Toni was enrolled at the Art Institute of Chicago to resume studies begun decades earlier at the Kunstgewerbeschule in Hamburg before their marriage. Decades later she would recount that the Dehns' contact at Black Mountain College was Erwin Straus [Harris, 1978]. Straus had coauthored a pair of papers [Straus and Wohlwill, 1924a,b] with Friedrich Wohlwill, a physician and neurologist at St. Georg Hospital in Hamburg and the brother-in-law of Max's sister Hedwig (Dehn) Wohlwill.

As reported in the Community Bulletin, Dehn's March 16th lecture told "the story of twenty-five thousand years of development in mathematics in a humorous, refreshing manner," that made "the subject quite interesting for even the layman."

The enthusiastic account, signed by student Flora Ricks, ends with the following elaboration of Dehn's theme:

> Animals, he pointed out, recognize units and pairs; and prehistoric man decorated his cave walls with geometrical forms. The development of symbols is a necessary preliminary to progress in Mathematics, and this was begun in about 2000 B.C., by the Babylonians. However, an essential method of solving problems, that of proof, was invented by the Greeks. It is interesting to note that although many fundamental theorems were proved thousands of years ago, it has not been until comparatively modern times that their importance has been recognized as the basis upon which all metrical geometry, prospective [sic] geometry and the mathematics of precision can be built.[2]

Brief as it is, the record suggests that Dehn's historical survey of mathematics impressed on the audience several features that distinguish his rather unusual perspective on the discipline, one that sees mathematics as: a basic capacity of human (animal) beings; rooted in a prehistory that coincides with the prehistory of art; a practice that, despite the importance of its symbolic notation, is defined by proof, not computation; and a science whose modern ideas rely crucially on insights that originated in antiquity. Implicit in this view of mathematics is that its history is a history of ideas. "Dehn saw the progress of mathematics in the conception of new ideas," his PhD students later recalled, "not in investigations into special cases and generalizations. His admiration was for creativity" [Magnus and Moufang, 1954, p. 224].[3]

Readers by now familiar with Dehn's contributions to the modern foundations of geometry and its exposition (cf. Chapters 2, 3 and 5) will infer what "fundamental theorems ... proved thousands of years ago" may have interested him, and through him, his listeners. In a 1928 address to a non-specialist audience at Frankfurt University, "The Mentality of the Mathematician," he had emphasized the belated appreciation of the Archimedean postulate, the "foundation stone of ... any ... rigorous theory that proposes to harmonize arithmetic and geometric phenomena" and the roots of projective geometry, which "deals with the remarkable and beautiful properties of figures that remain invariant under projection. But in antiquity the problem was not formulated in full generality. After the Renaissance, when interest in projective geometry was inspired by the use of perspective in painting, the problem was first treated in typically modern fashion ... " [Dehn, 1928]. At the time of his Black Mountain visit, Dehn had just reinscribed these ideas, among others, in English for a series that *The American Mathematical Monthly* editor hoped would "cover the whole history of mathematics in compact form"[Dehn, 1943, p. 357n]. In the month of March, 1944, he completed the fourth of four installments, to which Toni contributed illustrative maps of the ancient world.

Whether or not the *Monthly* series served as a draft of Dehn's guest lecture, it offers vivid examples of his philosophical and pedagogical approach to certain

---

[2]BMC Community Bulletin 1943-44 #24 p. 5, BMC Archives.

[3]Unless stated otherwise, translations are our own, with significant help from Peter Kalal, Jon Ording, and Google.

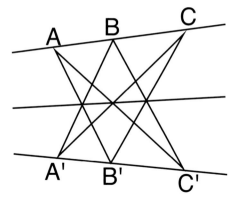

FIGURE 2. The theorem of Pappus: "Let A, B, C, be three points on one line, A', B', C', three points on another line, then the lines AB' and BA', BC' and CB', CA' and AC', respectively, meet in three points lying on one line" [Dehn, 1944b, p. 153].

key concepts in courses he would develop upon his appointment to the college.[4] Prominent among these is the idea of *configuration*, which Dehn attributes to a theorem of Pappus of Alexandria (Figure 2), circa 300 C.E.: "This theorem marks an event in the history of geometry. From the beginning geometry was concerned with measures: lengths of lines, areas of plane figures, volumes of bodies. Here we have for the first time a theorem which is established by the ordinary theory of measures but is itself free of all elements of measurement; it states the existence of a figure which is determined through the incidence of lines and points only. It is the first 'configuration' of projective geometry, and it was shown more than 1500 years later that this configuration alone is sufficient to build up projective geometry in the plane" [Dehn, 1944b, p. 153].

Dehn's correspondence in advance of his visit offers some clues about his interests and intellectual attraction to Black Mountain College at the time. In a letter to Erwin Straus dated February 9th, 1944, he proposed a few different subjects for his guest lecture, "I might speak about The Psychology of Mathematical Activities or about Common Roots of Mathematics & Ornamentics or Some Moments in the Development of Mathematical Ideas. Please let me know which subject you and your community prefer. Looking forward to getting your critical remarks on my papers."[5] Neither Dehn nor Straus identify in their correspondence the papers to which Dehn refers. They may have been installments from the *Monthly* but, given the topics he suggests, it seems more likely that he sought feedback from the professor of psychology and philosophy and his celebrated arts faculty colleagues on a yet more ambitious historical investigation of a very different sort.

Psychology of Mathematical Activity is the working title of an undated 226 page manuscript preserved at the Archives of American Mathematics. It begins with the story of its beginning:

---

[4]Not to mention his "humorous, refreshing manner," for example when he calls Archimedes "primarily a research man" whose pride was "perhaps not the sign of a philosophical mind" [Dehn, 1944a, p. 27].

[5]Dehn to Straus, February 9, 1944, Dehn Faculty File, BMC Archives.

> Some time ago I tried to present the roots that, deep in the human soul, are common to artistic and mathematical activity. Through this consideration I came to an arrangement that seemed quite useful, especially in the art of ornamentation. Since at that time I was only able to present these things very briefly, some only in outline, I wanted to try to give a systematic presentation of ornamentation on the basis of this arrangement, into which the known forms can be clearly integrated. But at the beginning of this work I came to the question: what *distinguishes* mathematical activity from that of the ornamental artist (*des Ornamentikers*)? And after that: what is characteristic of mathematical activity at all? And here, according to the reason for the question, the main emphasis should be placed on the processes in the soul. In the beginning this seemed like a side question that was easy to deal with, but soon I got caught up in ever new general and specific questions, and I was led to make a lengthy investigation, the results of which I would like to present here.[6]

Dehn cites his own work sparingly, but one infers that the preliminary effort from "some time ago" is the pair of articles "The Mathematical Ability in Humans" [Dehn, 1932], which presents itself as "the introduction to a history of mathematics" that aims "to link the science of mathematical science with the whole of human activity"[7] and the long paper "On Ornamentation" [Dehn, 1939]. Straus selected the last topic, "Some Moments in the Development of Mathematical Ideas," and suggested that there "might be an opportunity in the art class to speak on one of your other subject[s], namely 'Common Roots of Mathematics and Ornamentics.'"[8] His only further advice was that Dehn should travel from Annapolis to Black Mountain by Pullman rather than coach.

If Dehn did speak to an art class it is difficult to imagine one more sympathetic. Josef Albers (Figure 5) based the studio courses at Black Mountain on the foundational curriculum, or *Vorkurs*, that he had studied and taught at the Bauhaus. Albers characterized the first course, Drawing, as "a handicraft instruction, strictly objective, unadorned through style or mannerism" that began with "general technical exercises: measuring, dividing, estimating; rhythms of measure and form, disposing, modifications of form" [Albers, 1934, pp. 4-5]. Exercises often involved repetition with the aim of developing line control while also expanding student perceptiveness; "disposing" assignments demanded visualizing the line ahead of the pencil to achieve symmetry, balance, and regularity. Many of the results from this class and Basic Design coincide with the ornamental forms that preoccupied Dehn, including lattices, waves, spirals, and meanders (see Figure 3).

Dehn's 1939 study "On Ornamentation" pursued a quasi-axiomatic presentation that began with rudimentary motifs before synthesizing more complex types from these. At first he considers the markings found on Stone Age pottery, in particular, a row of equally spaced marks set horizontally around an earthenware

---

[6][Dehn, undated, p. 1], emphasis in the original. Section 4 below provides detailed information on the manuscript.
[7][Dehn, 1932, p. 1], translation from [Bergmann, 2012, p. 194].
[8]Straus to Dehn, February 15, 1944, Dehn Faculty File, BMC Archive.

FIGURE 3. Student drawings from Josef Albers's 1945 Basic Design Course at Black Mountain College by Lore Kadden Lindenfeld.

vessel. He observed that the execution of even such a humble pattern demanded that the artist coordinate three different instances of order: the equivalence of each mark to the next mark; the equivalence of each space between a pair of marks and the space between the next pair; and the proper choice of separation so as to evenly divide the vessel's circumference. "The solution," Dehn observes, "affords satisfaction as the solution of a reckoning assignment" [Dehn, 1939, p. 130].

By stacking rows so each mark is positioned between two neighboring points of the previous row the sense of a "rule" or "higher *rhythm*" emerges. Inevitably the artist is drawn to connect the dots, thereby dividing the surface into a lattice or "two-dimensional rhythm" that Dehn celebrated as a momentous event in the history of mathematics: "this new thing, *the unintended*, is already to be denoted as a *Theorem*... one of the first mathematical theorems ever encountered by a human" [Dehn, 1939, p. 135, emphasis in the original]. Dehn proceeded to more complex ornaments of checkerboards, meanders, spirals, ribbons, and so forth (e.g. see Figure 4), before concluding that,

> among all these forms, only a few principles must be acknowledged, basic tendencies of the human spirit which are inextricably bound up with music, architecture, and every artistic expression. These principles are mathematical in nature and mathematical propositions are discernible in many ornaments. Thus we are able through our observations to find a connection between the beautiful works of those who create sensuous forms and the manifold shapes with which the mathematician works in the abstract [Dehn, 1939, pp. 152-3].

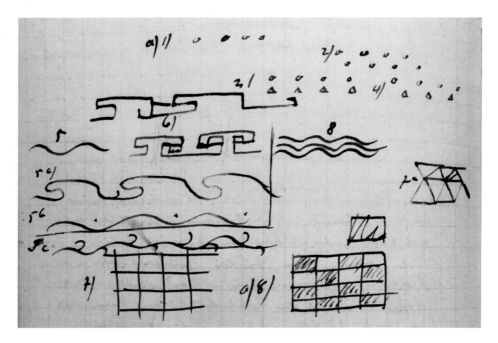

FIGURE 4. Elements of ornamentics in "Grundlagen der Geometrie" notebook, Max Dehn Papers.

Faculty appointment

The enthusiasm expressed in the Community Bulletin about Dehn's visit was mutual. According to his elder daughter Maria (Dehn) Peters, "when he was invited to visit Black Mountain College, he fell in love with it at once, and this love stayed with him" [Peters, n.d., p. 24]. The aims of Black Mountain College, as described by Anni Albers in the College Bulletin, mirror his vision of creative mathematical development: "Too often today education tends to develop receptive qualities and to neglect productive abilities. In most fields so much knowledge has to be acquired before new contributions can be made that a student is frequently confronted with results without being brought to understand the creative approach that led to the original discoveries. Naturally he cannot be expected to make new discoveries, but he can be brought to acquire an attitude that leads to discovery."[9]

Mathematics at Black Mountain College was taught by Theodore Dreier, who had left the Rollins physics department in 1933 to co-found the college. Dreier also served as the fledgling corporation's primary fundraiser, and by 1944 he was due a sabbatical. But internal disputes postponed Dreier's leave by a year. Dehn, however, had only to wait until the short days of December for the college's offer, such as it was. He negotiated a salary of $40 per month (today's equivalent of $680), including room and board.[10]

---

[9]Black Mountain College Bulletin 2:3, 1943.
[10]Dec 29, 1944 Special Meeting of the Board of Fellows, BMC Archives.

FIGURE 5. Max Dehn at right with colleagues (from far left) Elliot Merrick, Johanna Jalowetz, and Josef Albers at Black Mountain College, circa 1945. Photographer unknown.

His initial appointment as visiting professor allowed the possibility for him to leave mid-year for a permanent position should the opportunity present itself, but within months the faculty that constituted the governing Board of Fellows appointed Dehn Professor of Mathematics, a two-year position it renewed until his retirement in 1952. After Max had left the Illinois Institute of Technology for St. John's in 1943, Toni had kept their apartment in Chicago where she was employed at the Montgomery Ward department store as a commercial artist and designer. Largely for financial reasons, she remained in Chicago until March 1947 [Peters, n.d., p. 24a].

In his first term Dehn taught Introduction to Mathematics and Dialogues of Plato.[11] Courses in geometry and calculus followed later, balanced by more offerings in philosophy. Dreier returned from leave in fall 1946 to teach physics and analytic geometry. The absence of curricular distribution or attendance requirements at the college left instructors free to alter or entirely abandon any planned course of study, and few descriptive details accompany the provisional, often generic, course titles announced in the college bulletins and newsletters.[12] Dehn exercised these freedoms

---

[11] Course cards of students Ati Gropius and Betty Jennerjahn, BMC Archives.

[12] Course titles announced for mathematics vary, with introduction to mathematics appearing as Elementary mathematics, Introductory Mathematics, Basic Mathematics, or Introduction to Elementary and Higher Mathematics, whereas geometry courses appear alternately as Descriptive Geometry, Geometry without algebra, Geometry (Elementary, projective, and analytic), or Geometry for Artists. Calculus was simply Calculus, or possibly Advanced Mathematics. Outside mathematics, course listings included Problems in Philosophy, Ethics, Principles of Philosophy,

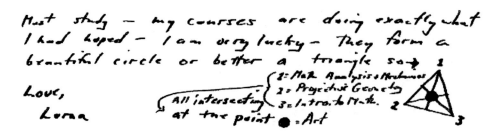

FIGURE 6. Detail of letter from Lorna to her mother, 1947, discussing math class and discussions with Dehn and Dreier.

to effect learning experiences that, while difficult to trace in official records, left lasting impressions on his students.

Lorna Blaine came to Black Mountain from New England in summer 1945 to study painting and stayed on as a regular student, eventually taking every mathematics course Dehn offered during her three years at the college. In an undated letter home from the college (Figure 6), she wrote, "my courses are doing exactly what I had hoped – I am very lucky – they form a beautiful circle or better a triangle." She diagrammed her curriculum as a projective configuration of six lines through four points labeled: "1 = Math. Analysis + Mechanics, 2 = Projective Geometry, 3 = Intro. To Math. All intersecting at the point • = Art."[13]

Harry Weitzer described himself as "a very square peg failing to fit the round hole provided by my Jewish middle class St. Louis family" before finding his way to Black Mountain to study art [Weitzer, n.d., p. 1]. The college assigned Dehn to be his academic advisor.

> My favorite of his classes was "Descriptive Geometry," which dealt with points in space and pure logical proofs that you knew where they were. We brought to class only a straight edge and pencil. We used no measurements, only lines and proofs. Once, standing on a chair (as his example required the reach), his chalk line wrapped neatly around a black painted pipe that crossed the board. Proud of his proof, he asked, "can you see the music?" To him, the harmony of mathematical relationships and those of music were of the same order [Weitzer, n.d., "Dehn" p. 1].

Weitzer also studied Philosophy with Dehn, who took a Socratic approach.

> We had no text, and his assignments were such as "Next time we will talk about Knowledge: think about it in the meantime." [...] There were about six or seven in our Philosophy class, which met once a week in his study. I remember the session with the assignment "Progress." We were all post-war cynics and thought we were mentally tough. One by one we crumpled under his questioning which had taken several hours. I remember Hank Bergman was the last to admit that there seemed to be

---

and Theory of Knowledge. Student transcripts ("course cards") that have so far been released from the BMC Archives confirm that most of these announced offerings were offered.

[13]Lorna Blaine letter, undated, BMC Archives. We thank Marjorie Senechal for bringing this letter to our attention.

at least some progress. In the silence that followed the "Ach So!", we began to turn the tables and tried to find out what Max thought about progress. Finally, he said that he had not been playing fair. At his age he could allow himself the luxury of doubting there was progress, but at our time of life that was something we could not afford [Lane, 1990, p. 192].

For students who sought to engage mathematics at a deeper level, Dehn led advanced tutorials. Black Mountain College was not accredited and relied on external examiners to evaluate students who applied for graduation after having completed their program of studies. In the two dozen years of its existence, sixty students graduated, and Dehn arranged examiners for two mathematics students: Emil Artin came from Princeton University to examine Peter Nemenyi, and Alfred Brauer from nearby Chapel Hill examined Trueman MacHenry.[14]

Dehn also held tutorials outside mathematics. Mildred Harding, whose husband John Adams taught anthropology, once joined a Greek tutorial. "From Max's, and Homer's, first words, I was enraptured," she remembered. "His eyes twinkling, Max read a few lines in Greek from the Calypso episode, confessed himself in love with Calypso, told us the story briefly, translated, then read a little more Greek, translated, paused for questions and comments" [Harding, 1986, p. 84]. Charles Olson wrote to fellow poet Robert Creeley, "I have again flung myself loose into those areas which are my concern: am studying Greek, for the 1st time, with an old German, Gottingen, named Dehn here (along with Con), and already know the first three lines of the ODYSSEY!"[15]

Geometry for Artists

Max Dehn left behind a sizeable record of at least one of his Black Mountain College courses. The folder labeled "Geometry for Artists Class BMC Winter & Spring 1948" in his archive contains eighty-one pencil drawings of remarkable precision and complexity. These are complemented by the notes of students Ruth Asawa, Albert Lanier, and Eine Sihvonen.[16] In graphite with occasional accents in colored pencil and ink, the diagrams range from simple sketches to elaborate compass and straight edge constructions with scores of lines. Captions are sparse but formal similarities across the drawings and the dates that Dehn and Asawa noted allow at least a reasonably consistent ordering.

Each day of the four month workshop was devoted typically to a single construction. It began in mid-February with a drawing of a sphere. A meridian line running through north and south poles projected to an ellipse. Students learned three ways to find the pair of tangent lines from an exterior point, or "pole," to a circle. A method that Dehn credited to Philippe de La Hire incorporated the polar, which is the line through the two points where the pair of tangents from the pole meet the circle. Pole and polar satisfy a reciprocal relationship that foreshadowed dualities and projective transformations soon to take center stage. In one of the

---

[14] Both graduates went on to complete doctoral studies in mathematics. See Chapter 11 for their stories.

[15] Olson to Creeley, September 30, 1951 in [Davidson, 1981, p. 211]. "Con" is short for Constance Wilcock, Olson's first wife.

[16] Asawa and Lanier's notes are preserved in the Ruth Asawa papers (M1585), Dept. of Special Collections and University Archives, Stanford University Libraries, Stanford, CA and the Asawa Estate, San Francisco, CA. The BMC Archives include a photocopy of Sihvonen's notes.

more elaborate assignments, students found the shortest paths between two points on projections of the globe. For his part, Lanier located approximate coordinates of Asawa's Los Angeles birthplace to which he joined those of Black Mountain (see Figure 7).

Circles disappear from the drawings by March as Dehn turned to projective geometry proper. Students constructed ten Desargues configurations of ten points and ten lines, starting each with a different choice of center (see Figure 8). A prize problem challenged students to demonstrate how to halve and double a line segment given only a parallel to the segment and an (unmarked) ruler. The ellipse returned as a section of a cone along with its conic cousins, the parabola and hyperbola, before spring break.

In April, the class worked through the theorems of Brianchon, Pascal, and Pappus. The balance leaned toward descriptive geometry with spirals, waves, helices, and double spirals in May. These and other curves appeared as planar projections from simple curves in the cylinder, cone, and sphere. The course concluded with the five regular solids and Euler's polyhedral formula. Dehn once referred to the tetrahedron, hexahedron, octahedron, dodecahedron, and icosahedron as "pure spatial ornaments," [Dehn, 1939, p. 144] but here, as if purity left them too plain, he wrapped their polygon faces with projections of a circle or trefoil knot (see Figure 9).

Taken together, the eighteen weeks of Geometry for Artists makes for an unusual course, by any standard. It is a blend of descriptive and projective geometry, applied and pure, demanding both dexterity and logical acumen. Dehn provided this brief course description when he offered the course for the last time, in spring 1952: "Special phenomena, projective configuration, structure of the theorems in geometry."[17] This leaves open the nature and purpose of the course, the picture of geometry it presented, and the sense in which it was a course "for artists."

Geometry for Artists does not appear to have been a course in perspective, as has been reported.[18] Central or one-point perspective drawings are constructed in relation to a fixed vantage point – the location of the observer's eye. The basic configuration of the center of projection, sight lines, picture plane and horizon appear briefly in the notes, but there is no evidence that Dehn instructed students in techniques of central perspective or that students made perspectival drawings.

As the distance between the center of projection and the picture plane increases without bound, sight lines align and central projection becomes orthogonal projection. This abstract limit is also known as parallel or right-angled projection. A magazine clipping among Dehn's Papers, "High noon on the Polynesian clock," shows a stand of palm trees in the midday sun, their shadows projected very nearly vertically to the ground beneath.[19]

The distinction between the two kinds of projection appears to have been Dehn's point of departure for the course. "Diff[erence] betw[een] artistic + math drawing" is the header of the first page of his notes, dated February 13th. A sequence of small sketches beneath illustrate the projection of a meridian ellipse

---

[17]Spring 1952 Bulletin, BMC Archives.

[18]Mary Emma Harris described it as "a course in perspective geometry that dealt with classical relationships such as the Golden Mean and optics" [Harris, 1987, pp. 109-10].

[19]"Drawings, sketches, undated," Dehn Papers.

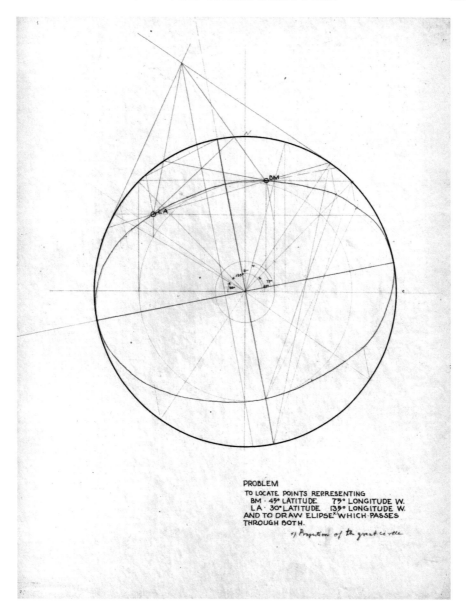

FIGURE 7. Albert Lanier, notes on construction of meridian ellipse connecting Los Angeles and Black Mountain, student notes from "Geometry for Artists," 1948. Courtesy of Ruth Asawa Lanier, Inc.

from a sphere, and the caption reads "circle projection of sphere at right angle = orthogonal projection." All the three-dimensional objects that Dehn's students constructed in Geometry for Artists (spheres, cylinders, cones, cubes, etc.) are rendered in parallel projection.

This classification leaves open the question of how and why Dehn conceived of his geometry course "for Artists."

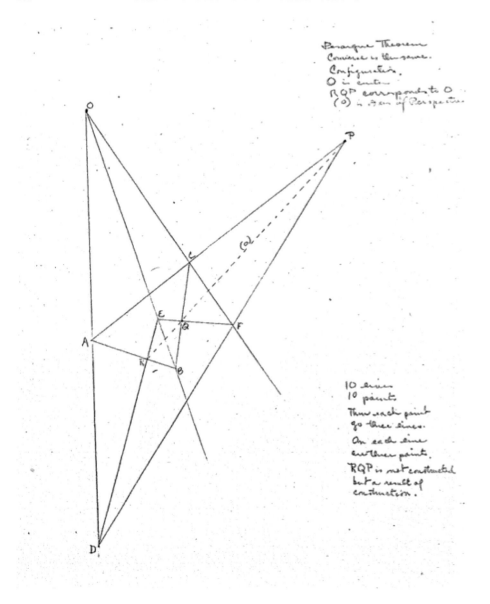

FIGURE 8. Desargues' theorem, student notes by Eine Sihvonen "Geometry for Artists," 1948.

To the extent that mathematics can be viewed as a craft, the ruler and compass are among its most elementary tools. Apart from the projective configurations, the constructions Dehn assigned could just as well have been furnished by any number of technical drawing manuals for artists.[20] More specifically, the "special phenomena" that he selected – circles, ellipses, spirals, helices, double spirals – adhere closely

---

[20]Dehn may have had in mind Albrecht Dürer's 1525 archetype, *A Painter's Manual* [Dürer and Strauss, 1977]. A literal translation of the original title is suggestive: *A Manual of Measurement with Compass and Straightedge* (*Unterweisung der Messung mit dem Zirkel und Richtscheit*).

to the collection of forms that students learned to draw in their studio art courses with Josef Albers. Albers stressed the importance of training the eye and hand through freehand drawing (he never taught perspective) and relied on unintuitive features of geometry, such as the hidden hexagonal cross-section of a cube. This example and other examples resurfaced in Geometry for Artists, where they were deduced rigorously. One could say, somewhat reductively, that Dehn's course was a rational recapitulation of his colleague's course: "Geometry for Albers."

Dehn's study "On Ornamentation" offers another sense in which to interpret Dehn's use of the term "Artists." He writes that "the domain of ornamentation is abandoned if the person seeks in the first place to represent, if ornament is subordinated to representation, if, one could say with some exaggeration, perspectival depiction is placed above symmetrical arrangement" [Dehn 1939 p. 150]. In this context, his February 13th class notes may imply that geometric drawings fall within the domain of ornamentation. Indeed, orthogonal projection, while still a form of representation, is distinguished from perspective projection by its symmetry: parallel lines project to parallel lines. In this way, we can view Geometry for Artists as a course that instructed the perspective artist in the ways of the ornamental artist, a more symmetric form of representation that resonated with a "higher rhythm."

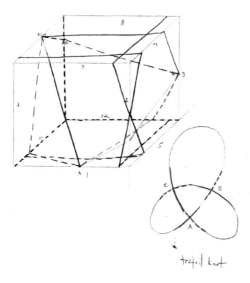

FIGURE 9. Ruth Asawa, notes on two projections of the trefoil to closed arcs on a cube, student notes from "Geometry for Artists," 1948. Courtesy of Ruth Asawa Lanier, Inc.

Some artists in the class, no doubt, failed to "see the music," but some did. Yet others found in Dehn's class a source of insight to which they would return after the end of term. In the fall of 1948, Lanier wrote to Asawa from San Francisco, where the two would eventually marry, raise a family, and develop their art and architecture practices: "I am [in] Dehn's math notes and I will make drawings of conic sections and famil[ies] of curves – Circles, Ellipses, Parabolas, Hyperbolas –

singly and together and repeated and in isometric and perspective – trying to learn to think in space more and then maybe to solids. This is a playful approach to geometry which is the foundation of new and more daring structural systems."[21] Years later, Asawa also found herself returning to her studies with Dehn, among others at Black Mountain College: "Teachers there were practicing artists, there was no separation between studying, performing the daily chores, and relating to many art forms. I spent three years there and encountered great teachers who gave me enough stimulation to last me for the rest of my life – Josef Albers, painter, Buckminster Fuller, inventor, Max Dehn, the mathematician, and many others. Through them I came to understand the total commitment required if one must be an artist."[22]

Dehn as working "artist"

Asawa's depiction of her teachers echoes founder John Andrew Rice's vision: "The law of a teacher at Black Mountain is to function as a working 'artist' in the teaching world, to be no passive recipient or hander-out of mere information, but to be and increasingly to become, productive, creative, *using* everything that comes within his orbit, including especially people."[23] If most people at the college were at least vaguely aware of how accomplished Dr. Dehn was, no one seems to have known what intellectual endeavors he pursued there. Erwin Straus, with whom he had begun a correspondence regarding his research and would have likely been his primary interlocutor had accepted a fellowship at Johns Hopkins just before Dehn's arrival.

One person with whom Dehn shared his research was Wolfgang Köhler. Köhler, a pioneer of Gestalt psychology, had directed the Psychological Institute at the University of Berlin (where Straus was a member of the medical faculty) before his open criticism of the Nazi regime and their interference with his work necessitated his accepting a position at Swarthmore College in Pennsylvania. It is unclear when and how the two first came into contact, but Dehn sent Köhler a sample of his investigations. In a March 24, 1945 letter, Köhler replied: "It is so gratifying to hear a mathematician talk about the human fundamentals of mathematical thought without any technical equipment."[24] He found the work to be reminiscent of existing studies in mathematical discovery, problem solving, and phenomenology, citing in particular Henri Poincaré, Karl Duncker and his book *On the Psychology of Productive Thinking*, and a "youthful publication" of Edmund Husserl, most likely *Philosophy of Arithmetic: Psychological and Logical Investigations*. The content of the letter and its placement in his archives implies that what Dehn sent Köhler was very likely an excerpt from his *Psychology of Mathematical Activity*.

The undated manuscript of *Psychology of Mathematical Activity* is divided across three folders in his archive: One folder is labeled "Untitled handwritten pages numbered 55-225 on the relationship between mathematics and ornamentation [?] described by Magnus." The second is labeled after the book's epigraph "*Animi sui complicatam notionem evolvere*" ("to unfold a complicated concept of the mind," from Cicero, *De Officiis* III.76) and contains thirty-six revised, typewritten pages

---

[21] Lanier to Asawa, November 30th, 1948, Asawa Papers.
[22] Website of the estate of Ruth Asawa, https://ruthasawa.com.
[23] Black Mountain College Bulletin, Spring 1952, emphasis in original.
[24] Köhler to Dehn, March 24, 1945, Dehn Papers.

and a handwritten page numbered 40. The third folder contains the intervening pages numbered 41-54, a two-page English outline of the book, and Köhler's letter. Given the limited scope of this chapter, we confine our study to that which can be gleaned from the most legible portion of the manuscript; the following partial synopsis derives from Dehn's typewritten outline and revisions.

Psychology of Mathematical Activity begins with an introduction to the "analytical method" with which it approaches psychological phenomena. Inspired by the axiomatic approach, it sets out to identify basic feelings that constitute the psychological prerequisites for mathematical activity. The first chapter concerns the "character" of mathematical statements. These are distinguished from chess problems, for example, by their infinite scope and the conviction with which they are held. Chapter Two takes up the process of comprehending elementary mathematical notions and results. Using the example of the algebraic identity $(a+b)^2 = a^2+2ab+b^2$, Dehn unpacks the mental processes involved with the use of symbols, sequences, numbers, and the "musical score" that counting and arithmetical operations entail. The third chapter turns to mathematical proofs. The essential feature of proof, he finds, is not the perfection of its reasoning but that its reasoning can be perfected. Beyond indirect proofs and proofs by induction, Dehn considers the role of memory, concentration, and the relationship between "mathematical reasoning and the reasoning behind cave paintings." The final chapter he outlined is on "productive activity." It is the most overtly psychological chapter as it addresses the motivations that drive mathematical activity. These include personal motives, such as "curiosity," "satisfaction derived from conquering difficulties," and "happiness found in creating harmonic structures," as well as social motives, such as "possessive spirit (priority)," "ambition," and "*Geltungstrieb*" – the drive for recognition.

Dehn introduces each psychological dimension of mathematical activity tentatively, cautioning readers against drawing any final conclusion let alone an epistemology. Nevertheless, through numerous examples, the *Psychology* contends that feelings play a constitutive function in mathematics that ought not be discounted. For instance, the following remarkable passage attests to the double-edged nature of "the feeling of absolute correctness" of a proof:

> The essential thing is the *feeling* that the proof is correct in and of itself, without the authority of any particular teacher or general tradition. This distinguishes the mathematician from the non-mathematician, who accepts such a fact even without proof because that is how it may be taught and learned. The mathematician, who has the feeling of absolute correctness, may yet be mistaken and consider a false proof correct. Indeed, [when this happens] he is almost always wrong because he is unaware of all the elements in the proof that depend on the collective in which he grew up, especially the language, but also more mathematical elements often taken for granted. [Dehn, undated, p. 33a, emphasis in the original]

The relevance of teaching and learning to Dehn's *Psychologischen Grundlagen*, as he referred to this project, is evident here and throughout the manuscript. In the introduction, he explains that consideration of psychological situations in learning mathematics is central to the method by which he achieves the results of his philosophical investigations. Preliminary as it is, our review of *Psychology of*

Mathematical Activity in the context of his teaching suggests that the converse may have been just as true: consideration of philosophical investigations were central to the method by which Dehn led students to achieve mathematical learning. When the University of Wisconsin invited him to teach in Madison for the fall of 1946, for example, he devised the course Concepts, Problems and Methods on the topics of: "Basic operations, the concept of numbers, and projective and metric geometry. Axiomatics. General methods, the *non-logical elements in Mathematics*."[25] Indeed, it would be difficult to know how to make sense of the surprising last topic without knowledge of Dehn's research at the time.

Dehn in the college community

In certain respects, the college offered Dehn a return to the communal atmosphere he had fostered as director of the Frankfurt Mathematics Seminar, with its reading seminar, hikes in the Taunus, and afternoon coffee at his and Toni's home. At Black Mountain, students accompanied Dehn on daily walks through forest paths above the college, stopping here and there to observe the root structure of a tree, a wild flower, a snake. On Sundays he organized outings in the farm truck to hike the ridge trails of Graybeard Mountain (Figure 10). Dorothea Rockburne was never formally one of his students, but she learned about mathematical principles in nature during walks with Dehn. She went on to a distinguished career as an artist and credits him for leading her to "an understanding of a more universal creative process" [Peifer, 2017, p. 1324].

The communal nature of Black Mountain College, in contrast to Frankfurt University, derived from a work program. All students and faculty were involved in the operation and maintenance of the campus. By the time Dehn arrived to the Lake Eden campus, faculty and students had cleared timber, cultivated a vegetable and livestock farm, and constructed the iconic Studies Building. Dehn enjoyed the honest work, too. Students remembered "hunting mushrooms" and "building a rough stone bridge with a world-class mathematician" [Lane, 1990, p. 148].

The faculty owned the college, and administrative decisions were made collectively, often in marathon meetings, sometimes ending in stalemate or resignations. According to historian Martin Duberman's chronicle of the college, Dehn was "a much-beloved figure" who found himself "the 'man in the middle' when new hostilities erupted within the community in the late forties" [Duberman, 1972, p. 229]. In a letter to Josef Albers, Dehn wrote, "I am always glad to do something for BMC which I consider – in spite of occasional trouble – a wonderful place where I can be together with young people without any institutional impediments. There, I can use what little abilities I have to transmit to them what I think is leading most surely towards a happy life. Not to forget the beauty of the surrounding nature which, I think, is of the greatest value to transform young and old people who live in it."[26]

Dehn's involvement with the college community extended beyond its classrooms, work program, and wooded trails. When the Alberses took extended leave in 1946, Dehn's niece Franziska Mayer managed the weaving studio in Anni's absence. Mayer had studied weaving in Stockholm before spending the war years in Newfoundland working for the International Grenfell Association. She was the

---

[25]Emphasis added. The syllabus is reproduced in Chapter 11.
[26]Max Dehn to Josef Albers, February 23, 1949, BMC Archive.

FIGURE 10. Max Dehn and students at top of Graybeard Mountain. Photograph by Margaret W. Peterson.

daughter of Dehn's sister Marie Dehn Mayer, who was killed in the Holocaust. Dehn organized a reduction in the college's meagre food budget in order to send four care packages to his friend and former student Willy Hartner in war-torn Frankfurt on Christmas 1947 [Remmert, 2016]. The Dehns were also frequent participants in the Asheville Chapter of the Southern Conference for Human Welfare, "a progressive group organized for the purpose of encouraging intelligent and effective citizenship."[27]

In the spring of 1948, the Alberses returned from their travels in Mexico and New Mexico where they had pursued their long-standing interest in pre-Hispanic art. Shortly afterward, prior tensions between established and newer faculty resurfaced. Amidst these hostilities, Dehn accepted a second invitation to teach at the University of Wisconsin for the 1948–49 academic year (see Chapter 11). Given the diminished size of the faculty then, Dehn agreed to return for monthly lectures on mathematics and philosophy.[28] Remarkably, given its financial precarity, the college paid for Dehn to commute from Madison by air.

Matters deteriorated further. Dehn's efforts to persuade Josef, whom he had nominated to the position of rector, were ultimately unsuccessful, and the Alberses resigned from Black Mountain College in spring 1949. Dehn was to return that summer to lead a course in "Mathematics for Artists" but taught in the Notre Dame Mathematics Teacher Training Program instead. It had been founded by Arnold Ross two years earlier and would later transform into the Ross Math Program for high school students, which continues "to instruct bright young students in

---

[27] Black Mountain College Bulletin 5:4, 1947.
[28] Josef Albers to the Dehns, October 2, 1948, Dehn Faculty File, BMC Archives.

the art of mathematical thinking" to this day.[29] "With particular reverence and appreciation," Ross noted "the summer term when the late Professor Max Dehn taught a course in Projective Geometry in our program."[30] In Dehn's absence, Buckminster Fuller organized the 1949 summer institute program.

Mary Emma Harris outlined the dire situation facing the college when Dehn returned:

> The survival of Black Mountain College depended not so much on outside events as on the ability of its faculty to redefine its educational goals and solve its administrative and financial problems. During the first sixteen years of its history, the college had evolved from a small experimental school offering a general curriculum into a unique creative community where the most vital work being done was in the arts. Still there were elements of both colleges in the community, and the dichotomy between what the college had set out to accomplish and what it had become was a source of conflict and confusion. Yet the critical years 1949 through 1951 were a period of indecision and internal conflicts, and by the fall of 1951 only a miracle could have saved the college [Harris, 1987, p. 168].

Dehn presided over the first faculty meeting of the 1951-52 academic year. He reopened a discussion of the college's educational aims, offering his own view that its purpose was simply to teach students "to think, to be tolerant, to understand patterns."[31] The discussion resumed the following month with poet Charles Olson and anthropologist John Adams. Dehn outlined how different disciplines cultivate the capabilities that he felt students needed most:

> Mr. Adams ... wonders if we shouldn't stress that students for the first year or two should take subjects stressing general disciplines; that as Dr. Dehn had pointed out, some subjects lend themselves more readily to basic disciplines ...
>
> Mr. Dehn said that he though[t] not only of math but of foreign languages, especially ancient lang[uages] where the structure is very strict, as a subject out of which knowledge of structure should come, then to go on to broaden knowledge; perhaps very strict work in music would do; perhaps biology since it is so broad and rich is not right for beginners but maybe for the second year; history is still more difficult and requires a very disciplined mind; then crowning all to grasp all together, philosophy ...
>
> Mr. Olson said that what makes the strata of the usual Amer[ican] education is not admitting that the individual is more complex than any curriculum... that he objected to a theory of chronological order of studies ...

---

[29] The Ross Mathematics Program website, https://rossprogram.org

[30] The Notre Dame Mathematics Teacher Training Program Summer 1959 report, p. 13 https://bpb-us-w2.wpmucdn.com/u.osu.edu/dist/e/5164/files/2014/12/1959-Training-Program-2k6l8yy.pdf.

[31] Minutes of the regular meeting of the Faculty and Corporation of BMC, October 5, 1951, BMC Archives.

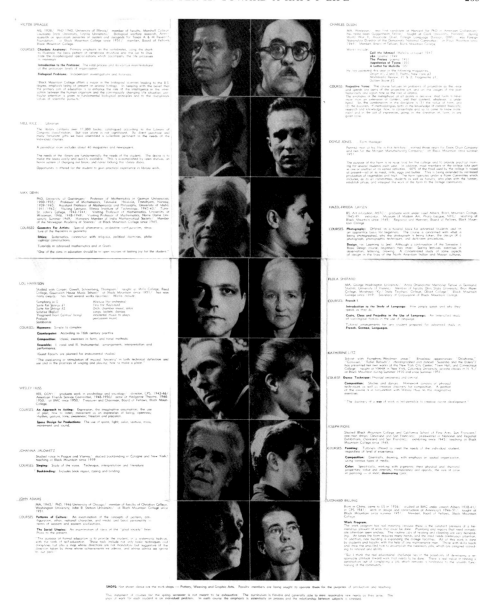

FIGURE 11. 1952 Spring Bulletin, overside broadside of courses including photographs and bios of faculty.

Mr. Adams said that Mr. Olson was objecting not to Mr. Dehn's method but to method.

Mr. Olson said that we gain by the fact that we do not make plans for a curriculum.[32]

---

[32]Minutes of the regular meeting of the Faculty and Corporation of BMC, November 21, 1951, BMC Archives.

Instead of resolving differences and committing themselves to one approach over the others, Olson suggested that the college project the multiplicity of educational perspectives of its faculty as a feature, rather than a problem. The majority, including Dehn, agreed and the spring 1952 bulletin displays the entire faculty of Black Mountain College in this spirit (Figure 11). Fourteen headshots sit one atop the other in two pillars, their credentials and course offerings flanking the composite faculty portrait like wings of a double totem pole. Beside his portrait, beneath course listings of Geometry for Artists and Ethics, Dehn distilled his psychological philosophy of education into one line: "One of the aims in education should be to open sources of lasting joy for the student."

Olson, who subsequently assumed leadership of Black Mountain, envisioned a plan for the college that would capitalize on the exceptional strengths of a diverse and professionally active faculty. The model was inspired by his and Dehn's monthly visits to the college during the 1948–49 year, when they would "descend on the place like arcangels and go away like banana skins" [Harris, 1987, p. 172]. For financial reasons, the college implemented a different program, even as both Olson and Dehn continued to plan scholarly itineraries beyond the Blue Ridge Mountains.

Dehn accepted an invitation to teach in Germany for the 1952-53 academic year. He was to give a seminar on "Mathematics in the Renaissance" at the University of Frankfurt in the fall and on "Historical presentation of the significant results, methods and concepts in mathematics" at the University of Göttingen in the spring [Remmert, 2016]. The correlation between the latter topic and his Madison course suggests that Dehn intended to further develop his psychological investigations.[33]

By this time, Dehn's German pension was finally transferred, and he and Toni were able to plan their retirement from the college. Through the Federal Housing Administration, they obtained a mortgage for a modest house on the outskirts of the town of Black Mountain.[34] In search of a successor, Dehn approached a number of contacts with the college's offer of emeritus status in exchange for room, board, and a light teaching load, but without success.[35] In a twist, the faculty organized a special meeting without Dehn and decided unanimously to confer the honor to *him*, which they did at the close of the spring term.

Dehn appreciated the privilege and agreed to serve the college in an advisory capacity, but within three weeks, on June 27th, he died. The previous day he had tried to preserve a stand of flowering trees in the vicinity of a beloved picnic site from inadvertently being felled in an ill-conceived plan to raise funds through the sale of timber. His effort that day or the early arrival of midsummer heat or both were blamed for the release of the blood clot that lodged itself in one of his lungs. Coincidentally, on the same day, June 26, 1952, Ruth Asawa gave birth to her and Albert Lanier's second son. In honor of their teacher they named him Hudson Dehn Lanier.[36]

---

[33]That the course was planned for Göttingen may also explain why Dehn, a polyglot who delivered his lectures in the local language, made his typed revisions in German.

[34]Dehns to Asawa and Lanier, June 12, 1952, Asawa Papers.

[35]Faculty meeting minutes 1952. Dehn had recently proposed Dirk Struik.

[36]Toni Dehn to Ruth Asawa and Albert Lanier, October 6, 1952, Asawa Papers.

Dehn as late modernist

What initially attracted us to research Max Dehn at Black Mountain College was the possible impact that his interactions with its celebrated student and faculty artists had on their art work. The more we uncovered about his endeavors as a "teaching 'artist'" however, the more we felt compelled to concentrate our attention on his own creative work, despite the growing realization that its depth and complexity exceeded the scope of our remit. By reconstructing the syllabus for Geometry for Artists and analyzing his *Psychology of Mathematical Activity* and contribution to the intellectual life of the college, we have aimed to dispel the tacit notion that Dehn's Black Mountain appointment was some form of retirement; to attract further study of his late work; and to provide context with which to meaningfully evaluate what effect he may have had on his students and colleagues.

To isolate and pursue lines of influence is a complicated affair of disentanglement to be sure. Many of the artists at Black Mountain who were receptive to Dehn had encountered mathematical sources prior to his arrival just as he had been preoccupied with art before Straus invited him. The meander that threads through notes and writings of Josef and Anni Albers and Dehn, for example, represents a common yet independent interest in the origins of design that predates their meeting by many years. One also has to guard against facile equivalences, to which formalist approaches to material culture seem especially vulnerable. That Charles Olson chose to name his 1950 manifesto "Projective Verse" does not by itself imbue the essay with mathematical meaning – it warrants further examination. As Dehn cautioned in his study of ornamentation, "in this analysis, the emotional origin must always be considered lest we interpret our findings as having more mathematics in mind than the maker of ornaments intuited" [Dehn, 1939, p. 129].

In closing, we confine ourselves to highlight a single, albeit broad, resonance between the work of Dehn and his Black Mountain colleagues. In her essay "Imaginary Landscape" for the exhibition catalog, *Leap Before You Look: Black Mountain College 1933–1957*, art historian Helen Molesworth observes that despite the differences great and small that separated the visual, performing, and literary artists at the college, "nearly all were committed, in profound ways, to the exploration of form" [Molesworth and Erickson, 2015, p. 51]. In three brief case studies, she recounts how painter Josef Albers, composer John Cage, and poet Charles Olson innovated their respective domains through structural experimentation in the use of color, time, and breath, adding (p. 57):

> They also explored the in-between – Albers, in the relativity of color's perception; Cage, in the parity of sound and silence; and Olson, in the movements connecting writer and listener. Indeed, both Albers and Cage shared Olson's sense that "it is not the things in themselves but what happens between things where the life of them is to be sought." It is precisely the space between their formal rigor and their openness to perception and its contingencies that places their practices in the nascent shift from modernism to postmodernism, and turned Black Mountain College into the unlikely wellspring of avant-garde culture.

Viewed from outside the discipline, all of mathematics may appear to be a formalist exploration of figures of one kind or another. Yet even the casual observer of Dehn's mathematics teaching record at the college would recognize the priority he placed on figures of a particular sort – projective configurations. As structures, configurations are defined in terms of the incidence of points and lines "free of all elements of measurement," making them intrinsically relational or schematic. The content of some configurations is even preserved when the roles of point and line are exchanged, as in the case of Pappus's theorem (Figure 2). In this sense, configurations are a vivid mathematical expression of "not the things in themselves but what happens between things."

Projective configurations, Dehn would be the first to acknowledge, were neither new nor a creation of his own. But the elementary modern geometry course he developed around them was a pedagogical experiment that derives from his unique philosophy of mathematics. Dehn was a modernist in the sense that he understood mathematics as "an autonomous body of ideas, having little or no outward reference, placing considerable emphasis on formal aspects of the work" [Gray, 2008, p. 1]. At the same time, he warned against the "false method of merely formal generalization" and "Descartes' delusion" that one could "comprehend the whole world by means of pure reason" [Dehn, 1928, pp. 22-3]. The basic assumption of his meditation on the origin of mathematics is that "it arises from the methods of logical inference – both conscious and *unconscious* – that allow us to 'manufacture' hitherto unknown facts" [Dehn, 1932, p. 131, emphasis added]. Through his investigation of non-logical elements he sought to reduce the sensations in mathematical activity to a collection of particularly simple elements, "basic feelings." As such, his *Psychology of Mathematical Activity* represents a departure from mathematical modernism no less profound than that of the work of his anti-establishment artist students and colleagues at Black Mountain. One could say, by analogy with Molesworth's assessment of Albers, Cage, and Olson, that Dehn explored the psychological between the logical.

In some respects, Dehn's *Psychology* anticipates later studies of the social and psychological underpinnings of mathematics. For example, Brian Rotman has characterized "mathematical thought as a kind of waking dream" in which the mathematician imagines an idealized "subject" who instructs an idealized "agent" in ways reminiscent of Dehn's description of the psychic or "soul" processes involved in mathematical activity [Rotman, 1993, p. xii]. Whereas Rotman reinscribes the body into mathematical practice through semiotics, cognitive scientists George Lakoff and Rafael Núñez have sought more direct connections between biology and the nature of mathematical ideas, arguing, rather less tentatively than Dehn, that "mathematics as we know it arises from the nature of our brains and our embodied experience" [Lakoff and Núñez, 2000, p. xvi]. Of course, Dehn was not a semiotician or a cognitive scientist, and their aims differed from his.

There is a notable contrast between the abundance of Dehn's Black Mountain College activity and its scarcity in published accounts. To be fair, much of the literature on the college is the product of investigations into the arts at the college, and the visual record of mathematics and science at the college is thoroughly upstaged by Buckminster Fuller. On the mathematical side, the only hint that Dehn had work in progress is a reference to "largely unpublished studies of ornamentation" in the memorial by Magnus and Moufang. The obituary that Willy Hartner wrote for his friend in *Die Frankfurter Allgemeine Zeitung* is no less opaque: "He has worked at six or seven smaller colleges over the past ten years, the longest at the progressive Black Mountain College in North Carolina, where, besides mathematics, he studied Plato, the history of philosophy, especially Oriental philosophy, gradually adopting many of its ideas into his own philosophy of life." Perhaps Dehn was too conscientious to share his incomplete manuscript with mathematicians. We hope that this preliminary study initiates deeper investigations that will bring forward, among other things, more of his unpublished *Psychology*. In 1947 Dehn reviewed Swiss mathematician Andreas Speiser's *Die Matematische Denkweise* for the *Monthly*. His review of the book opens with the following passage, which we expect will likely serve just as well to describe Dehn's own contribution.[37]

> In this book we have the philosophy of a mathematician. It is written with the enthusiasm of a distinguished mathematician who penetrates the arts and the world in his peculiar way. It will transmit, I imagine, this enthusiasm to every mathematician who is not only a craftsman but possessed by the sacred fire as the poet and philosopher ought to be. [Dehn, 1947]

References

Josef Albers. Concerning art instruction. *Black Mountain College Bulletin*, June (2), 1934.

Josef Albers. *Interaction of Color*. Yale University Press, 1963.

Birgit Bergmann. *Transcending Tradition: Jewish Mathematicians in German Speaking Academic Culture*. Springer Science & Business Media, 2012.

Michael Davidson. Charles Olson & Robert Creeley: The Complete Correspondence. 1981.

Max Dehn. Über die geistige Eigenart des Mathematiker. *Frankfurter Universitaetsreden*, (28), 1928.

Max Dehn. Das Mathematische im Menschen. *Scientia*, 26 (52), 1932.

Max Dehn. Ueber Ornamentik. *Norsk Matematisk Tidsskrift*, 21: 121–153, 1939.

Max Dehn. Mathematics, 600 B.C.–400 B.C. *The American Mathematical Monthly*, 50 (6): 357–360, 1943.

Max Dehn. Mathematics, 300 B.C.–200 B.C. *The American Mathematical Monthly*, 51 (1): 25–31, 1944a.

Max Dehn. Mathematics, 200 B.C.–600 A.D. *The American Mathematical Monthly*, 51 (3): 149–157, 1944b.

Max Dehn. Die Matematische Denkweise by Andreas Speiser. *The American Mathematical Monthly*, 54 (7): 424–426, 1947.

Max Dehn. Manuscripts 'animi sui complicatum' [sic] typed ms and carbon copy. each w/ annotations. ca 60 pages total – on psychology of mathematics, undated, Psychology. Max Dehn Papers, Archives of American Mathematics, The Dolph Briscoe Center for American History, The University of Texas at Austin.

---

[37] We wish to thank Marjorie Senechal for inviting us to the Mathematisches Forschungsinstitut Oberwolfach Mini-Workshop: Max Dehn: his Life, Work, and Influence which prompted this collaboration. The second author received support for this research while a visitor to the Program in Interdisciplinary Studies at the Institute for Advanced Study in Princeton, NJ. We are grateful to the following members of the Dehn family: Antonio Alcala, Joanna Beresford Dehn, Dr. Matthias Brandis, Max Dehn Jr., Enrique Mayer, Renata Mayer Millones, Maria Mayer Scurrah, and Peter Spencer.

Martin Duberman. *Black Mountain: An Exploration in Community*. E.P. Dutton & Co, New York, 1972.

Albrecht Dürer and Walter S. Strauss. *The Painter's Manual: A Manual of Measurement of Lines, Areas, and Solids by Means of Compass and Ruler Assembled by Albrecht Dürer for the Use of All Lovers of Art with Appropriate Illustrations Arranged to be Printed in the Year MDXXV*. Abaris Books, 1977.

Jeremy Gray. *Plato's Ghost: The Modernist Transformation of Mathematics*. Princeton University Press, Princeton, 2008.

Mildred Harding. My Black Mountain. *The Yale Literary Magazine*, 151 (1): 76–89, 1986.

Mary Emma Harris. Toni Dehn interview, 8 April 1978. Collection of Joanna Dehn Beresford.

Mary Emma Harris. *The Arts at Black Mountain College*. MIT Press, Cambridge, MA, 1987.

George Lakoff and Rafael Núñez. *Where Mathematics Comes From*. Basic Books, New York, 2000.

Mervin Lane. *Black Mountain College: Sprouted seeds: An Anthology of Personal Accounts*. Univ. of Tennessee Press, 1990.

Wilhelm Magnus and Ruth Moufang. Max Dehn zum Gedächtnis. *Mathematische Annalen*, 127 (1): 215–227, 1954.

Helen Anne Molesworth and Ruth Erickson. *Leap Before You Look: Black Mountain College, 1933-1957*. Yale University Press, 2015.

David Peifer. Dorothea Rockburne and Max Dehn at Black Mountain College. *Notices of the AMS*, 64 (11), 2017.

Maria Dehn Peters. Letter to Joanna Dehn Beresford, n.d. Partially dated November 1, 1997, and June 8, 1998, Courtesy Dehn Family Collection.

Volker Remmert. Max Dehn: 1878-1952, 2016. Lecture at Oberwolfach Research Institute for Mathematics.

Brian Rotman. *Ad Infinitum... The Ghost in Turing's Machine*. Stanford University Press, 1993.

Carl Ludwig Siegel. On the history of the Frankfurt mathematics seminar. *The Mathematical Intelligencer*, 1 (4): 223–230, 1979.

E.W. Straus and F. Wohlwill. Der Hitzschlag. *Spezielle Pathologie und Therapie innerer Krankheiten*, 2: 445–454, 1924a.

E.W. Straus and F. Wohlwill. Nichteitrige Entzündungen des Centralnervensystems. *Spezielle Pathologie und Therapie innerer Krankheiten*, 2: 455–464, 1924b.

Harry Weitzer. Six BMC Stories, n.d. Box 19, Black Mountain College Project Collection, Western Regional Archives, North Carolina.

Selected Published Titles in This Series

46 Jemma Lorenat, John McCleary, Volker R. Remmert, David E. Rowe, and Marjorie Senechal, Editors, Max Dehn: Polyphonic Portrait, 2024

45 Craig A. Stephenson, Periodic Orbits: F. R. Moulton's Quest for a New Lunar Theory, 2021

44 Christopher D. Hollings and Reinhard Siegmund-Schultze, Meeting under the Integral Sign?, 2020

43 Sergei S. Demidov and Boris V. Lëvshin, Editors, The Case of Academician Nikolai Nikolaevich Luzin, 2016

42 David Aubin and Catherine Goldstein, Editors, The War of Guns and Mathematics, 2014

41 Christopher Hollings, Mathematics across the Iron Curtain, 2014

40 Roman Duda, Pearls from a Lost City, 2014

39 Richard Dedekind and Heinrich Weber, Theory of Algebraic Functions of One Variable, 2012

38 Daniel S. Alexander, Felice Iavernaro, and Alessandro Rosa, Early Days in Complex Dynamics, 2011

37 Henri Poincaré, Papers on Topology, 2010

36 Éric Charpentier, Étienne Ghys, and Annick Lesne, Editors, The Scientific Legacy of Poincaré, 2010

35 William J. Adams, The Life and Times of the Central Limit Theorem, Second Edition, 2009

34 Judy Green and Jeanne LaDuke, Pioneering Women in American Mathematics, 2009

33 Eckart Menzler-Trott, Logic's Lost Genius, 2007

32 Jeremy J. Gray and Karen Hunger Parshall, Editors, Episodes in the History of Modern Algebra (1800–1950), 2007

31 Judith R. Goodstein, The Volterra Chronicles, 2007

30 Michael Rosen, Editor, Exposition by Emil Artin: A Selection, 2006

29 J. L. Berggren and R. S. D. Thomas, Euclid's *Phaenomena*, 2006

28 Simon Altmann and Eduardo L. Ortiz, Editors, Mathematics and Social Utopias in France, 2005

27 Miklós Rédei, Editor, John von Neumann: Selected Letters, 2005

26 B. N. Delone, The St. Petersburg School of Number Theory, 2005

25 J. M. Plotkin, Editor, Hausdorff on Ordered Sets, 2005

24 Hans Niels Jahnke, Editor, A History of Analysis, 2003

23 Karen Hunger Parshall and Adrian C. Rice, Editors, Mathematics Unbound: The Evolution of an International Mathematical Research Community, 1800–1945, 2002

22 Bruce C. Berndt and Robert A. Rankin, Editors, Ramanujan: Essays and Surveys, 2001

21 Armand Borel, Essays in the History of Lie Groups and Algebraic Groups, 2001

20 Kolmogorov in Perspective, 2000

19 Hermann Grassmann, Extension Theory, 2000

18 Joe Albree, David C. Arney, and V. Frederick Rickey, A Station Favorable to the Pursuits of Science: Primary Materials in the History of Mathematics at the United States Military Academy, 2000

17 Jacques Hadamard, Non-Euclidean Geometry in the Theory of Automorphic Functions, 2000

16 P. G. L. Dirichlet and R. Dedekind, Lectures on Number Theory, 1999

For a complete list of titles in this series, visit the
AMS Bookstore at **www.ams.org/bookstore/hmathseries/**.